A series of student texts in

Contemporary Biology

General Editors:
Professor Arthur J. Willis
Professor Michael A. Sleigh

A Biologist's Guide to Principles and Techniques of Practical Biochemistry

Third Edition

Edited by
Keith Wilson

B.Sc., Ph.D.
Head of Division of Biological and Environmental Sciences,
The Hatfield Polytechnic

and
Kenneth H. Goulding

M.Sc., Ph.D.
Head of School of Applied Biology,
Lancashire Polytechnic

Edward Arnold
A division of Hodder & Stoughton
LONDON NEW YORK MELBOURNE AUCKLAND

First published in Great Britain 1975
Reprinted 1976, 1979
Second edition 1981
Reprinted with corrections 1983
Reprinted 1984
Third edition 1986
Reprinted 1988, 1989

Distributed in the USA by Routledge, Chapman and Hall, Inc.
29 West 35th Street, New York, NY 10001

British Library Cataloguing in Publication Data

A Biologist's guide to principles and
techniques of practical biochemistry.—
3rd ed.—(Contemporary biology)
1. Biological chemistry—Technique
I. Wilson, Keith, *1936*– II. Goulding,
Kenneth H. III. Series
576.19′2′028 QP519.7

ISBN 0-7131-2942-5

Typeset in 10/11pt Times Compugraphic by Colset Private Ltd, Singapore. Printed and bound in Great Britain for Edward Arnold, the educational, academic and medical publishing division of Hodder and Stoughton Limited, 41 Bedford Square, London WC1B 3DQ by Richard Clay Ltd., Bungay, Suffolk.

Preface to the third edition

The decade that has passed since the first edition of this book was published has seen a quantum leap in our understanding of gene structure, gene expression and gene manipulation and has coincided with the emergence of biotechnology. The impact of the new technology associated with these developments has been to create great hopes for the diagnosis and control of numerous human genetic disorders, and for the introduction of commercially attractive characteristics into animal, plant and microbial cells. The developments have been made possible by the isolation and purification of numerous enzymes associated with nucleic acid metabolism, by the refinement and application of existing analytical techniques and the development of new ones. One of the greatest impacts has been made by the development of relatively simple procedures for the production of monoclonal antibodies which can be used in the detection and assay of specific proteins.

All of the principles and techniques associated with these new methodologies have quickly found their way into undergraduate curricula. Practical exercises based on them form an increasing component of courses in biochemistry, microbiology, genetics, plant physiology and immunology. We have attempted to respond to these developments by updating all chapters and by including appropriate new ones in this third edition. Chapter 1 has been expanded to cover a consideration of the rationale and methodology involved in *in vitro* and *in vivo* biochemical experimentation including cell and tissue culture, cryopreservation and the approaches to metabolic investigations. The chapter also considers the importance of mutants in biochemical studies and the applications of light and electron microscopy. Chapter 3 is a new chapter on Enzyme Techniques which covers the basic principles of enzymology, enzyme and substrate assays and ligand binding techniques. Chapter 4 on Immunochemical Techniques was first introduced in the second edition, but has been expanded to give appropriately greater consideration to monoclonal antibodies. Chapter 5 on Techniques in Molecular Biology is the second totally new chapter and considers the principles behind the recent developments in nucleic acid isolation, analysis and structure determination and of genetic manipulation including the isolation of specific genes, the production of gene libraries and gene cloning.

The new additions to the book have been made at the expense of the chapter on Manometric Techniques, the essential outlines of which are now included in Chapter 1, and as a result of an increase in the total size of the book. The dilemma faced by all authors of undergraduate texts of balancing comprehensive cover against a reasonable text length and viable retail price is a difficult one to solve, but we hope that the moderate increase in the size of this new edition will be compensated by its wider appeal.

The general approach to the preparation of all the chapters remains unchanged from that of previous editions. Our aim was not to produce a comprehensive text for the specialist, but a general and, where necessary, simplified account for those students who have recourse to use some of the techniques during their undergraduate or postgraduate careers. Greatest attention has been given to those techniques which generally feature prominantly in undergraduate practical classes and less detailed coverage to other techniques which would be referred to in lectures and tutorials but which students are less likely to encounter in the laboratory. The main principles of the techniques and their associated instrumentation are discussed and reference given to their main applications and limitations. The book is intended for students on degree courses and Higher National Certificate and Higher National Diploma courses of BTEC in the biological, medical, paramedical and veterinary sciences in which biochemistry is an important component. It may also be of value to students on M.Sc. or other post-graduate courses who will be encountering biochemical techniques for the first time.

The third edition has been produced without the involvement of Bryan Williams who made such an invaluable contribution to previous editions. We are pleased to welcome Stephen Boffey as a first time contributor to the new edition. We would like to thank our two colleagues Dr Donald Bailey and Dr Michael Trevan for permission to reproduce their electron micrographs in Fig. 1.1. We are again indebted to the staff of Edward Arnold (Publishers) Limited, and particularly Nancy Loffler, for their continued enthusiastic support and helpful advice. We also gratefully acknowledge the unique scientific and linguistic skills of those people who have been responsible for the translation of previous editions into German, Italian, Russian and Spanish. We continue to welcome constructive comments and criticisms from all those who use the book as part of their studies.

Lancashire Polytechnic
The Hatfield Polytechnic
1986

Kenneth H. Goulding
Keith Wilson

Contents

Contributors

Stephen Boffey, B.Sc., Ph.D.
Senior Lecturer in Biochemistry, The Hatfield Polytechnic, Hatfield
David H. Burrin, B.Sc., Ph.D.
Principal Lecturer in Biochemistry, Coventry (Lanchester) Polytechnic, Coventry
Michael G. Davis, M.Sc., Ph.D.
Senior Lecturer in Biochemistry, The Hatfield Polytechnic, Hatfield
Kenneth H. Goulding, M.Sc. Ph.D.
Head of School of Applied Biology, Lancashire Polytechnic, Preston
Alwyn Griffiths, B.Sc., Ph.D.
Senior Lecturer in Biochemistry, Oxford Polytechnic, Oxford
M. Rosalind Jenkins, M.Sc., Ph.D.
Senior Lecturer in Biochemistry, The Hatfield Polytechnic, Hatfield
Ivor Simpkins, B.Sc., Ph.D.
Principal Lecturer in Biochemistry, The Hatfield Polytechnic, Hatfield
Keith Wilson, B.Sc., Ph.D.
Head of Division of Biological and Environmental Sciences, The Hatfield Polytechnic, Hatfield

Abbreviations and SI units

The following abbreviations have been used throughout this book without definition:

AMP	Adenosine 5′-monophosphate
ADP	Adenosine 5′-diphosphate
ATP	Adenosine 5′-triphosphate
DDT	2,2-bis-(*p*-chlorophenyl)-1,1,1-trichlorethane
DNA	Deoxyribonucleic acid
EDTA	Ethylenediaminetetra-acetate
FAD	Flavin adenine dinucleotide
FMN	Flavin mononucleotide
mol. wt.	Molecular weight
NAD^+	Nicotinamide adenine dinucleotide (oxidised)
NADH	Nicotinamide adenine dinucleotide (reduced)
$NADP^+$	Nicotinamide adenine dinucleotide phosphate (oxidised)
NADPH	Nicotinamide adenine dinucleotide phosphate (reduced)
Pi	inorganic phosphate
PPi	inorganic pyrophosphate
RNA	Ribonucleic acid
s.t.p.	Standard temperature and pressure
Tris	2-amino-2-hydroxymethyl propane-1,3-diol
e.m.f.	Electromotive force
e^-	Electron

Basic SI Units (Système International D'Unités)

Physical Quantity	*Name of SI Unit*	*Symbol*
Length	metre	m
Mass	kilogramme	kg
Time	second	s
Electric current	ampere	A
Thermodynamic temperature	kelvin	K
Amount of substance	mole	mol

Derived Units

Defined in terms of basic SI Units or other derived units.

Physical Quantity	*Name of Unit*	*Symbol*
Energy	joule	J
Force	newton	N
Pressure	pascal	Pa
Power	watt	W
Electric charge	coulomb	C
Electric potential difference	volt	V
Electric resistance	ohm	Ω
Frequency	hertz	Hz
Magnetic flux density	tesla	T
Area	square metre	m^2
Volume	cubic metre	m^3
Velocity	metre per second	$m\ s^{-1}$
Acceleration	metre per second squared	$m\ s^{-2}$
Density	kilogramme per cubic metre	$kg\ m^{-3}$
Electric field strength	volt per metre	$V\ m^{-1}$
Concentration	mole per cubic metre	$mol\ m^{-3}$
Magnetic field strength	ampere per metre	$A\ m^{-1}$
Dipole moment	coulomb metre	C m
Entropy	joule per kelvin	$J\ K^{-1}$

Volume

The SI unit of volume is the cubic metre, m^3. The litre has been redefined as being exactly equal to the cubic decimetre. Although the term litre still remains in common usage, it is recommended that both the litre and fractions of it (e.g. millilitre) are abandoned in exact scientific work.

$$1 \text{ litre (l)} = 1\ dm^3 = 10^{-3}\ m^3$$
$$1 \text{ millilitre (ml)} = 1\ cm^3 = 10^{-6}\ m^3$$
$$1 \text{ microlitre } (\mu l) = 1\ mm^3 = 10^{-9}\ m^3$$

Powers of Units – Prefixes

Multiple	*Prefix*	*Symbol*
10^9	giga	G
10^6	mega	M
10^3	kilo	k
10^2	hecto	h
10	deca	da

10^{-1}	deci	d
10^{-2}	centi	c
10^{-3}	milli	m
10^{-6}	micro	μ
10^{-9}	nano	n
10^{-12}	pico	p
10^{-15}	femto	f

Conversion Table for Common Units to SI Equivalents

Unit	*SI equivalent*
ångström (Å)	100 pm = 10^{-10} m
atmosphere (standard) (760 mmHg at s.t.p.)	101 325 Pa
calorie	4.186 J
centigrade (°C)	(t°C + 273) K
Curie, Ci	3.7×10^{10} s^{-1}
erg	10^{-7} J
gauss (G)	10^{-4} T
micron, μ	1 μm
millimetre mercury (mmHg)	133.322 Pa
pound-force/sq in (lb f in^{-2}) (p.s.i.)	6894.76 Pa
ln x	2.303 $\log_{10} x$

Values of some physical constants in SI Units

Gas constant (R)	8.314 J K^{-1} mol^{-1}
Planck constant (h)	6.63×10^{-34} J s
Molar volume of ideal gas at s.t.p.	22.41 dm^3 mol^{-1}
Faraday constant (F)	9.648×10^4 C mol^{-1}
Speed of light in a vacuum (C)	2.997×10^8 m s^{-1}

1
General principles of biochemical investigations

1.1 Introduction

Biochemistry is concerned with understanding and exploiting the chemical unity and diversity of living organisms and seeks to correlate chemical structure with biological reactivity at the molecular, subcellular, cellular and organism level.

Biochemical studies have substantiated the cell as the fundamental unit of life, since it alone possesses all the characteristics for independent energy transformation and replication. A unifying feature of all cells is that they contain many common chemical constituents, common metabolic pathways, and common mechanisms of cellular regulation. For example, only twenty different amino acids are found in proteins, and the membranes obtained from different organelles or even species are similar in phospholipid composition. There is overall similarity also in the chemical structure and function of enzymes and in the various metabolic pathways associated with the synthesis and degradation of carbohydrate, lipid, protein and nucleic acids. These unifying principles have assisted in the development of theories on biochemical evolution and phylogenetic interrelationships between organisms (*comparative biochemistry*) and they form the basis for a mode of biochemical deduction, based on extrapolation of results obtained in one species (usually of lower phylogenetic order) to another. Thus micro-organisms, animal tissue cultures or laboratory animals are frequently used for monitoring the biochemical, physiological, pharmacological or toxicological responses to foreign exogenous compounds (xenobiotics) as a prelude to their use in humans. However, this approach must be treated with caution since biological variation between cell types or species is possible and there may be gross physiological differences, particularly between unicellular and multicellular species.

Metabolism depends on the enzymatically coupled turnover of a relatively few energy-rich *group transfer* molecules (such as certain acyl phosphates, nucleoside diphosphates and triphosphates, and enoyl phosphates) and strongly reducing substances, generated in catabolism, (such as reduced pyridine nucleotides and flavin nucleotides) being used to overcome thermodynamic barriers in biosynthesis. Nutritional classification is based on both the external source of electrons for reduction purposes and the energy source. Organisms which rely on inorganic electron donors are said to be *lithotrophic* whilst those which rely on organic sources are said to be *organotrophic*. To each of these classes may be added the prefix *photo* if energy is provided by light within the visible and far-red region of the spectrum or *chemo* if energy is provided by oxidation of either organic or inorganic compounds.

Eukaryotes exhibit a much narrower range of nutritional types than prokaryotes but display great heterogeneity in differentiation of cells, tissues and organs which perform especial physiological functions. Each different cell type in a multicellular organism must reflect accompanying biochemical and physiological differences operating within these cells and invoke mutual cooperation of cells in physiological processes. A large part of developmental biochemistry is concerned with elucidating, at the molecular level, the mechanisms of selective gene expression leading to differentiation.

Essentially two types of biochemical investigation are possible. The *in vivo* technique uses intact, whole organisms (plants or animals) or, alternatively, parts of animals subjected to perfusion techniques, to maintain as far as possible the integrity of tissues. The advantage of the *in vivo* method is that artefacts are reduced but often it does not permit precise analysis to be undertaken because of permeability barriers, the complexities of metabolism associated with the multicellular state and mutual cell interference. The *in vitro* method involves the incubation of biologically-derived material in artificial physical and chemical environments. The term is equally applicable to enzyme preparations, to isolated organelles, to intact microorganisms and to excised parts of animals or plants. Conditions may be chosen to promote a limited degree of growth, differentiation and development as for instance in *cell, tissue and organ culture* of animals and plants. The specific advantage of cell and tissue culture methods is that they reduce the physiological and biochemical constraints imposed by contiguous cells. The approach has found widespread application in the biosciences. In its most fundamental sense, cell culture facilitates the investigation of the developmental potential (*totipotency*) of cells, i.e. the capacity, within the limits of its genetic constitution, of one cell to form any other type of cell, given an appropriate artificial chemical and physical environment. A general criticism of experimentation *in vitro* is that extrapolation to the situation *in vivo* may be unjustified, i.e. that the methodology *in vitro* is the study of artefacts.

Biochemical investigations frequently require the purification of a particular compound from a complex mixture. In *analytical separations*, the objective is to identify and estimate small amounts of the compounds and frequently it is not necessary to recover the compound after the separation process. In *preparative separations* the main aim is to isolate and recover as large an amount as possible of the compound in a high degree of purity, in order, subsequently, to study its chemistry and/or its biological properties. Whether an analytical or preparative approach is being adopted may well dictate the choice of separation and purification techniques, mainly because a preparative approach requires much larger amounts of starting material and need not employ techniques which give a high percentage recovery.

Analytical separations may be used on a qualitative basis as a short cut to identifying the components present in a mixture. If a compound is suspected to be present in a mixture, then if its behaviour during analytical separation is identical with the behaviour of a known compound (a *standard*), the two compounds are likely to be identical. In assessing the behaviour of the standard and the unknown as many high resolution techniques as possible should be used in which the standard should undergo exactly the same treatment, preferably at the same time, as the unknown compound. This technique removes the need for analysing the compound by physical methods, after separation, in order to establish its structure.

It is not always necessary to carry out preliminary isolation and purification of molecules before estimating their concentration in biological material if they possess some unique property which may be used as a basis for their quantitative determination. Such unique features are an exemplification of biological specificity and are illustrated in the assay of enzymes in crude extracts, in intact cells and in cellular organelles. Some spectrophotometric methods by which enzyme assays may be carried out are given in Sections 3.4 and 8.2.3. In cases where such specificity is not exhibited, then some degree of purification of the molecule is required before it can be analysed. Any purification procedures adopted will inevitably involve some loss of material. It is essential that the number of purification stages should be kept to a minimum and therefore the techniques employed should be those which are capable of giving the greatest resolution with small amounts of material. Some of the techniques, described in later chapters, such as chromatography (Chapter 6) and electrophoresis (Chapter 7) are particularly suitable.

High resolution methods of separation normally have a limited capacity and cannot be used to separate large amounts of material. The normal procedure in preparative separations, therefore, involves a first stage with a high separation capacity but which may have low resolution. Techniques such as precipitation, dialysis, adsorption, liquid-liquid chromatography and partition column chromatography, countercurrent distribution and continuous electrophoresis, all of which are discussed in subsequent chapters, are important because, although they may not achieve as good a resolution as other methods, they can be scaled-up to cope with large amounts of material. Some of the high resolution methods, for example ion-exchange, exclusion and affinity chromatography, preparative GLC and HPLC and preparative gel electrophoresis may then be employed for the final purification steps. Sequential methods used for separation should be based on as many different properties of the molecule as possible since comparatively little further purification is likely to be obtained by using another method which exploits the same physical property for separation.

When confronted with the problem of choosing a particular method (or methods) to purify a compound, it is obvious that although several techniques may be theoretically applicable, in practice certain techniques may be superior. Several factors must be taken into consideration when deciding which techniques to adopt. These include:

(i) whether the separation is to be carried out on an analytical or preparative basis so that the resolution and/or capacity of the techniques are considered;
(ii) the physical properties of the compound, for example solubility, volatility, molecular weight and charge properties;
(iii) the stability of the compound during a particular treatment;
(iv) a consideration of comparable successful separations documented in the literature;
(v) the availability of equipment and expertise;
(iv) the cost of the procedure.

1.2 pH and buffers

1.2.1 Effect of pH on biological processes

Organisms and cells can generally withstand large variations in the pH of their external environment. In contrast, cellular processes are sensitive to pH changes and take place in a medium the pH of which is carefully regulated. There may, however, be localised intracellular pH variations, for example at membrane surfaces. The majority of intracellular processes occur at a pH maintained near neutrality. This neutral pH is generally the one at which the various metabolic processes occur at their maximum rate. The hydrolases of lysosomes, however, have their maximum activity at a pH in the region of 5, a pH which prevails following the death of the cell. The gastric juice of mammals is quite exceptional in having a pH of approximately 1, a pH at which the enzyme pepsin, which initiates the digestion of dietary proteins in the stomach, has its maximum activity.

The control of a virtually constant pH in biological systems is achieved by the action of efficient buffering systems whose chemical nature is such that they can resist pH changes due to the metabolic production of acids such as lactic acid and bases such as ammonia. The major buffering systems found in cellular fluids involve phosphate, bicarbonate, amino acids and proteins.

The sensitivity of biological processes to pH may be due to one of several possible reasons. The process may be catalysed by hydrogen ions or may involve a hydrogen ion as a reactant or product. Alternatively, a pH change may alter the distribution of a compound or ion across a membrane possibly by altering the permeability of the membrane. Membranes, like many other biological structures and molecules, possess ionisable groups whose precise state of ionisation influences their molecular conformation and thus their biological activity. This is particularly true of proteins and thus of enzymes. Some proteins rely on a slight change of the pH of their environment to complete their biological function. In the case of haemoglobin, whose primary function is to transport oxygen from the lungs to the tissues, a slight decrease in the pH of the tissues due to carbon dioxide and hydrogen ion production as a result of the active respiration of the tissue, helps to facilitate the unloading of the oxygen when and where it is required. The unloading of oxygen is accompanied by the uptake of protons by the haemoglobin, thus simultaneously helping to buffer the system.

In vitro studies of metabolic processes require the use of buffers which may not be physiological. Deliberate changes in pH, however, may help in the analytical study of certain groups of molecules, e.g. amino acids, proteins and nucleic acids, by such techniques as electrophoresis and ion-exchange chromatography.

1.2.2 The pH-dependent ionisation of amino acids and proteins

The state of ionisation of a weak electrolyte is dependent upon the prevailing pH and the numerical value of its ionisation constant. For weak acids which ionise according to the equation:

$$\underset{\text{acid}}{RCOOH} \rightleftharpoons \underset{\text{conjugate base}}{RCOO^-} + H^+$$

the ionisation constant K_a, is given by the expression

$$K_a = \frac{[RCOO^-][H^+]}{[RCOOH]}$$

In the case of weak bases such as amines, which ionise according to the equation:

$$\underset{\text{base}}{RNH_2} + H_2O \rightleftharpoons \underset{\text{conjugate acid}}{RNH_3^+} + OH^-$$

the ionisation constant may be expressed in terms of a K_b value

$$K_b = \frac{[RNH_3^+][OH^-]}{[RNH_2][H_2O]}$$

or more commonly in terms of the K_a value of the conjugate acid

$$K_a = \frac{[RNH_2][H^+]}{[RNH_3^+]}$$

In such cases, the product of K_a and K_b for a weak base is equal to K_w, the ion product of water.

In practice, since K_a values are numerically very small, it is customary to use pK_a values, where pK_a = $-\log_{10} K_a$. It follows that for weak bases pK_a + pK_b = 14. The precise way in which the state of ionisation of a weak electrolyte varies with pH is given by the *Henderson–Hasselbalch equation*. For a weak acid, this takes the form:

$$\mathrm{pH} = \mathrm{p}K_a + \log_{10} \frac{[\text{conjugate base}]}{[\text{acid}]}$$

or

$$\mathrm{pH} = \mathrm{p}K_a + \log_{10} \frac{[\text{ionised form}]}{[\text{unionised form}]} \qquad 1.1$$

In the case of a weak base, expressing the equation in terms of the ionisation of the conjugate acid we get:

$$\mathrm{pH} = \mathrm{p}K_a + \log_{10} \frac{[\text{base}]}{[\text{conjugate acid}]}$$

or

$$\mathrm{pH} = \mathrm{p}K_a + \log_{10} \frac{[\text{unionised form}]}{[\text{ionised form}]} \qquad 1.2$$

It can be appreciated from these equations that weak acids will be predominantly unionised at low pH values and ionised at high pH values. The exact opposite is the case for weak bases; at low pH values the conjugate acid will predominate and at high pH values the unionised free base will exist. This sensitivity to pH of the state of ionisation of weak electrolytes is important physiologically, for example in the absorption by passive diffusion from the gastrointestinal tract of weak electrolytes and their excretion by the kidney, as well as in *in vitro* investigations employing such techniques as electrophoresis and ion-exchange chromatography.

The α-amino acids which are incorporated into proteins are all weak electrolytes and, with the exception of proline, conform to the general formula $RCH(NH_2)COOH$. Since they possess both an amino group and a carboxylic acid group, they are ionised at all pH values, i.e. a neutral species represented by the general formula does not exist in solution irrespective of the pH. This can be seen as follows:

$$\begin{array}{ccccc} \mathrm{R} & & \mathrm{R} & & \mathrm{R} \\ | & & | & & | \\ \mathrm{CH{-}\overset{+}{N}H_3} & \underset{}{\overset{\mathrm{p}K_{a_1}}{\rightleftharpoons}} & \mathrm{CH{-}\overset{+}{N}H_3} & \overset{\mathrm{p}K_{a_2}}{\rightleftharpoons} & \mathrm{CH{-}NH_2} \\ | & & | & & | \\ \mathrm{COOH} & & \mathrm{COO^-} & & \mathrm{COO^-} \end{array}$$

net positive charge — zero net charge 'zwitterion' — net negative charge

increasing pH →

Thus at low pH values an amino acid exists as a cation and at high pH values as an anion. At a particular intermediate pH the amino acid carries no net charge and is called a *zwitterion*. It has been shown that in the crystalline state and in solution in pure water, such amino acids exist predominantly as this zwitterionic form. This confers upon them physical properties characteristic of ionic compounds, i.e. high melting point and boiling point, water solubility and low solubility in such organic solvents as ether and chloroform. The pH at which the zwitterion predominates in aqueous solution is referred to as the *isoionic point* since it is the pH at which the number of negative charges on the molecule produced by protolysis is equal to the number of positive charges acquired by proton acceptance. In the case of amino acids this is equal to the *isoelectric point* (pI) since the molecule carries no net charge and is therefore electrophoretically immobile. The numerical value of this pH is related to the acid strength by the equation:

$$\mathrm{pH} = \mathrm{pI} = \frac{\mathrm{p}K_{a_1} + \mathrm{p}K_{a_2}}{2}$$

In the case of glycine, $\mathrm{p}K_{a_1}$ and $\mathrm{p}K_{a_2}$ are 2.3 and 9.6 respectively so that the isoionic point is 6.0. At pH values below this, the cation and zwitterion will co-exist in a ratio determined by the Henderson–Hasselbalch equation, whereas at higher pH values the zwitterion and anion will co-exist.

For a so-called acidic amino acid such as aspartic acid, the ionisation pattern is different owing to the presence of a second carboxyl group.

$$\begin{array}{ccccccc}
\mathrm{COOH} & & \mathrm{COOH} & & \mathrm{COO^-} & & \mathrm{COO^-} \\
| & & | & & | & & | \\
\mathrm{CH_2} & & \mathrm{CH_2} & & \mathrm{CH_2} & & \mathrm{CH_2} \\
| & & | & & | & & | \\
\mathrm{CH{-}\overset{+}{N}H_3} & \underset{2.1}{\overset{\mathrm{p}K_{a_1}}{\rightleftharpoons}} & \mathrm{CH{-}\overset{+}{N}H_3} & \underset{3.9}{\overset{\mathrm{p}K_{a_2}}{\rightleftharpoons}} & \mathrm{CH{-}\overset{+}{N}H_3} & \underset{9.8}{\overset{\mathrm{p}K_{a_3}}{\rightleftharpoons}} & \mathrm{CH{-}NH_2} \\
| & & | & & | & & | \\
\mathrm{COOH} & & \mathrm{COO^-} & & \mathrm{COO^-} & & \mathrm{COO^-}
\end{array}$$

cation	zwitterion	anion	anion
(1 net positive charge)	pH 3.0 (isoionic point)	(1 net negative charge)	(2 net negative charge)

In this case, the zwitterion will predominate in aqueous solution at a pH determined by $\mathrm{p}K_{a_1}$ and $\mathrm{p}K_{a_2}$.

In the case of lysine, which is a basic amino acid, the ionisation pattern is different again and its isoionic point is determined by $\mathrm{p}K_{a_2}$ and $\mathrm{p}K_{a_3}$.

$$\begin{array}{ccccccc}
\mathrm{\overset{+}{N}H_3} & & \mathrm{\overset{+}{N}H_3} & & \mathrm{\overset{+}{N}H_3} & & \mathrm{NH_2} \\
| & & | & & | & & | \\
\mathrm{(CH_2)_4} & & \mathrm{(CH_2)_4} & & \mathrm{(CH_2)_4} & & \mathrm{(CH_2)_4} \\
| & & | & & | & & | \\
\mathrm{CH{-}\overset{+}{N}H_3} & \underset{2.2}{\overset{\mathrm{p}K_{a_1}}{\rightleftharpoons}} & \mathrm{CH{-}\overset{+}{N}H_3} & \underset{9.0}{\overset{\mathrm{p}K_{a_2}}{\rightleftharpoons}} & \mathrm{CH{-}NH_2} & \underset{10.5}{\overset{\mathrm{p}K_{a_3}}{\rightleftharpoons}} & \mathrm{CH{-}NH_2} \\
| & & | & & | & & | \\
\mathrm{COOH} & & \mathrm{COO^-} & & \mathrm{COO^-} & & \mathrm{COO^-}
\end{array}$$

cation	cation	zwitterion	anion
(2 net positive charges)	(1 net positive charge)	pH 9.8 (isoionic point)	(1 net negative charge)

As an alternative to possessing a second amino or carboxyl group, an amino acid side chain may contain a quite different chemical group R which is also capable of ionising at a characteristic pH. Such groups include a phenolic group (tyrosine), guanidino group (arginine), imidazolyl group (histidine) and sulphydryl group (cysteine) (Table 1.1). It is clear that the state of ionisation of the main groups of amino acids will be grossly different at a particular pH. Moreover, even within a given group there will be minor differences due to the precise nature of the R group. These differences are exploited in the electrophoretic and ion-exchange separation of mixtures of amino acids such as those present in a protein hydrolysate.

The ionisation of protein molecules is qualitatively similar to that of amino acids, but quantitatively different due to the very large number of ionisable groups present. Proteins are formed by the condensation of the α-amino group of one amino acid with the α-carboxyl of the adjacent amino acid. With the exception of the two terminal amino acids, therefore, the α-amino and carboxyl groups are all involved in peptide bonds and are no longer ionisable in the protein. Amino and carboxyl groups in the side chain are, however, free to ionise and of course there may be hundreds of these. Some of them may be involved in electrostatic attractions which help to stabilise the three-dimensional structure of the protein molecule. Proteins frequently fold in such a manner that the majority of these ionisable groups are on the outside of the molecule where they can interact with the surrounding medium. The relative numbers of positive and negative groups in a protein molecule naturally influence its physical behaviour. The histones have a predominant number of cationic groups, whereas other proteins may have either roughly equal numbers of anionic and cationic groups or an excess of anionic groups.

Table 1.1 Ionisable groups found in proteins.

Amino acid group	pH-dependent ionisation	Approximate pK_a
N-terminal α-amino	$-\overset{+}{N}H_3 \rightleftharpoons NH_2 + H^+$	8.0
C-terminal α-carboxyl	$-COOH \rightleftharpoons COO^- + H^+$	3.0
Asp-β carboxyl	$-CH_2COOH \rightleftharpoons CH_2COO^- + H^+$	3.9
Glu-γ carboxyl	$-(CH_2)_2COOH \rightleftharpoons (CH_2)_2COO^- + H^+$	4.1
His-imidazole	$-CH_2$-imidazolium (HN, $\overset{+}{}$, NH) $\rightleftharpoons$ $-CH_2$-imidazole (N, NH) $+ H^+$	6.0
Cys-sulphydryl	$-CH_2SH \rightleftharpoons -CH_2S^- + H^+$	8.4
Tyr-phenolic	$-C_6H_4-OH \rightleftharpoons -C_6H_4-O^- + H^+$	10.1
Lys-ε amino	$-(CH_2)_4\overset{+}{N}H_3 \rightleftharpoons -(CH_2)_4NH_2 + H^+$	10.8
Arg-guanidino	$-NH-C(=\overset{+}{N}H_2)-NH_2 \rightleftharpoons -NH-C(=NH)-NH_2 + H^+$	12.5

The isoionic point of a protein and its isoelectric point, unlike that of an amino acid, are generally not identical. This is because, by definition, the isoionic point is the pH at which the protein molecule possesses an equal number of positive and negative groups formed by the association of basic groups with protons and dissociation of acidic groups respectively. In contrast, the isoelectric point is the pH at which the protein is electrophoretically immobile. In order to determine this, the protein must be dissolved in a buffered medium containing low molecular weight anions and cations which are capable of binding to the multi-ionised protein. Hence the observed balance of charges at the isoelectric point would partly be due to there being more mobile anions (or cations) than bound cations (or anions) at this pH. This could mask an imbalance of charges on the actual protein.

In practice, protein molecules are always studied in buffered solutions; hence it is the isoelectric point which is important. It is the pH at which, for example, the protein has minimum solubility since it is the point at which there is greatest opportunity for attraction between oppositely charged groups of neighbouring molecules leading to aggregation and easy precipitation.

Aggregation and consequently precipitation of protein molecules can also be brought about by the addition to the protein solution of such salts as ammonium sulphate (Section 3.2.3). The protein molecules and the inorganic ions compete for the water molecules for their hydration. At a characteristic salt concentration, protein–protein interactions become predominant over protein–water interactions and aggregation and precipitation occur. This technique is widely used in the preliminary purification of proteins. Differences in the isoelectric point of different protein molecules are also exploited in their separation by electrophoretic and ion-exchange chromatographic means.

1.2.3 Buffer solutions for biological investigations

A buffer solution is one which resists a change in hydrogen ion concentration on the addition of acid or alkali. This resistance is called the *buffer action*. The magnitude of the buffer action is called the *buffer capacity* (β), and is measured by the amount of strong base required to alter the pH by one unit:

$$\beta = \frac{\mathrm{db}}{\mathrm{d(pH)}}$$

where d(pH) is the increase in pH resulting from the addition db of base.

In practice, buffer solutions usually consist of a mixture of a weak acid or base and its salt, e.g. acetic acid and sodium acetate (on the Brönsted and Lowry nomenclature, a mixture of a weak acid and its conjugate base).

In a solution of a weak acid (RCOOH) and its salt ($RCOO^-$), added hydrogen ions are neutralised by the anions of the salt, which therefore act as a weak base, and conversely, added hydroxyl ions are removed by neutralisation of the acid. It is clear from this that the buffer capacity of a particular acid and its conjugate base will be a maximum when their concentrations are equal, i.e. when $pH = pK_a$ of the acid. Buffer capacity also depends upon the total concentration as well as the ratio of acid and salt - the greater the total concentration, the greater the buffer capacity. The usual concentration of acid and salt in buffer solutions is of the order of 0.05–0.20 M and generally the mixtures possess acceptable buffer capacity within the range $pH = pK_a \pm 1$.

The criteria for buffers suitable for use in biological research may be summarised as follows. They should:

(i) possess adequate buffer capacity in the required pH range;
(ii) be available in a high degree of purity;
(iii) be very water soluble and impermeable to biological membranes;
(iv) be enzymatically and hydrolytically stable;
(v) possess a pH which is minimally influenced by their concentration, temperature and ionic composition or salt effect of the medium;
(vi) not be toxic or biological inhibitors;
(vii) only form complexes with cations that are soluble;
(viii) not absorb light in the visible or ultraviolet regions.

Needless to say, not all buffers which are commonly used meet all these criteria. Thus, phosphates tend to precipitate polyvalent cations and are often metabolites or inhibitors in many systems and Tris is often toxic or may have inhibitory effects. However, a number of zwitterionic buffers of the HEPES and PIPES type fulfil the requirements and are often used in tissue culture media containing sodium bicarbonate or phosphate solutions as nutrients. They suffer from the disadvantage that they interfere with Lowry protein determinations (Section 3.2.2).

Physiologically one of the most important groups of buffers is the proteins. By virtue of their large numbers of weak acidic and basic groups in the amino acid side chains, proteins have a very high buffer capacity. Haemoglobin is mainly responsible for the buffer capacity of blood.

Some of the commonly used buffers are listed in Table 1.2. To obtain buffer solutions covering an extended pH range, but which are derived from the same ions, mixtures of different systems may be used. Thus the McIlvaine buffers cover the pH range 2.2 to 8.0 and are prepared from citric acid and disodium hydrogen phosphate.

1.3 Physiological solutions

1.3.1 Introduction

Physiological solutions used in either *in vivo* or *in vitro* experimentation should closely resemble the properties of the intact tissue in question. The design of effective physiological solutions is often based on the results of chemical investigations of cellular fluids and on the responses of cells to

Table 1.2 pK_a values of some acids and bases which are commonly used as the basis for buffer solutions.

Acid or base	pK_a (at 25°C)
Acetic acid	4.75
Barbituric acid	3.98
Carbonic acid	6.10, 10.22
Citric acid	3.10, 4.76, 5.40
Glycylglycine	3.06, 8.13
*HEPES	7.50
Phosphoric acid	1.96, 6.70, 12.30
Phthalic acid	2.90, 5.51
†PIPES	6.80
Succinic acid	4.18, 5.56
Tartaric acid	2.96, 4.16
‡Tris	8.14

*HEPES = N-2-hydroxyethylpiperazine-N′-2-ethanesulphonic acid
†PIPES = piperazine-N-N′-bis (2-ethanesulphonic acid)
‡Tris = 2-amino-2-hydroxymethylpropane-1, 3-diol

nutrient solutions. Physiological solutions may be broadly classified according to their use for either short-term incubations or for supporting cell and tissue culture. In either case, however, it is essential that the suspending medium is *isotonic* with the tissue (i.e. possesses the same osmotic pressure) and is buffered to maintain metabolic integrity.

1.3.2 Microbial media

The composition of the medium for incubation of microorganisms *in vitro* is determined largely by their nutritional classification i.e. whether a chemolithotroph, chemoorganotroph, photolithotroph or photoorganotroph, and whether or not they are wild-type or mutant cell lines. Typically, a *minimal* medium for the growth of a fully biosynthetically competent (*prototrophic*) chemoorganotroph would contain salts of Na^+, K^+, Ca^{2+}, Mg^{2+}, NH_4^+ or NO_3^-, Cl^-, HPO_4^{2-}, and SO_4^{2-} and a simple carbon source such as glucose. A chemoorganotroph mutated such that synthesis of alanine is prevented would be an alanine *auxotroph* and as such must be supplied with alanine in the culture medium. Different auxotrophic mutants may be exploited in the elucidation of metabolic pathways. Complex organic supplements may sometimes be added to accelerate the growth rate of chemoorganotrophs in culture or when the nutritional requirements are ill-defined. The gelling agent agar, may be added to solidify the medium to facilitate the surface growth of microorganisms.

1.3.3 Higher plant media

The mineral element composition of a nutrient solution supporting the growth of plants in hydroponic incubution (Section 1.4.2) is similar to that used for microorganisms except that it usually also contains additional essential microelements such as Mn^{2+}, B^{3+}, Zn^{2+}, Cu^{2+} and Mo^{2+}.

A variety of different types of medium is available for the proliferation of plant cells in culture. In addition to a balanced mixture of macro and micro-elements, carbon sources are provided since higher plant cells are normally photosynthetically incompetent in culture. Nitrogen is supplied as either NO_3^- or NH_4^+ or as an organic nitrogen supplement such as urea or glutamine. Vitamin inclusions are normally from the B group. Most non-tumour cultures require auxin-like regulators which promote cell expansion, such as naturally occurring indole-3-acetic acid (IAA) or synthetic compounds such as naphthalene acetic acid (NAA) or 2,4-dichlorophenoxyacetic acid (2,4-D). Cytokinins, included to stimulate cell division, may include naturally occurring zeatin or synthetic analogues such as 6-furfurylaminopurine (kinetin). The induction and maintenance of growth depends primarily on an appropriate balance of hormones rather than on their absolute concentrations. The majority of media currently used are chemically-defined but some may include complexes such as coconut milk or yeast extract.

1.3.4 Animal media

Numerous physiological salt solutions have been developed for short-term work with animal tissues, e.g. Tyrode's, Young's, Locke's, Meng's and Da Jalon's, many being derived from Ringer's solution which was one of the first to be formulated. Two of the most common are Krebs-Ringer bicarbonate, the bicarbonate solution having been designed, originally, to approximate to the ionic composition of mammalian serum. Typically such salt solutions contain NaCl, KCl, $MgSO_4$, $CaCl_2$, $NaHCO_3$, and KH_2PO_4 in various amounts and may be gassed with O_2–CO_2 mixtures.

Media supporting the long term maintenance of animal cells in culture must contain a balanced salt solution, vitamin supplements, a carbon source and amino acids. Protein inclusions may include plasma or serum which are complex physiological mixtures of ill-defined composition. Hormone supplements are not as important for animal cells as for plant tissue culture. Antibiotics such as streptomycin and penicillin, antifungal substances such as griseofulvin and antimycoplasmal agents such gentimycin may be included to prevent microbial contamination.

1.3.5 Media for organelle isolation and tissue homogenisation

In addition to a buffered, balanced salt solution, media usually contain a carbon source. Sucrose is commonly included but it interferes with certain enzyme assays and for this reason is often replaced with mannitol which is non-metabolisable by plant and animal cells.·There are many individual recipes used to preserve organelle integrity and to prevent enzyme inactivation, and these differ in details such as carbon source, the presence or absence of chelating agents such as EDTA, and sulphydryl protecting-agents such as reduced glutathione or 2-mercaptoethanol. Non-ionic media are sometimes preferred to saline solutions which often agglutinate polymorphonuclear leucocytes and liver organelles. Citric acid has been used in the isolation of nuclei and chromosomes because of its ability to inhibit the activity of neutral deoxyribonucleases. Solutions of glycerol and ethyleneglycol have been used in the isolation of nuclei and the carbowaxes (ethyleneglycol polymers) for the preparation of plastids from plant cells.

One of the factors which makes enzyme assays more difficult in plant extracts is that large amounts of phenols are often released by the cells during the homogenisation process. When oxidised to quinones via peroxidase and polyphenol oxidase these phenols form hydrogen bonds with the carbonyl groups of peptide bonds in proteins, and thus inactivate many enzymes. To prevent this inactivation polyvinylpyrrolidone (PVP) is often added to plant extracts because it forms an insoluble complex with the phenols which can then be removed by filtration. Non-aqueous media may also be used in the isolation of organelles. The suspending fluid is usually a blend of a light and a heavy organic solvent such as ether-chloroform or benzene-carbon tetrachloride. The density of the medium can be varied so that the required organelles either float or sediment from the remainder of the homogenate in the subsequent centrifugation stages. Use of non-aqueous fractionation procedures has been made in the preparation of chloroplasts and haemosiderin granules from spleen. The technique does have some disadvantages in that morphological alterations may occur in some tissues and most enzymes are inactivated by organic solvents.

1.4 Whole organism techniques

1.4.1 Animal studies

The use of laboratory animals for biochemical experiments *in vivo* is very extensive, since alternative model techniques such as cell culture are often considered physiologically too simplistic to meet the needs of the investigation. Animals are used for fundamental research, for medical and clinical research, for veterinary and agricultural research, in the manufacture of vaccines, antibodies and hormones, in testing the potency of drugs or other biological products where chemical determination is not feasible, and also very widely in toxicological testing. Much animal work is directed towards the study of the metabolism of xenobiotics (drugs, food additives) and extrapolation of results to humans as a prerequisite to clinical trials. Mice, rats and guinea-pigs are most commonly used because of their lower cost and ease of handling, but rabbits, cats, dogs, marmosets, monkeys and baboons are also used. The animals are bred specifically for experimental purposes and in the UK their husbandry is subject to scrutiny by the Home Office Inspectorate, who issue licences for work involving live vertebrates. Age, sex, nutritional state, strain, circadian rhythms and stress state of animals, each have to be closely regulated to minimise variability in results. Experiments should be performed, therefore, on small groups of animals, as physiologically similar as possible, and the collated results analysed statistically. When metabolic responses to xenobiotics are being monitored, the results obtained must be compared with a control group which has received a *placebo* (a completely inert substance such as lactose). In man or a laboratory animal the fate of an administered compound, the xenobiotic, can be traced by monitoring the concentrations of the compound and/or its metabolites in blood, urine, faeces, bile, expired air, sweat or saliva. Excretion patterns of small laboratory animals are best studied by placing the animal in a metabolic cage for the duration of the experiment. Such cages allow urine and faeces to be collected separately and expired carbon dioxide to be trapped. If ^{14}C-labelled compounds are used, the cage is supplied with CO_2-free air and the expired $^{14}CO_2$ absorbed in sodium hydroxide solution. The assay of $^{14}CO_2$ expired allows an estimate of the extent of complete degradation of the compound to be made. Since many excretion products in urine and faeces are conjugated (i.e. linked to polar molecules such as glucuronic acid, sulphate and glycine) it is essential that samples are hydrolysed, enzymatically or chemically, before the free metabolites are extracted and identified by established analytical techniques such as GLC and HPLC.

The technique of whole body autoradiography (Section 9.2.4) has proved very useful as a method of visualising the distribution of radioactive materials administered to small animals. Experimental information on distribution, relative accumulation in tissues, rates of excretion and ability to cross biological membranes is obtainable by this technique. At a suitable time interval after injection with a radioactively-labelled compound, the animal is killed by anaesthetisation and rapidly cooled in an acetone-solid CO_2 mixture at $-78°C$ or in liquid N_2. The frozen animal is embedded in an aqueous solution of resin (gum acacia) at low temperature. After the resin has set, the animal may either be machined to a suitable level with a cutter attached to an electric drill, or serial sections made using a microtome which has a special tungsten carbide blade. The prepared animal surface is placed in close contact with X-ray film at a low temperature for one to two weeks after which time the film is developed. (Fig. 9.11).

The fate of a compound in a living laboratory animal may also be studied through the vascular perfusion of an organ such as the liver or kidney. This procedure involves infusing the compound through a fine hollow needle inserted into the artery carrying blood to the organ and performing subsequent analysis on the blood being transported from that organ by the corresponding vein.

The cost and ethics of whole animal experimentation is leading to increasing effort being directed towards finding alternatives to whole animal experiments and, coincidentally, to reducing the number of animals used for essential experimentation. The main approach being increasingly adopted is to change to *in vitro* methods where appropriate. These methods include the use of isolated vascularly perfused organs (e.g. liver, heart, lungs), isolated organs or tissue slices (e.g. stomach, intestine, pancreas, liver, kidney and eye) and the use of either isolated single cells (e.g. red blood cells, adipocytes, muscle cells, neurones, hepatocytes) or subcellular organelles.

1.4.2 Plant studies

Traditionally higher plants have been used extensively for the study of the biochemistry of such processes as photosynthesis, photorespiration, respiration and nitrogen assimilation which all have a general relationship to plant productivity. The physiology of plants is complex, however, and very little is known in biochemical terms of the factors which control growth and development. The range of secondary plant products is very extensive indeed and is currently being investigated as an existing and potential source of economically important products such as pharmaceuticals, pigments, perfumes, agrochemicals and enzymes. Again, very little is known of how secondary metabolism is regulated in higher plants, and this is greatly restricting attempts to employ genetic engineering practices using genes which code for useful plant products. Suppression of photorespiration and integration of nitrogen fixing *Nif* genes into crop plants, using genetic engineering techniques would, if achievable, be two examples of innovations having revolutionary significance to world agriculture.

Methods of studying the metabolism of whole plants depend largely on the degree of organisation of the plant. Higher plants, for whole plant studies, may be cultivated on a small scale in sterile compost mixtures or in nutrient solutions (*hydroponics*) inside glasshouses or special growth chambers where some of the physical environmental factors such as temperature, light and relative humidity can be controlled. Test compounds may be administered either to the roots by adding the compound to the growth medium, or by foliar application by painting or spraying the compound onto the foliage, or by injection into the plant. The subsequent distribution of the compound and any of its metabolites within the plant may then be monitored by examination of the different parts of the plant. Certain gross physiological changes can be conveniently measured with intact plants, for example, rates of CO_2 fixation may be measured *in vivo* using infra-red gas analysis (Section 8.3) or the effects of potential growth regulators on visible features such as leaf expansion may be monitored. A major difficulty in studying plant metabolism is that plant organs are not as highly differentiated as those of animals and consequently are generally less convenient as a concentrated source of enzymes or metabolic processes.

1.5 Organ and tissue slice techniques

1.5 Perfusion of isolated organs

In this technique an organ such as liver, kidney or heart is removed from an animal and maintained at constant temperature in suitable apparatus. It is not always necessary to remove the organ completely from the animal since perfusion may be carried out on an organ exposed within an anaesthetised animal, in which case the nerve supply and perhaps part of the vascular system may still be intact. In order to study the metabolism of a compound by the organ, the compound is added to the perfusion fluid which is normally infused into the organ through the artery, and analysis made of the fluid emerging through the vein. The metabolic fate of the perfused compound in the organ may then be determined. The perfusing fluid may be passed once only through the organ or may be recirculated several times before finally being analysed. The perfusing fluid may be passed through the organ by gravity feed or by using a small pump. The latter method is obligatory when the perfusing fluid is to be recirculated. In some perfusion systems the flow of fluid through the organ is produced in pulses (*pulsatile flow*) rather than at a continuous pressure. Pulsatile flow resembles the *in vivo* situation more closely than non-pulsatile flow since it parallels the type of flow produced by the heart.

The effect of a compound on a tissue may be determined by studying the mechanical response of the isolated tissue to the compound. The compound is added to the liquid bathing the tissue and the response of the tissue followed by clamping one end of the tissue and attaching the other end to a recording lever which moves in response to any tissue movement. In this way a permanent record can be made of the movement observed. Response of isolated organs to nanogramme quantities of active substances has been obtained by this technique.

1.5.2 Slice techniques

This method consists essentially of the preparation of slices of a particular tissue which are thin enough both to enable oxygen to reach the innermost layers of the slice and for adequate removal of waste products by diffusion. Slices 0.5 to 5 mm thick generally meet the above requirements whilst allowing the proportion of disrupted cells to intact cells to remain small.

The organ under study should be removed immediately after the animal is killed in order to minimise any post-mortem changes. Cutting of the slices is achieved by using razor blades, or for more uniform preparations, by using a microtome. The tissue slices are transferred to a vessel containing a suitable suspending medium (Section 1.3). The metabolism of the slices may then be studied and the effects of added compounds on the metabolic processes determined. The use of tissue slices is suited to manometric studies (Section 1.10) but for studies under aerobic conditions, because slices are not thin enough, about 95% oxygen must be used in the gas phase in order to achieve aeration of the innermost cells of the slices. An obvious disadvantage of this technique, however, is that by trying to ensure adequate diffusion of oxygen to the innermost cells of the slices, toxic concentrations of oxygen may be in contact with the outer layers of cells.

1.6 Cell and tissue culture

1.6.1 Microbial culture

The term microorganism usually includes bacteria, fungi, yeasts, unicellular algae, filamentous algae and protozoa. When grown axenically in culture for experimental or industrial purposes such organisms are effectively whole organisms and, as such, are simpler than animal or plant cells and tissues grown in culture. A large number of biochemical and physiological studies aimed at understanding reactions and processes in the *whole organism* have therefore been conducted using bacteria (e.g. *Escherichia coli*, *Bacillus subtilis*), yeasts (e.g. *Saccharomyces cerivisiae*, *Candida albicans*) and algae (e.g. *Chlorella vulgaris*, *Chlamydomonas dysosmos*). The structural simplicity of bacteria in particular is a great advantage in such work, especially in studies of self-assembly processes and on morphogenesis (e.g. the life cycle of lysogenic bacterial viruses), and on DNA replication, RNA transcription and translation and control of gene expression, since their haploid nature facilitates the use of mutants to elucidate these processes. Types of mutation, their induction, selection and application are dealt with more fully in Section 1.7.

The extrapolation to higher organisms of results obtained using microbial systems is not without its pitfalls but, as mentioned in Section 1.1, biochemistry is the unifying theme of the Biological Sciences. Information obtained using microorganisms provides a basis for comparison with higher organisms notwithstanding inherent difficulties in extrapolation. The basic techniques of microbial cell culture are now being applied, with some modification, to the culture of animal and plant cells and tissues, this in itself being a major advance since such systems have many fundamental and industrial applications (Sections 1.6.2 and 1.6.3).

Besides being used in biochemical and physiological studies microbial cultures have many industrial applications as sources of, for example, alcohol, amino acids, antibiotics, coenzymes, organic acids, polysaccharides, solvents, sterols and converted sterols, sugars, surfactants and vitamins. The ability to genetically engineer microbial cells, as discussed in Chapter 5, is leading to a rapid extention of this list to include foreign compounds (i.e. compounds not naturally produced by the microbe itself). These include protein products from higher organisms (e.g. insulin and human growth factor) since the DNA coding for their synthesis can be isolated and introduced (*engineered*) into a bacterial host which then begins to produce the protein.

Microbial degradation of waste products, particularly those from the agricultural and food industries, is another important industrial application of microbial cells in culture, especially in the degradation of toxic wastes (e.g. pesticides) or in the bioconversion of wastes to useful end products (e.g. manure conversion to methane by methanogenic bacteria including conversion of cellulosic wastes to ethanol by *Trichoderma reesi*). Biomass production as *single cell protein* (SCP) by growing microorganisms on various substrates is a further important use of microbial cultures (e.g. growth of *Methylosinus* on methanol in the ICI Pruteen system and growth of *Fusarium graminearum* on glucose in the production of Mycoprotein). Microorganisms also have value in replacing animals in preliminary toxicological testing of xenobiotics.

The nature of microbial growth media has been discussed previously (Section 1.3.2). Microbial growth can be in either *closed* or *open systems*. In closed systems a finite amount of medium is supplied and growth continues until one factor in the medium becomes limiting. This may be a nutrient, including oxygen, or a toxic by-product of growth. In open systems of so-called *continuous culture*, outflow from the system of both cells and medium is balanced by an equivalent volume of fresh medium which is added.

Closed systems, which involve a single aliquot of growth medium are called *batch cultures*, but if fresh medium is added periodically, thereby increasing the total culture volume and stimulating further growth, they are called *fed-batch* cultures. The objective of fed-batch culture is to control important nutrients within a narrow concentration range whilst open, continuous culture systems enable both nutrients, waste products and biomass to be controlled by varying the *dilution rate* (i.e. the rate at which culture is removed and replaced). Open systems are referred to as either *chemostats* when growth rate is controlled at submaximum levels by the rate of dilution of the medium containing a single limiting nutrient or *biostats* where the cell population is maintained within narrow predetermined limits by monitoring some physiological property of the culture (e.g. concentration of outlet gas or cell population) which acts as a signal for dilution and wash out of excess cells. The specific advantage of continuous cultures over batch cultures is that they facilitate growth under *steady-state* conditions in which there is a tight coupling between biosynthesis and cell division. Thus all facets of metabolism are proceeding in an equilibrium set by the dilution rate employed (i.e. the supply of the limiting nutrient). It follows that if the rates of biosynthesis and cell division are constant and equal, the mean composition of cells in the culture does not vary with time and growth is said to be balanced. This is extremely useful in many biochemical studies and is in contrast to the situation in batch or fed-batch culture where cellular composition and state varies throughout the growth cycle.

1.6.2 Animal cell and tissue culture

The proliferation of animal cells in culture has many applications which include their use as model systems for biochemical, physiological and pharmacological studies and the production of growth factors, blood factors, monoclonal antibodies, interferons, enzymes, vaccines and hormones. A brief outline of some of the specific uses of particular animal cell and tissue cultures is given in Table 1.3. The distinction between simple techniques of incubation *in vitro*, such as perfusion and cell and tissue culture is to some extent arbitrary since conditions chosen for cell culture may not permit cell

Table 1.3 Some examples of the applications of animal cell and tissue culture.

Cell type	Process investigated
Monocytes and macrophages	Pinocytosis and phagocytosis
Blood lymphocytes	Karyotype analysis for detection of genetic defects in humans
Normal and transformed fibroblasts	Surface adherence properties of normal and malignant cell membranes
Kidney tubule epithelial cells	Differentiation of monolayers; electrical and vectorial transport of solutes; monoclonal antibody production
Myeloma cells and β-lymphocytes	Purification and characterisation of specific membrane proteins e.g. α- and β-adrenergic receptors, dopamine receptors
Kidney epithelial cells	To investigate relationship between membrane polarity and budding properties of enveloped RNA viruses
Transformed leucocytes, fibroblasts and either lymphocytes or lymphoblastoid cells	Cells are infected with Sendai virus to produce α-, β- and γ-interferon respectively
Transformed Hela cells, mammalian cells	Radiation therapy and the design of radiosensitisers and radioprotectors
Mouse fibroblasts	Acute and chronic toxicity testing and metabolism of xenobiotics; vaccine production
Primary monkey kidney cells	Production of poliovaccines; hormone secretion
Fibroblasts, mammalian brain cells	To identify chemicals capable of inducing chromosome aneuploidy

division. However, the maintenance of viable cells for periods in excess of 24 hours generally leads to them being referred to as cells in culture.

The major problem associated with the isolation of free cells and cell aggregates from organs is that of releasing the cells from their supporting matrix without affecting the integrity of the cell membrane. The earliest approaches to this problem employed mechanical techniques such as forcing the tissue through cheese cloth or silk or shaking the tissue with glass beads in an appropriate buffer. These procedures inevitably caused considerable cell damage and resulted in a low cell yield which raised the practical problem of determining whether or not the isolated cells were representative of those originally present in the tissue. The use of a biochemical approach, rather than a mechanical one, has largely overcome these problems. In the case of hepatocyte isolation, the use of collagenase and hyaluronidase in a calcium-free medium to digest the matrix has enabled cells with a viability in excess of 95% to be isolated on a routine basis. The liver may either be removed, thinly sliced and then incubated with the enzymes, or be cannulated *in situ*, perfused with an oxygenated, calcium-free medium containing the enzymes, and then removed and broken up with a blunt spatula. This latter technique requires greater practical expertise but appears to give a greater and more reproducible cell yield and viability. In all cases, viability is generally assessed by the ability of the cells to exclude the dye trypan blue, but other procedures based upon respiration or protein synthesis are probably better.

Cells obtained directly from embryonic, adult or tumour cells are termed *primary* cultures. Free cells and small cell aggregates, released by enzymic digestion of the original excision, when dispersed into appropriate culture medium attach themselves to the surfaces of the culture vessel and may then either remain viable without dividing or divide more or less rapidly according to the tissue type. Primary cells of non-tumorous tissue develop as monolayers and division ceases when all the available surface is covered i.e. when confluence is attained and the dense packing of cells restricts cell movement. This is known as *contact inhibition*.

Primary cultures, which are capable of cell division, consist of a relatively heterogeneous mixture of cells of differing physiological state. Cells will enter into senescence and death phase if critical nutrients become limiting, or toxic products accumulate in the medium. Some primary cultures (e.g. human fibroblasts) are capable of regrowth when cells are reinnoculated into fresh medium and may be subcultured several times before becoming moribund. Although of limited viability, these cultures usually show a high degree of fidelity to the *in vivo* state, with cells retaining both the original diploid karyotype and state of differentiation as when first isolated. Such cultures are often termed *cell strains* and are very useful as model systems and for proliferation of virus for vaccine production (Table 1.3). The proliferation of mixed populations of differentiated cells is usually referred to as *organ culture*.

Differentiation or *transformation* of primary cell cultures may arise spontaneously on subculture of cells of certain types of primary cell culture (e.g. baby hamster kidney (BHK) cells) to give rise to so-called *established cell lines*. Transformation is a physiological adaptation and does not result from mutation although aneuploidy is frequently associated with the phenomenon. Transformed cells show a much altered physiology compared with normal cells being capable of indefinite subculture, more rapid growth, not showing contact inhibition and generally exhibiting similar physiological characteristics as neoplastic, tumour cells. Transformation can be induced in primary cultures by the same chemicals or viruses that may produce tumours *in vivo*, e.g. 3,4-benzpyrene, nitrosomethyl urea or Rous sarcoma virus. Injections of transformed cells into laboratory animals will also induce tumour formation *in vivo*. The close similarity between transformed and tumour cells is often exploited experimentally, with transformed cells being used instead of tumorous cells because of their tendency to be, cytologically, more homogeneous.

Animal cells may be broadly subdivided into those which remain viable only when attached to a solid substrate (e.g. primary cultures, normal diploid fibroblast cell strains) and those that will proliferate in fine suspension (e.g. human tumour, HeLa cell lines, and hybridomas). The material to which cells stick must be nontoxic, sterilisable and preferably transparent so that the cells may be observed microscopically. Many surfaces have been used successfully including plastic, glass, teflon tubing and DEAE-Sephadex. Cells may be detached from such surfaces by mild trypsin digestion.

Small scale culture of *anchorage-dependent cells* has traditionally employed a variety of modified glass slide chambers, tubes, Petri dishes and prescription bottles. Growth of cells under these conditions is very inefficient since diffusion gradients are set up within static medium layers and waste products remain close to the surfaces of the cells. This problem is avoided by using slowly rotating roller tubes or bottles which ensure that monolayers are alternately exposed to air and medium. The surface area to volume ratio (S/V) of each roller tube remains comparatively small however and low cell yields of the order of 3×10^7 cells are obtained from a surface area of 500 cm^2. Improvements to the basic design of roller bottles intended to improve S/V have included the addition of internal plastic spirals or plates. Roller bottles providing culture surfaces of up to 1500 cm^2 have been used commercially for producing virus for vaccine production. This method is very expensive in labour and equipment since the system has inherent problems associated with efficient monitoring and control of the growth of the cell population, with medium refeeding and also cell harvesting. The introduction of charged dextran microcarrier beads, suspended in liquid medium to support the growth of anchorage dependent cells, significantly increases S/V ratios. This has facilitated the development of stirred tank vessels incorporating facilities both for monitoring the environment of the cells and for automatic refeeding in fed-batch processes. It has also eliminated expensive and bulky controlled environment bioreactors.

Cells amenable to *suspension culture* may be grown either in low density static culture, where cells settle to the bottom of the container or in high density in medium agitated by rocking, shaking, spinning or aerating the culture vessel. Mammalian cells are readily damaged by liquid shear forces such as those generated in pumps, narrow pipes, centrifuges and on vigorous agitation of their growth medium. This fragility has to be overcome in the design of large-scale bioreactors for suspensions in which medium is agitated not with blade impellers but usually by either the rapid vertical displacement of a horizontal perforated metal disc or by the circular movement of flexible sheets attached to a vertically orientated drive shaft. Cells may also be protected from shear forces by being encapsulated within semipermeable membranes e.g. polymer reinforced calcium alginate. Such encapsulation enables higher cell numbers to be obtained than in free suspensions and the capsules can be easily recovered by gravity. A disadvantage of encapsulation is the increased operating costs. Other major barriers involved in the scale-up of mammalian cell culture are oxygen transfer limitations, accumulation and removal of toxic waste products, the cost of medium constituents (e.g. serum), and the maintenance of sterility and safety of vessels.

1.6.3 Plant cell and tissue culture

This term refers to the *in vitro* proliferation of parts of mosses, liverworts and vascular plants. The technique has evolved from being mainly a model system for studying plant development, to fulfil a role in practically all aspects of the physiology and biochemistry of higher plants (Table 1.4). The general principles of plant cell and tissue culture are similar to those operating for

Table 1.4 Some examples of the applications of plant cell and tissue culture.

System	Process investigated
1. Protoplasts	
Nicotiana tabacum	Cell wall regeneration; clonal propagation of mutant plants; interspecific hybrids; uptake of *Rhizobium*
Solanum tuberosum	Somaclonal variation in cultivar Russet Burbank; genetic transformation using Ti plasmid; transformation by co-cultivation of protoplasts with isolated *Agrobacterium tumefaciens* cells
Arabidopsis thaliana + *Brassica campestris*	Intergenetic hybridisation
Glycine max	Isolation of viable bacteroids from root nodule protoplasts
2. Cell suspensions	
Acer pseudoplatanus	Control of cell division and expansion in single cells, batch and continuous cultures, and monitoring the associated biochemical changes
Daucus carota	Demonstration of totipotency in higher plants
Nicotiana tabacum	Industrial fermentation of biomass for tobacco industry. Selection of nitrate reductase-less (NR^-) mutants by means of their resistance to chlorate. Selection and regeneration, and sexual transmission in regenerants of resistance to the herbicide picloram
Digitalis purpurea	Biotransformation of digitoxin to digoxin
3. Callus	
Phaseolus vulgaris	Physiology and biochemistry of differentiation
Vicea faba	Cytogenetic studies indicating chromosome instability
Elaeis guineesis	Cloning of oil-palm plantlets and crop improvement
4. Anther and microspores	
Nicotiana tabacum	Embryogenesis from cultured anthers
Brassica napus	Embryogenesis from cultured microspores
5. Embryos	
Papaver somniferum	Plant breeding; direct pollination of excised ovules to overcome prezygotic barriers to fertility
6. Meristem	
Solanum tuberosum	Elimination of potato virus *X* by thermotherapy followed by plantlet regeneration
Chrysanthemum morifolium	Radiation induced mutation breeding

microorganisms and animal cells. A wide variety of explant has been successfully proliferated *in vitro* following surface sterilisation in appropriate disinfectants, e.g. 10% v/v calcium hypochlorite solution, and copious washing in sterile distilled water. Examples include single cells, callus (i.e. multicellular aggregates of mainly undifferentiated parenchymatous cells), cell suspensions, mixtures of single and cell aggregates dispersed in agitated liquid medium, protoplasts, (i.e. plasmalemma-bound cytoplasm released by cellulase digestion of the cell wall), microspores and ovules, anthers, root and shoot meristems and even complete organs.

Plant cells are not as amenable to manipulation in suspension as are microbial cells, but they may be multiplied in batch culture, fed-batch and continuous cultures in reactors which are similar in principle to the fermenters used for microbial and animal cell suspensions. Particular problems associated with the suspension culture of higher plant cells are the slow doubling time (usually about two days), inability of cells to disaggregate, partial anaerobiosis in culture, the susceptibility of cell walls to shear forces in large fermenters and ploidy instability on prolonged subculture.

Higher plant cells may become transformed in an analogous sense to animal cells in culture. In the case of plant cells this transformation may arise *in vivo* as a result of a sexual cross, e.g. *Nicotiana glauca* × *Nicotiana langsdorfii*, or viral infections e.g. wound tumour virus, or bacterial infections e.g. *Agrobacterium tumefaciens*. Such transformed cells in culture afford a useful model system for comparison with healthy tissue, since they show a greater biosynthetic capacity, in particular showing prototrophy for growth regulators. The especial value of bacterial and viral transformations is that they afford a means of introducing and integrating foreign DNA into host chromosomes following uptake of the DNA vector, e.g. cloned *Agrobacterium* tumour-inducing (Ti) plasmid, responsible for tumour formation, into a host protoplast (Table 1.4 and Section 5.8.4).

The potentialities for somatic hybridisation in higher plants, involving chromosome transfer following protoplast fusion techniques, are also being realised (Table 1.4). Protoplast fusion is promoted by polyethylene glycol (PEG). Selection procedures may involve micromanipulation of fused protoplasts, observed microscopically to contain different plastids, or by selection of genetic markers using suitable media and/or environmental conditions. The emphasis of somatic hybridisation studies in higher plants currently differs from that in animals (Section 4.2.2) in being more readily directed towards regeneration of novel species particularly in cases of species or genera that are normally sexually incompatible.

1.6.4 Cell sorting

The physical isolation of populations of particular cells in bulk from a mixed population or individual cells from a mixed population is often essential to the cell biologist. Cells may often be separated in bulk on the basis of biological survival in an adverse environment exploiting the capacity of a particular cell for division. Survival, and therefore selection, may be based on many factors including resistance to either an infection or a cytotoxin, due to the absence of receptors, prior infection which has led to an immune response or to cell transformation.

Bulk cell separation may also be achieved by exploiting the different sedimentation characteristics of cells by differential centrifugation, density gradient centrifugation, isopycnic centrifugation and continuous flow centrifugation (Chapter 2). The different surface charge properties of cells facilitate their electrophoretic separation particularly by continuous flow electrophoresis and isoelectric focusing (Chapter 7). Other surface properties of cells may be important in their bulk separation by affinity chromatography, (Section 6.7.3), by agglutination involving attachment of polyvalent ligands (immunoglobulins and lectins) (Section 4.8), and by phase partitioning (Section 6.3.2). Phase partitioning is also known as *cell surface chromatography* and frequently employs separation of cells into PEG or dextran on a multiple counter current distribution basis, separation being dependent on the nature of exposed polar groups of phospholipids and membrane carbohydrates of animal cells. A final method of bulk separation involves the simple approach of *sequential filtration* through meshes of differing pore size to separate cells on the basis of cell size, shape and deformability.

Individual cells may be isolated by similar selection procedures to those mentioned above particularly when resistance is based on a rare mutation. Microscopy and micromanipulation may be employed as in somatic cell hybridisation studies.

More sophisticated flow sorting methods are being developed to sort out individual cells from a mixed population. One of the most important of these methods is *fluorescence activated cell sorting* (FACS) which measures the fluorescence emitted by individually-labelled cells in a mixed cell population in a *cytofluorimeter* (also termed a *cytofluorograph*). FACS separates cells which contain preselected fluorescence characteristics by enclosing them inside charged droplets which can be deflected into reservoirs or slides under the influence of an applied electrostatic field according to the polarity and quantity of charge. Unselected cells held in uncharged droplets fall into another reservoir. Fluorescence may be generated from dyes specific to cell components (e.g. propidium iodide for DNA) or to antibodies to surface antigens. The instrument uses an argon-laser beam to excite fluors of fluorescently-labelled cells being passed in single file through a laminar flow chamber. A suspension of single cells is a prerequisite and consequently tissue and tumour specimens raise the difficulty of disruption without loss of cellular markers and integrity. Scattering of the monochromatic light, which is a function of cell size and viability is measured to one side of the incident beam with a photomultiplier tube. The data from cytofluorimeters is generally displayed on a VDU as histograms of fluorescence intensity, as a function of the number of cells with that intensity, in order to depict subpopulations of cells.

If the fluorescence intensity, size and viability of a given cell meets the criteria preselected by the FACS operator, a droplet containing the cell is charged and recovered. The procedure is very sensitive and cytofluorimeters capable of measuring a single virus particle and less than 100 fluorophores per particle have been developed. Automation allows up to 5000 cells to be sorted per second into viable sub-populations.

A new approach to adherence methods of cell separation, known as *magnetophoretic separation* involves the use of magnetic microspheres bearing conjugated ligands. So far the microspheres used have consisted of magnetite (Fe_3O_4) cores surrounded by polyacrylamide and/or agarose layers to which fluorescent labels and protein are chemically attached. Cells are allowed to mix with these microspheres in suspension and the mixture is then passed through a magnetic field. Those cells attached to the ferromagnetic beads are attracted to the poles of the magnet and washed free of the remaining cells. Application of this technique to neuroblastoma cells with expression of surface gangliosides, has allowed a purification greater than 98%, with viability of cells remaining unaffected.

1.6.5 Cell counting

Biochemical measurements are often related to growth cycle phases or to the important physiological parameter of the developmental state of the organism concerned, results being expressed mainly in relation to cell number, fresh or dry weight, cell volume or per unit of nitrogen. Where cell numbers are used, these must be determined by a suitable counting method. Non-motile bacteria are conveniently counted by observing colony formation of cell clones on the surface of a solid medium following innoculations of serial dilutions of the original sample.

Before cell counting can be undertaken in suspension, multicellular aggregates must be digested gently either enzymatically or chemically e.g. with a dilute solution of chromium trioxide. A known volume of cell suspension may then be introduced to a counting chamber and cell numbers determined microscopically. The *haemocytometer* is the most commonly used counting chamber for blood cells, microorganisms and dispersed animal cells. Some counting chambers for algae, plankton and surface-aggregated animal cells in culture bottles use inverted microscopes in which the light source shines down through the counting chamber and the microscope optics are arranged to view the sample from underneath the chamber.

Counting of cells under the microscope is slow, time consuming, tedious and may be inaccurate especially at low cell densities. *Electronic particle counters* resolve these difficulties, many being fully automated and include facilities for estimating mean cell sizes. The *Coulter Particle Counter* operates on the following principles. A current flows between two electrodes immersed in a liquid electrolyte. One of the electrodes is inside a glass counting tube and the other is outside the tube, but immersed in the sample which is suspended in the same electrolyte. A small orifice in the counting tube ensures the flow of current. A known volume of sample, usually 0.5 cm^3, is automatically drawn into the tube via the orifice and any particles present increase the resistance between the two electrodes as they pass through the orifice. This results in a pulse appearing on an oscilloscope screen and a single count recording on a digital scaler. The size of each pulse is directly proportional to the size of the particle. The pulses are very small and amplification is necessary to accurately record the pulse produced. The degree of amplification depends on the size of the particles being counted and must be determined for each kind of sample. Similarly, the size of the orifice tube must be chosen with care. An orifice of 100 μm diameter is appropriate for blood cells and most bacteria, whereas 70 μm is better for viruses. Pulse height analysis resolves cell numbers and biovolume for subpopulations. Mean cell size is determined by taking cell counts at increasing pulse height analyser settings and calculating the mean threshold setting when half the total population is counted and half is not. Various standard sized particles such as pollen grains or, more usually, latex particles are used to calibrate the instrument so that a particular mean pulse height analyser setting can be related to a particular particle volume. From a knowledge of this data the mean cell size can be determined. Multiplying this by the total cell count gives the total cell biovolume. Errors in sizing may be introduced by osmotic effects on cells by the suspending electrolyte whilst errors in counting may be introduced by cell settling, by debris in suspension and partial blocking of the orifice. The technique is very suitable for relatively simple populations but not for natural populations of, for example, phytoplankton where the range of sizes of organisms precludes effective choice of orifice.

1.6.6 Cryopreservation

Repeated subculture of cell lines may result in genetic aberration and hence loss of biosynthetic capacity or morphogenetic potential. Various strategies are possible to alleviate these difficulties including methods for slowing down growth in stock cultures, or freeze-drying for certain microbial cultures. *Cryopreservation* in liquid nitrogen is the preferred method for most eukaryotes and is based on the principle that the only reactions that occur at −196°C, the boiling point for liquid nitrogen, are ionisations due to background irradiation, which are very rare. Biochemical reactions cease at about −130°C. Various experimental protocols have been devised, often empirically, for different tissue types.

Successful cryopreservation requires that viability is maintained after the sample has been held for indefinite periods in liquid nitrogen. The rate of freezing is very important since too rapid freezing causes large crystals of ice to form intracellularly. Such crystals, on thawing, lyse cell membranes. Slow freezing rates favour the deposition of extracellular ice at subzero temperatures, provided the internal concentration of solutes is sufficiently high to make the internal water potential lower than that externally. This process may be assisted by adding a penetrating *cryoprotective agent* such as glycerol to the medium supporting the specimen. When ice has formed extracellularly water is withdrawn from the cells because the vapour pressure density of ice is lower than that of the predominantly aqueous cytoplasm. Water moves out until the vapour pressure densities are equilibrated. During this process, the protoplast shrinks and plant and microbial cells may become plasmolysed. Alternatively cryoprotective agents such as polyethylene glycol (PEG) and polyvinyl pyrrolidone (PVP) may be used, which do not penetrate the cells, to ensure that cell dehydration proceeds smoothly during slow freezing. If the dehydration is too rapid, denaturation of proteins may occur due to localised high salt deposits, which could in turn result in membrane damage. At a critical temperature corresponding to the eutectic point for the cytoplasm, the cell contents freeze, without ice crystals forming.

Slow cooling rates are best achieved with samples held in ampoules within commercial units, which cool at rates of 0.1°C to 10°C min^{-1}. Such units consist of containers enclosing a low melting point liquid alcohol, cooled by a refrigeration coil. Stepwise cooling, which may be equally effective, involves holding the specimen for prescribed periods at a given temperature in improvised units. When the temperature reaches −50°C to −70°C for the improvised units, or −100°C for controlled freezing, the ampoules are transferred directly to liquid nitrogen.

Thawing of samples is best achieved rapidly by removing them to a water bath held at 30°C to 40°C for a few minutes, whilst applying gentle agitation. Washing to remove the cryoprotectant should not be necessary as cryoprotectants should be non-toxic. Damage to tissue during an experimental protocol may be estimated by employing light or electron microscopic techniques (Section 1.9). Viability may be assessed either by monitoring respiration, by cytological staining (e.g. exclusion of Evans Blue Stain from viable cells) or by regrowth in appropriate media.

1.6.7 Culture collections

The increasing importance of culture techniques for experimental purposes relies heavily on reference collections of plant, animal and microbial cell lines being readily available, so as to enable coincidental comparative studies to be made in different laboratories. Special characteristics of particular cell lines may include biochemical markers, karyotype analyses, nutritional mutants, drug resistance or sensitivity, tumour production, cellular inclusions (e.g. microbial contaminants, phage or other episomes) or secretory products (e.g. hormones or immunoglobulins). An example of a culture collection is the National Collection of Type Cultures, Central Public Health Laboratory, Colindale Avenue, London.

1.7 Mutants in biochemistry

1.7.1 Introduction

Much of the biochemistry associated with molecular biology is concerned with defining the structural and metabolic consequences of *mutations* i.e. altering the base sequences of DNA. Our knowledge of the principles of linkage and genetic mapping of the genetic code and the control of transcription and translation in prokaryotic and eukaryotic organisms is based largely on the exploitation of mutant organisms. The particular value of mutations, if expressed, is that they form the basis of a comparative study with the normal (*wild-type*) cells.

1.7.2 Mutant classification

In chemical terms there are several classes of mutation, which may either occur naturally or be induced by treating cells with physical or chemical mutagens e.g. X-rays, ultraviolet radiation, hydroxylamine, alkylating agents such as ethylmethanesulphonate or acridine dyes such as proflavin. The introduction of restriction endonucleases and genetic recombination *in vitro* (Section 5.12) is facilitating *directed mutagenesis*, the intentional alteration of a gene at a specific location for a specific purpose. A *point mutation* is said to occur when a single base in DNA is changed and a double mutation when there are two such changes. *Deletions* and *deletion substitutions* refer either to removal or exchange base pairing in larger parts of a gene.

Microorganisms have traditionally been used for studies in biochemical genetics because they are predominantly haploid, have short generation times, can be grown in/on defined media, making mutant isolation easier, and can be propagated as large biomasses of biochemically uniform cell populations in industrial-scale fermenters. The genetic transfer systems of microorganisms are also simpler than those of eukaryotes and are now relatively well defined both in terms of genetic mapping and in complementation tests.

Largely as a result of microbial studies, mutations may be further classified according to the conditions under which mutation is expressed. So-called *non-conditional mutants* display the mutant phenotype under all (natural) conditions. Examples of non-conditional mutants include *auxotrophy* where mutation results in an enzyme deficiency or inactivity which may be rectified by supplements of the product of those defective enzymes, provided they can be transported into the cell. *Prototrophs* are non-conditional mutants in which auxotrophs *gain* a biosynthetic function leading to autonomy from a nutritional supplement. Regaining a wild-type phenotype in this way is called *reversion* and is usually associated with point mutations. A *resistance mutation* confers increased resistance to an antimetabolite or environmental stress. Antimicrobial (drug) resistance in bacteria, yeast or plasmids is an example of resistance mutation often of single gene origin.

A *conditional* mutant is one that does not always display the mutant phenotype. *Suppressor-sensitive mutations* (*sus*) do *not* display the mutant characteristic in the presence of the gene product from a wild-type suppressor gene. Hence the mutation expression is being suppressed. For example, a bacteriophage carrying a *sus* mutation will produce progeny when it infects a host cell expressing the su^+ gene (the so called *permissive condition*) but will not do so when the host cell lacks the suppressor gene (su^-), under so called *non-permissive conditions*. Another good example of a conditional mutant is *temperature-sensitive mutant* in which a physiologically active protein is rapidly inactivated above or below a critical temperature. The particular value of conditional mutants for biochemical studies is that they are conditionally or potentially lethal and enable investigations to be made on enzymes which cannot be corrected by nutritional supplements. For example, mutants of DNA and RNA polymerases or amino acyl tRNA synthetases cannot be corrected by simple nutritional supplements to the medium as these enzymes are involved in the metabolism of complex macromolecules.

1.7.3 Mutant selection

The application of mutants in genetical, biochemical or physiological studies presupposes that such mutants can be selected from a mixed population. *In vitro* techniques are widely exploited for such selection procedures. Selection of biochemical auxotrophs is usually based on incorporating effective concentrations of antibiotics (e.g. penicillin or streptomycin) into minimal media of bacterial suspensions to kill dividing, wild-type cells. Auxotrophs present will be incapable of growth but remain viable and may be recovered after centrifugation, and washing to remove excess penicillin. Surviving cells can then be incubated on a supplemented medium. The nature of the supplement to minimal medium will biochemically characterise the auxotroph. *Replica plating* from complete medium to minimal medium using a sterile velvet pad is another potential means of detecting auxotrophic colonies on subsequent transfer to complete medium.

In eukaryotes there is no effective replica plating technique. Thus selection for auxotrophs is currently mostly based either on the addition of an antimetabolite to minimal medium to kill wild-type dividing cells, leaving the auxotrophs to be protected and recovered by transfer to nutrient enriched media, or by a laborious total isolation approach.

1.8 Cell fractionation

1.8.1 Introduction

Cell fractionation experiments must proceed in two successive stages, namely *homogenisation* and *separation*. Homogenisation is essentially a disorganisation stage in which a tissue is converted into a so-called *homogenate* whilst separation reintroduces a new degree of order into the system by grouping together components of the homogenate which possess certain common physical properties, for example size and/or density. The ideal isolation procedure would yield the desired intracellular components as they exist in the cell, i.e. in an unchanged morphological and metabolic state, and in quantitative yield. Most cell fractionation techniques fall short of this ideal. For example fractionation methods that preserve morphology may destroy activity and methods preserving activity often result in loss of morphological structure. Thus a degree of compromise is nearly always necessary when using cell fractionation techniques.

The design and object of an experiment often limit the selection of tissue. Tissues and cells from different species may differ in composition, fragility and density and this variation may favour a particular isolation technique. The liver, for example, is an ideal choice of tissue for the study of mitochondrial function because liver cells contain large numbers of mitochondria, whereas the thymus is a tissue often used for nuclear isolations because the nuclei of thymocytes comprise nearly 50% of the cell mass. Tissues in the main, however, are heterogeneous with regard to cell shape and size and subcellular fractions isolated from homogenates of animal tissues invariably reflect this heterogeneity. Chemical analysis of isolated fractions can therefore yield only an average representation of the composition of the fraction. Animal organs differ also in their blood and connective tissue content. Generally, the more connective tissue found in an organ, the greater is the problem of tissue homogenisation and hence the yields of subcellular components are lower.

Homogenisation of tissues results in the loss of morphological and biochemical detail. This may not be important if homogenisation is intended as a preliminary step in the isolation of a compound from a tissue. If metabolic processes are to be studied, however, it is essential to preserve as much of the morphological and biochemical integrity of the tissue as possible. The unfortunate feature of homogenisation techniques is that their use in tissue fractionation studies is still empirical. The aim of the process is to disrupt the tissue and break open the boundary (cell wall and/or membrane) of the cells to release the cellular contents. Many different devices and techniques have been utilised in an attempt to accomplish this, but the principles on which they operate are not always clearly understood. Homogenisation in the absence of a satisfactory theoretical basis and of adequate standardisation of the methods, tends largely to be more of an art than a science. In view of the wide variations between different tissues, both in fragility of certain cellular organelles and in the resistance of cells and tissues to disruption, the homogenisation of each type of material poses a separate problem which can be solved only by trial and error.

The major use of homogenisation is its role as the preliminary stage for the separation of cellular components, hence allowing the intracellular localisation of metabolic processes to be deduced. Homogenates have also been used in their own right to study the uptake and metabolism of added compounds. This is a useful technique if difficulty is experienced in getting compounds into intact cells because of membrane permeability problems.

1.8.2 Methods of disrupting tissues and cells

Elevated temperatures denature most enzymes and consequently cell disruption procedures are best carried out around 0°C preferably in purpose built low temperature laboratories (*cold rooms*). Cell disruption may often be obtained simply by enzymic digestion of walls or membranes using appropriate combinations of cellulases, chitinases, lipases and proteases (e.g. collagenase, trypsin, hyaluronidase), by bursting cells by osmotic shock from exposure to alternating freezing and thawing, or by using organic solvents such as toluene. Usually, however, more vigorous techniques have to be applied. Animal cells are susceptible to mild stresses whereas the presence of cell walls in bacteria and plants makes disruption more difficult. Physical methods may be grouped according to whether disruption is produced by shearing forces between cells and solids or is produced in cell suspensions by liquid shear.

Solid Shear Methods Mechanical shaking with abrasives may be achieved in Mickle shakers which oscillate suspensions vigorously (300 to 3000 times min^{-1}) in the presence of small glass beads of 500 nm diameter. This method may result in organelle damage. More controlled solid shearing operates in the Hughes Press where a piston forces either moist cells together with abrasives, or a deep frozen paste of cells, through a 0.25 mm diameter slot in a pressure chamber. Pressures up to 55×10^6 Pa (8000 p.s.i.) may have to be achieved to lyse bacterial preparations.

Liquid Shear may be produced in cell suspensions either in *blenders* by high speed reciprocating or rotating steel blades or, in *homogenisers*, by the upward and downward movement of a plunger and ball. The cutting blades of blenders are inclined at different angles to permit efficient mixing of the vessel contents and work in association with a specially designed stainless steel cup which has indented walls to aid maximum solution turbulence. Homogenisers are usually operated in short high speed bursts of a few seconds to minimise local heating. The technique is not suitable for microorganisms.

Homogenisers consist either of a hand operated (Dounce or Tenbroeck) or power driven (Potter-Elvejham) pestle made of Pyrex glass, Teflon or Lucite, and a glass tube. The tube is fixed and the pestle rotates to establish liquid shear forces, these being greatest at the surface of the pestle and least at the walls of the tube. The clearance between the pestle and the tube is kept between accurate limits because the rate of shear depends on the radii of the pestle and tube, and the rate of rotation of the pestle.

The shearing forces established using homogenisers are too small to disrupt intact plant and microbial cell membranes, but even very gentle liquid shear using simply repeated pipetting through a Pasteur pipette will disrupt some animal cells such as polymorphonuclear leucocytes.

Disruption by High Pressure Extrusion The most frequently used equipment for disruption of microbial cells is the French pressure cell in which pressures of 10.4×10^7 Pa (16 000 p.s.i.) are commonly employed. A pressure cell consists of a stainless steel chamber which opens to the outside by means of a needle valve. The cell suspension is placed in the chamber with the needle valve in the closed position. After inverting the chamber, the valve is opened and the piston pushed in to force out any air in the chamber. With the valve closed again, the chamber is restored to its original position, placed on a solid base and the required pressure exerted on the piston by a hydraulic press. When the required pressure has been attained, the needle valve is opened fractionally to slightly release the pressure, and as the cells expand they burst. The valve is kept open while the pressure is maintained so that a trickle of smashed cells is collected.

Disruption by Ultrasonic Oscillation High frequency ultrasonic oscillations have been found useful for cell disruption. The method by which ultrasonic waves break cells is not fully understood, but it is known that high transient pressures are produced when suspensions are subjected to ultrasonic vibration. The main disadvantage of the technique is that considerable amounts of heat are generated.

In order to minimise heat effects specially designed glass vessels are used to hold the cell suspension. Such designs allow the cell suspension to circulate away from the ultrasonic probe to the outside of the vessel where it is cooled as the flask is suspended in ice. Some flasks, however, have built in cooling units, but localised heating effects are not fully eradicated.

1.9 Microscopy

1.9.1 Introduction

Biochemical analysis is frequently accompanied by light and electron microscopic examination of tissue, cell or organelle preparations to evaluate the integrity of samples and to correlate structure with function. Microscopy serves two independent fuctions of enlargement (*magnification*) and improved resolution (the rendering of two objects as separate entities). Light microscopes employ optical lenses to sequentially focus the image of objects whereas electron microscopes use electromagnetic lenses. Light and electron microscopes may work either in a *transmission* or *scanning* mode depending on whether the light or electron beam either pass through the specimen and are diffracted or whether they are deflected by the specimen surface. Polarised light microscopes detect optically active substances in cells, e.g. particles of silica or asbestos in lung tissue, or starch granules in amyloplasts. Light microscopes in *phase contrast* mode are often used to improve image contrast of unstained material, e.g. to test cell or organelle preparations for lysis. Changes in phase of emergent light are caused mainly by either diffraction or by changes in the refractive index of material within the specimen, or even by differences in thickness of the specimen. At their point of focus the converging light rays show interference resulting in either increases or decreases in the amplitude of the resultant wave (*constructive* or *destructive interference* respectively), which the eye detects as differences in brightness.

Microscopes using visible light will magnify approximately 1500 times and have a resolution limit of about 0.2 μm whereas a transmission electron microscope is capable of magnifying approximately 200 000 times and has a resolution limit for biological specimens of about 1 nm. The excellent resolving power of transmission electron microscopy (TEM) is largely a function of the very short wavelength of electrons accelerated under the influence of an applied electric field. (An accelerating voltage of 100 kV produces a wavelength of 4×10^{-3} nm.) Scanning electron microscopes (SEM) use a fine beam of electrons to scan back and forth across the metal-coated specimen surface. The secondary electrons that are generated from this surface are collected by a scintillation crystal which converts each electron impact into a flash of light. Each light flash inside the crystal is amplified by a photomultiplier and used to build up an image on a fluorescent screen. The principal application of SEM is the study of surfaces such as those of cells. The resolution limit of a scanning electron microscope is about 6 nm. Figure 1.1 illustrates the application of TEM and SEM in biological investigations.

1.9.2 Ion probe analysis

Electron microscopes may be equiped to perform *X-ray spectrochemical analysis*. When a specimen is irradiated by an electron beam, an electron may be displaced from an inner to outer orbital. When this happens the vacated orbital may be infilled by an electron from a higher energy orbital with the emission of an X-ray photon characteristic of the difference in energy levels of the two orbitals. The binding energy of the electrons in an orbital is related to the charge on the nucleus; hence each element produces its own characteristic emission spectrum. Emitted photons are usually detected by *energy-dispersive analysis*, using lithium drifted solid-state detectors. Each photon emitted reacts with silicon atoms to produce an electrical pulse proportional to the energy of the photon. An electronic pulse height analyser and microcomputer are used to process the spectral data which is normally displayed on a video monitor. X-ray spectrochemical analysis permits the measurement of ion distrubutions *in situ* in SEM and TEM by combining the high spatial resolution of electron microscopy with the ability to determine subcellular elemental composition. Areas as small as 100 nm^2 can be analysed under optimal conditions and detection of any element with an atomic number greater than 10 is possible. Measurement is made in terms of total concentration of the element rather than free ion activity and in this sense the method is comparable to flame photometry (Section 8.7) though with a higher sensitivity.

Another highly sensitive method for the analysis of elements in biological materials uses a high energy proton beam to excite characteristic X-rays. This technique is known as *proton induced X-ray emission* (PIXE) and has a

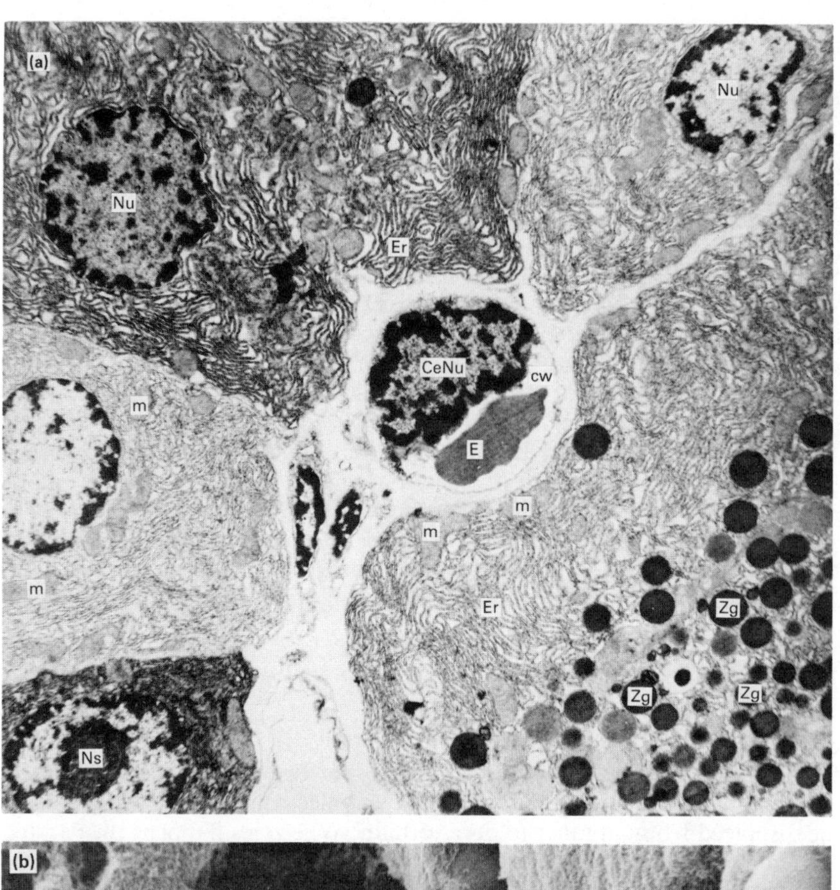
(a)
Nu
Nu
Er
CeNu
cw
m
E
m
m
m
Er
Zg
Zg
Zg
Ns

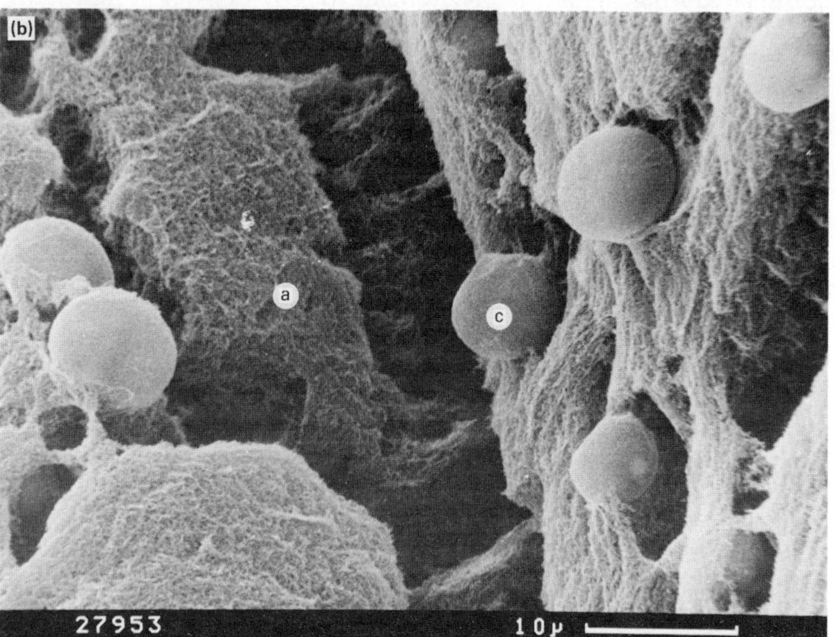
(b)
a
c
27953
10µ

spatial resolution of 1 μm with normally an analytical sensitivity of 10 ppm, although, under optimal conditions, a sensitivity of one part in 10^7 can be achieved for many elements. The principle of the technique closely resembles that of X-ray spectrochemical analysis, although in PIXE, X-rays are produced by collisions between protons rather than between electrons and the target atoms. The technology associated with electron focusing cannot be directly applied to the PIXE method as the lenses are too weak to cope with high energy protons. Consequently, strong focusing magnetic quadruple lenses have been developed. An advantage of PIXE is that high energy protons penetrate much further than electrons, being less easily deflected by atomic collisions. Thus, the resolution of the proton beam is maintained through thicker specimens.

1.9.3 Preparation of specimens for microscopy

Transmission microscopy requires thin sample specimens, i.e. squashes, smears, hanging drops or very thin sections. Preservation and integrity of cellular components requires initial *tissue fixation* which may be achieved either by rapid freezing or by chemical treatment to stabilise and crosslink protein and lipid components of membranes. Fixation with formaldehyde (principally used in light microscopy) and glutaraldehyde (alone or in combination with formaldehyde for electron microscopy), is the result of the formation of methylene bridges with side chain amino groups of proteins whilst osmium tetroxide, a common fixative in electron microscopy, mainly crosslinks with unsaturated fatty acid side chains (Fig. 1.2). Fixed tissue is then subjected to sequential processes of staining and sectioning.

Samples for ion probe analysis (Section 1.9.2) are prepared either by chemical fixation or, more successfully, by ultra-rapid freezing methods, such as immersion in liquid propane or nitrogen slush, which serve to immobilise ions. Where frozen tissue is used, sections must be extremely thin and may require the use of an ultra-cryomicrotome.

Histological stains are used to produce contrast which aids resolution.

Fig. 1.1 **(a)** Transmission electron micrograph of a transverse section of capillary in the exocrine pancreas of the rat ($\times$ 9750). The relationship between the high concentration of rough endoplasmic reticulum and a considerable quantity of protein synthesis and storage is shown in the cell at bottom right of the photomicrograph. Substrate for such synthesis is supplied by the blood via capillaries such as that in the centre of the micrograph. The tissue was fixed in buffered paraformaldehyde and gluteraldehyde, secondarily fixed in osmium tetroxide, then stained with lead citrate and uranyl acetate solutions. **Ce Nu,** capillary endothelial cell nucleus; **cw,** capillary wall; **E,** erythrocyte in capillary lumen; **Er,** rough endoplasmic reticulum; **m,** mitochondrion; **Ns,** nucleolus; **Nu,** nucleus; **Zg,** zymogen granule. **(b)** Scanning electron micrograph of cells of *Chlorella* sp. Immobilised cells were prepared by suspending cells in 5% w/v solution of sodium alginate and dropping the mixture from a pippette into a 0.1M solution of calcium chloride. Beads of insoluble calcium alginate, 2–3 mm in diameter, form instantaneously. For electron microscopy, the beads were taken through a series of increasing concentrations of acetone (to a maximum of 100%), dried in a critical point dryer and coated with gold. **c,** *Chlorella* cell; **a,** alginate.

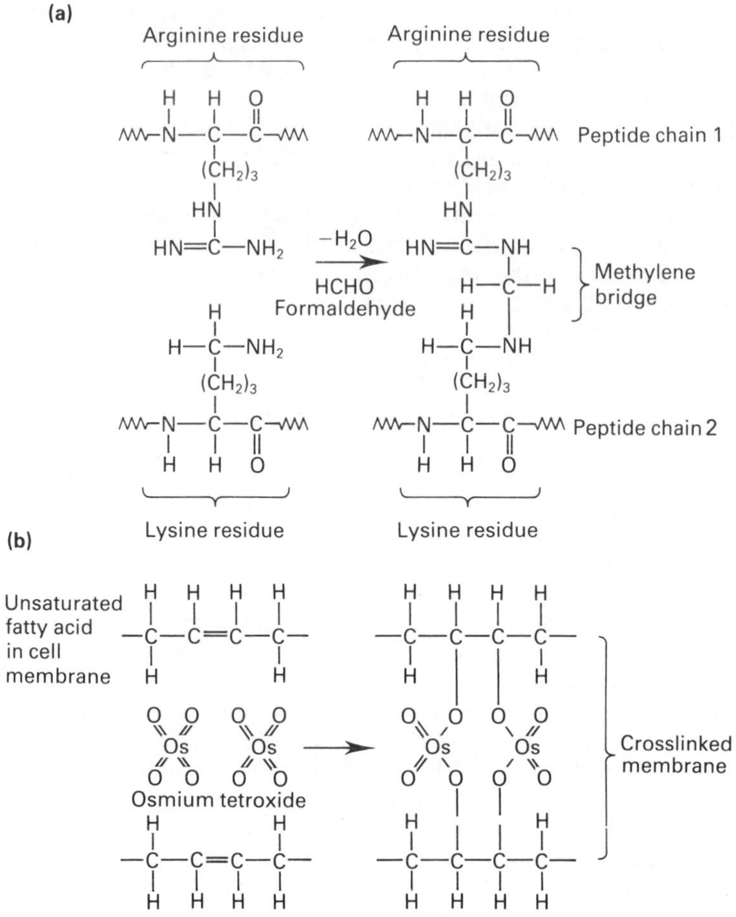

(a)

Arginine residue Arginine residue

Peptide chain 1

(CH₂)₃ HN HN=C—NH₂

$-H_2O$

HCHO
Formaldehyde

Methylene bridge

Peptide chain 2

Lysine residue Lysine residue

(b)

Unsaturated fatty acid in cell membrane

Osmium tetroxide

Crosslinked membrane

Fig. 1.2 Crosslinking by some chemical fixatives: **(a)** formaldehyde; **(b)** osmium tetroxide.

Many stains used in light microscopy rely on anionic and cationic reactions with intracellular ampholytes, and as such their efficiency is profoundly influenced by pH. Cytoplasmic components are mostly cationic in the slightly acid pH range, inferring preferential binding to anionic (acidic) stains such as eosin. Chromatin and DNA at pH 6.0 are anionic and thus bind to cationic stains such as methylene blue. Constrast in material for TEM is improved by incorporating heavy metal salts into the specimens to induce a greater extent of electron absorption. This can be achieved by exploiting the binding of uranium to nucleic acids and proteins and of lead to lipids.

Fixed tissue, when not frozen, needs supporting before sections are cut for microscopic study. *Embedding media* such as waxes and epoxyresins (araldite or epon) are immiscible with both water and alcohol which necessitates the initial dehydration and equilibration of fixed tissues by passing them through solutions containing increasing concentrations of ethanol followed by transfer to xylene or propylene oxide, before infiltration with an appropriate embedding medium in its liquid phase. Sections, 10 μm thick when cut from tissue frozen at −20°C in a refrigerated microtome, called a *cryostat*, are good for rapid examination by light microscopy but suffer damage if thawed. Wax sections, 5 μm thick, also for light microscopy are floated on warm water, placed on glass slides, dried and subsequently dewaxed and stained for classical light microscope examination. Ultra-thin sections for TEM (less than 100 nm) are cut with *ultramicrotomes* using either fractured glass or diamond as the cutting edge, from an area of block face trimmed to about 0.1 mm^2. Section *ribbons* are floated onto water and mounted on fine copper grids for staining and examination. Cell organelles are examined in the electron microscope in isolated, intact form rather than in thin sections.

In *negative staining*, a heavy metal stain, often phosphotungstic acid, is allowed to dry in a puddle around the surfaces of isolated cell particles supported on a thick carbon or plastic film. The stain molecules deposit into surface crevices in the specimen during drying and produce a ghost image in which the specimen appears light against a dark background, often outlining details very clearly. The contrast and surface details of isolated specimens are also frequently increased by the technique of *shadowing*. A thin layer of a heavy metal such as platinum is deposited on the surface of the specimen from one side. The effect in the electron microscope is as if a strong light is directed toward the specimen from one side, placing surface depression in deep shadow. *Freeze-fracture techniques*, in which frozen samples are cleaved with a knife along fracture planes in membranes, utilise this shadowing technique to produce replicas of broken cellular material which show the membrane surface structure on, and within, cell organelles.

1.9.4 Cytochemistry

Cytochemical techniques may be used to identify specific chemical components, especially enzymes, in cells by direct microscopic observations of tissue sections *in situ*. The technique is based on the colorimetric detection of the components or of their metabolic products. The products of most enzyme reactions are soluble and must be made insoluble before visualisation can be achieved. The formation of insoluble electron dense precipitates suitable for electron microscopic studies involves such so-called capture mechanisms. Table 1.5 gives some specific examples of enzyme cytochemical techniques. Cytochemical techniques have been very useful in locating the enzymes associated with oxidative phosphorylation, β-oxidation, fatty acid synthesis etc.

Immunocytochemical microscopy exploits the capacity of cell constituents to act as antigens and to bind specifically to antibodies produced against them. The immunoglobulin antibody is isolated and purified chromatographically from the innoculated animal's blood (say a rabbit) and this is used to

Table 1.5 Some examples of the cytochemical assay of enzymes by light microscopy (LM) and electron microscopy (EM).

Enzyme	Cellular function	Localisation methods
Acyl transferases	Catalyse transfer of acyl group through CoA	Localisation depends on production of free sulphydryl group and can involve (i) production of electron dense ferrocyanide precipitate, or (ii) incubation in presence of cadmium/lanthanum ions to form insoluble mercaptides
Cytochrome oxidase	Terminal cytochrome in electron transfer chain involved in the transfer of electrons from flavoproteins to O_2	'Nadi' reaction. Naphthol plus aromatic diamine mixed in the presence of cytochrome *c* gives coloured indophenol. Modification for EM involves the formation of an indoaniline product which yields a coordination polymer
DNase and RNase	Release nucleotides from DNA and RNA	Lead precipitation, plus an extra initial step e.g. hydrolysis of nucleotide by phosphatase
Esterases	Hydrolyse a range of carboxylic acid esters	For EM, thioacetic acid is used as substrate. If incubation medium includes lead nitrate, an electron dense lead sulphide is produced. Alternatively, for LM and EM, azo dye methods can be used
Sulphatases	Catalyse the hydrolytic cleavage of sulphate esters	For LM, naphthol sulphates substituted as substrates and liberated naphthols coupled to diazonium salts give an insoluble dye product. (Azo dye method.) For EM, heavy metal trapping agents used e.g. lead and barium

produce a second antibody in another animal (usually a goat). A fluorescent dye, e.g. fluorescein is attached to the goat anti-rabbit antibody which may be observed to detect conjugation between rabbit antibodies and the host antigens as a characteristic fluorescence under an ultraviolet microscope. This indirect method, which has widespread pathological application, has the advantage that the goat anti-rabbit antibody may be used with different rabbit antibodies.

Fluorescent analogue cytochemistry (FACS) is a relatively new method which permits a study of the spatial organisation of cellular components in living cells by covalently labelling functional molecules or organelles with fluorescent probes and reincorporating these fluorescent analogues into cells, e.g. the molecular organisation of the cytoskeletal protein actin, labelled with 5-iodoacetamidofluorescein, injected into cells by hypo-osmotic shock treatment may be followed using fluorescence microscopy.

Fluorescent analogues and fluorescent antibodies may both be used in cell sorting methods (Section 1.6.4).

1.10 General approaches to metabolic investigations

Certain generalities may be applied to the way in which metabolic pathways are elucidated and, once they are known, to determine whether or not a particular organelle, cell, organ or organism is exploiting a specific pathway. These approaches are mainly, although not exclusively, applied to *in vitro* preparations and thus may also be used, along with microscopy, to evaluate the integrity of organelles etc. Experiments may be performed in which all intermediates and enzymes involved in a pathway are first identified and quantified by chromatographic, spectroscopic or other means. When tissue extracts are supplemented with these intermediates, the rate of processes in the pathway would be expected to accelerate in a predictable manner due to faster turnover. Changes in metabolic rate may also be achieved by altering environmental conditions e.g. glucose breakdown via the Embden-Meyerhof Pathway in facultative anaerobes, such as the yeast *Saccharomyces cerevisiae*, is accelerated by switching from aerobic to anaerobic conditions. Such metabolic perturbations closely relate to changes in kinetic activities of key regulatory enzymes as they respond to the binding of different positive or negative effector ligands which induce conformational (allosteric) changes in the enzymes (Section 3.3.2). The identification and characterisation of rate-limiting enzymes is thus an important component in the elucidation of metabolic pathways, particularly with respect to how the pathways are regulated. In general terms, there must be a positive correlation between the rate of the overall physiological process and the kinetic and regulatory properties of key enzymes. Addition of *enzyme inhibitors* will cause intermediates to accumulate thereby aiding in their identification.

Isotopic tracers are a very powerful tool for identifying the metabolic fate of precursors and for following metabolic turnover (Section 9.5.1). *Time course studies* and *pulse chase* techniques are the two main approaches employed in metabolic studies in which isotopically-labelled compounds are identified and recovered. Often different elements in compounds may be labelled and this aids in discriminating biosynthetic origins in pathways, e.g. in amino acid biosynthesis. The great advantage of radio-labelling is in its sensitivity and specificity.

Preliminary evidence for identifying whether or not a particular pathway is operating may be inferred from yield (biomass production) studies on microorganisms growing on a particular substrate, or by measurements of gaseous exchange processes. The value of this approach relies on prior knowledge of stoichiometry and of the pathways for dissimilation.

Manometry may be used to measure either uptake or evolution of both CO_2 and O_2 whereas O_2 and CO_2 electrodes simply monitor changes in O_2 and CO_2 levels respectively albeit with greater sensitivity. Manometry also has the distinctive feature of allowing the simultaneous determination of O_2 and CO_2 exchange and has the advantage that the magnitude of the exchange is independent of the partial pressure of the gas at the beginning of the experiment. Manometric studies are carried out in a small flask attached to some form of manometer which measures changes in the amount of gas in the flask. In all types of manometer, the flasks, immersed in a water bath with a temperature control of ± 0.5°C, are shaken mechanically at rates of 100 to 120 oscillations per minute to ensure that respiratory gas exchange is not limited by diffusion of gas into the liquid phase. The total volume of liquid in the flask should generally not exceed 4 cm^3 because of gas diffusion limitations. The two principal types of manometer are the *Warburg constant volume manometer* (Fig. 1.3) and the *Gilson constant pressure manometer* (Fig. 1.4).

The biological applications of manometry are extensive. *Respiratory quotients* (*RQ*), defined as the relationship between the volume of CO_2 produced and the volume of O_2 consumed during respiration i.e.

$$RQ = \frac{CO_2 \text{ evolved}}{O_2 \text{ absorbed}}$$

may give an indication of the nature of the endogenous substrate being metabolised. The complete oxidation of a simple carbohydrate gives an *RQ* of 1, whereas an average fat gives a value of approximately 0.7 and a protein of 0.8. Deviations from these values are sometimes obtained in practice since some CO_2 may be incorporated into cellular material, lowering the volume of CO_2 evolved. The rate at which an organism or tissue consumes O_2 or evolves CO_2 is expressed by a *metabolic quotient* Q_x, where x is the gas being measured. Thus Q_{O_2} is defined as the volume of O_2 taken up per mg dry weight of biological material per hour. In some cases it may be expressed in terms of mg of nitrogen (generally determined by the Kjeldahl method, Section 3.2.2) or mg protein or DNA. In such cases it is expressed as follows:

$$Q_{O_2}(N) = \text{mm}^3\text{O}_2 \text{ mg tissue N}^{-1}\,\text{h}^{-1}$$

Metabolic quotients may also be determined in atmospheres of different or varying gas composition, for example pure N_2, in which case an additional suffix is added to the quotient, thus:

$$O^{N_2}_{CO_2}(N) = \text{mm}^3\text{CO}_2 \text{ mg tissue N}^{-1}\,\text{h}^{-1} \text{ in an atmosphere of N}_2 \text{ gas.}$$

If there is a possibility of confusion, the quotients may be indicated as positive or negative, indicating the production or removal of metabolite respectively.

Manometric techniques have been applied in studies on tissue slices and homogenates with attendant problems of homogeneity of gas supply and artefacts, in mitochondrial studies for the study of respiratory control, and the effect of inhibitors on mitochondrial respiration (although the

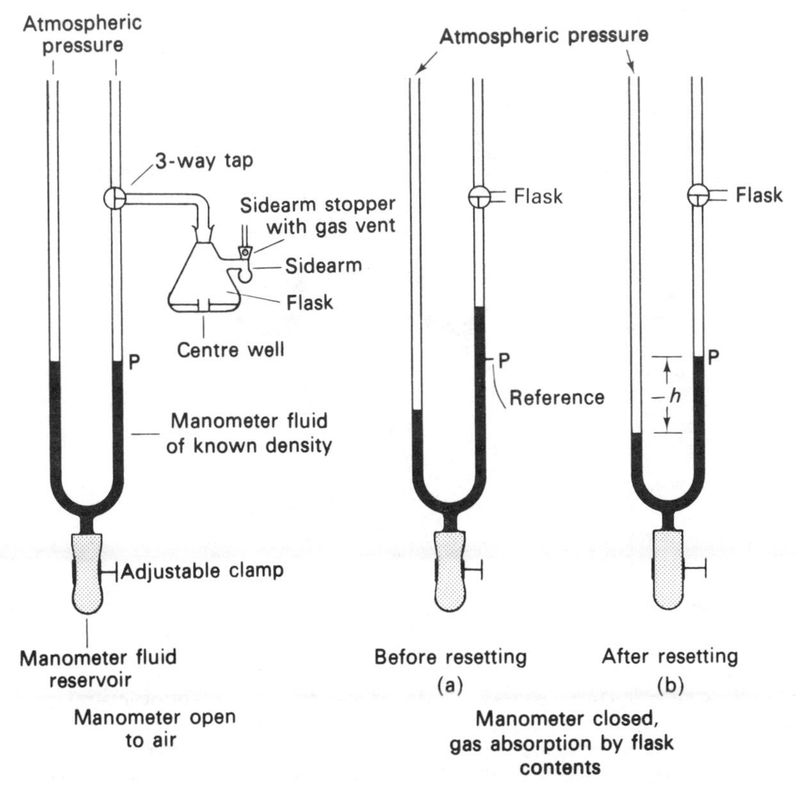

Fig. 1.3 Diagrammatic representation of a Warburg manometer. CO_2 gas is being absorbed in the experimental flask and the resulting decrease in pressure forces the fluid level in the right-hand limb to rise and in the left-hand limb to fall. At regular time course intervals the meniscus of the fluid in the left-hand limb is returned to the reference point P by withdrawing fluid into the reservoir using the adjustable clamp, thereby measuring the resultant decrease in pressure at constant volume as $-h$ mm. The change h is related to the quantity X of gas evolved at s.t.p. by the equation:

$$X = h \left[\frac{V_g \dfrac{273}{T} + V_f \alpha}{P_o} \right]$$

where V_g = volume of the gas space in the flask including that of the capillary from the flask to the reference mark P (mm^3); V_f = volume of liquid in the flask (mm^3); α = solubility of the gas in the liquid in the flask; T = temperature of the water bath in °K; P_o = standard pressure expressed in mm of manometric fluid. Since, for a given experimental flask being used to study a particular reaction under defined conditions, all values within the brackets in the equation are constant, $X = kh$ where k is the flask constant. Flask constants for particular manometers are usually supplied by manufacturers or can be obtained by a suitable form of calibration.

technique is less sensitive than the O_2 electrode, Section 10.5). When used for photosynthetic studies it is necessary to carry out a control determination in darkness and to keep the partial pressure of one gas constant during the experiment. This may be achieved by maintaining a constant partial

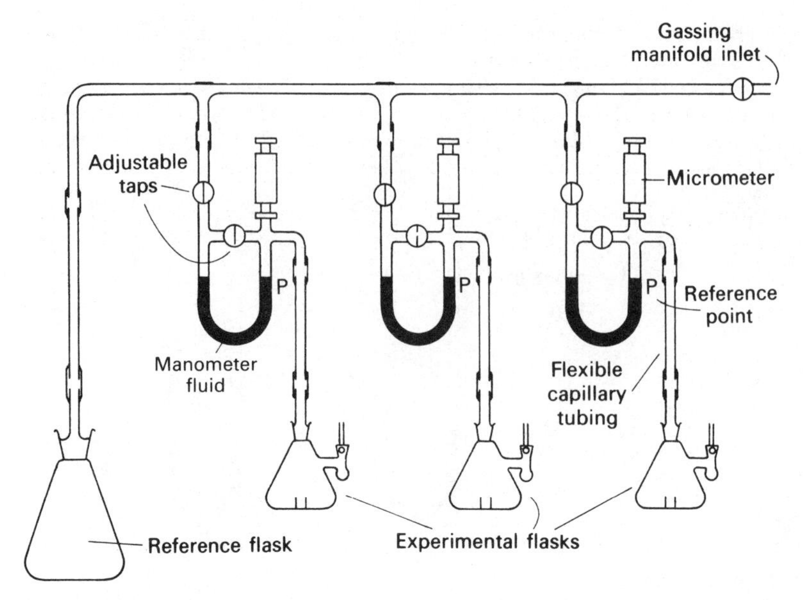

Fig. 1.4 Diagrammatic representation of the Gilson constant pressure manometer. Up to 14 experimental flasks and small U-tube capillaries are connected to the same reference flask by a gassing manifold. This arrangement allows all flasks to be gassed simultaneously. The reference flask, which is much larger than the experimental flasks, eliminates errors due to changes in barometric pressure and water bath temperature. Attached to each U-tube is a micrometer calibrated in microlitres, which is used to adjust the level of fluid in the tube to a reference mark prior to taking readings, thereby ensuring that all gas volume changes are recorded at constant pressure. Changes in gas volumes within the experimental flask are thus measured directly, thereby obviating the need to calibrate the flask and U-tube. Volume changes must however be adjusted to s.t.p. allowing for the vapour pressure of water in the flask.

pressure of CO_2 by using carbonate-bicarbonate buffers or by removing all the O_2 by chemical means. This has limited value however since it is known that the rate of CO_2 exchange is dependent on the O_2 content of the atmosphere.

1.11 Suggestions for further reading

Altman, P.L. and Katz, D.D. (1976). *Cell Biology*. The Federation of American Societies for Experimental Biology. (A good reference text to the general area of cell biology.)

Freifelder, D. (1983). *Molecular Biology*. Jones & Bartlett Publishers Inc., Boston. (A comprehensive introduction to prokaryotes and eukaryotes.)

Hall, J.L. (1978). *Electron Microscopy and Cytochemistry of Plant Cells*. Elsevier/North Holland Biomedical Press, Oxford.

Morris, J.G. (1974). *A Biologists Physical Chemistry*, 2nd edition. Edward Arnold, London. (A detailed discussion of pH, acid and base strength, buffer action and other physico-chemical principles underlying biological phenomena.)

Smyth, D.H. (1978). *Alternatives to Animal Experiments*. Scolar Press in association with the Research Defence Society.
Stanbury, P.F. and Whitaker, A. (1984). *Principles of Fermentation Technology*. Pergamon Press, Oxford. (An excellent introduction to fermentation processes.)
Thomas, E. and Davey, M.R. (1975). *From Single Cells to Plants*. Wykeham, London. (An intermediate level monograph on plant tissue and cell culture techniques.)
Umbreit, W.W., Burris, R.H. and Stauffer, J.F. (1972). *Manometric and Biochemical Techniques*, 5th edition. Burgess Publishing Company, Minneapolis. (A comprehensive text on manometric techniques.)

2
Centrifugation techniques

2.1 Introduction

Centrifugation separation techniques are based upon the behaviour of particles in an applied centrifugal field. The particles are normally suspended in a specific liquid medium, held in tubes or bottles, which are located in a rotor. The rotor is positioned centrally on the drive shaft of the centrifuge. Particles which differ in density, shape or size can be separated since they sediment at different rates in the centrifugal field, each particle sedimenting at a rate which is proportional to the applied centrifugal field.

Centrifugation techniques are of two main types. *Preparative centrifugation* techniques are concerned with the actual separation, isolation and purification of, for example, whole cells, subcellular organelles, plasma membranes, polysomes, ribosomes, chromatin, nucleic acids, lipoproteins and viruses, for subsequent biochemical investigations. Very large amounts of material may be involved when harvesting microbial cells from culture media, plant and animal cells from tissue culture or plasma from blood. Relatively large amounts of cellular particles may also be isolated in order to study their morphology, composition and biological activity. It is also possible to isolate biological macromolecules, such as nucleic acids and proteins, from preparations which have received some preliminary purification by, for example, fractional precipitation (Section 3.2.3). In contrast, *analytical centrifugation* techniques are devoted mainly to the study of pure or virtually pure macromolecules or particles. It is primarily concerned with the study of the sedimentation characteristics of biological macromolecules and molecular structures, rather than with the actual collection of particular fractions. It requires only small amounts of material and utilises specially designed rotors and detector systems to continuously monitor the process of sedimentation of the material in the centrifugal field. Such studies yield information from which the purity, molecular weight and shape of the material may be deduced. Since preparative centrifugation techniques are more commonly used in undergraduate studies, this chapter will concentrate on these techniques and deal only briefly with analytical centrifugation techniques.

2.2 Basic principles of sedimentation

The rate of sedimentation is dependent upon the applied *centrifugal field* (G) being directed radially outwards, which is determined by the square of the angular velocity of the rotor (ω, in radians per second) and the radial distance (r, in cm) of the particle from the axis of rotation, according to the equation:

$$G = \omega^2 r \qquad 2.1$$

Since one revolution of the rotor is equal to 2π radians, its angular velocity, in radians per second, can readily be expressed in terms of revolutions per minute (rev min^{-1}), the common way of expressing rotor speed being:

$$\omega = \frac{2\pi \text{rev min}^{-1}}{60} \qquad 2.2$$

The centrifugal field (G) in terms of rev min^{-1} is then:

$$G = \frac{4\pi^2(\text{rev min}^{-1})^2 r}{3600} \qquad 2.3$$

and is generally expressed as a multiple of the earth's gravitational field (g = 980 cm s^{-2}), i.e. the ratio of the weight of the particle in the centrifugal field to the weight of the same particle when acted on by gravity alone, and is then referred to as the *relative centrifugal field* (RCF) or more commonly as the 'number times g'.

Hence
$$\text{RCF} = \frac{4\pi^2(\text{rev min}^{-1})^2 r}{3600 \times 980} \qquad 2.4$$

which may be shortened to give:

$$\text{RCF} = (1.11 \times 10^{-5})(\text{rev min}^{-1})^2 r \qquad 2.5$$

When conditions for the centrifugal separation of particles are reported, therefore, rotor speed, radial dimensions and time of operation of the rotor must all be quoted. Since biochemical experiments are usually conducted with particles dissolved or suspended in solution, the rate of sedimentation of a particle is dependent not only upon the applied centrifugal field but also upon the density and size of the particle, the density and viscosity of the medium in which it is sedimenting and the extent to which its shape deviates from spherical. When a particle sediments it must displace some of the solution in which it is suspended, resulting in an apparent upthrust on the particle equal to the weight of liquid displaced. If a particle is assumed to be spherical and of known volume and density, then the net force (F) it experiences when centrifuged at an angular velocity of ω radians per second is given by:

$$F = \frac{4}{3}\pi r_p^3(\rho_p - \rho_m)\omega^2 r \qquad 2.6$$

where $\frac{4}{3}\pi r_p^3$ = volume of a sphere of radius r_p

ρ_p = density of the particle
ρ_m = density of the suspending medium
r = distance of the particle from the centre of rotation

Particles, however, generate friction as they migrate through the solution. If a particle is spherical and moving at a known velocity, then the frictional force opposing motion is given by *Stokes' law*:

$$f_o = 6\pi\eta r_p v \qquad 2.7$$

where f_o = frictional coefficient for a spherical particle
η = viscosity coefficient of the medium
v = velocity or sedimentation rate of the particle

A particle of known volume and density, and present in a medium of constant density, will therefore accelerate in a centrifugal field until the net force on the particle equals the force resisting its motion through the medium, i.e.

$$\mathrm{F} = f_o \qquad 2.8$$

or

$$\frac{4}{3}\pi r_p^3(\rho_p - \rho_m)\omega^2 r = 6\pi\eta r_p v \qquad 2.9$$

In practice the balancing of these forces occurs quickly with the result that the particle sediments at a constant rate. Its rate of sedimentation (v) is then given by:

$$v = \frac{\mathrm{d}r}{\mathrm{d}t} = \frac{2}{9}\,\frac{r_p^2(\rho_p - \rho_m)\omega^2 r}{\eta} \qquad 2.10$$

It can be seen from equation 2.10 that the sedimentation rate of a given particle is proportional to its size, and to the difference in density between the particle and the medium. It will be zero when the density of the particle and the medium are equal; it will decrease when the viscosity of the medium increases, and increase as the force field increases. However, since the equation involves the square of the particle radius, it is apparent that the size of the particle has the greatest influence upon its sedimentation rate. Particles of similar density, but only slightly different in size can therefore have large differences in their sedimentation rate. Integration of equation 2.10 yields equation 2.11 which gives the sedimentation time for a spherical particle in a centrifugal field as a function of the various variables and in relation to the distance of travel of the particle in the centrifuge tube.

$$t = \frac{9}{2}\,\frac{\eta}{\omega^2 r_p^2(\rho_p - \rho_m)}\ln\frac{r_b}{r_t} \qquad 2.11$$

where t = sedimentation time in seconds
r_t = radial distance from the axis of rotation to liquid meniscus
r_b = radial distance from the axis of rotation to bottom of tube

It is thus clear that a mixture of heterogeneous, approximately spherical particles can be separated by centrifugation on the basis of their densities and/or their size, either by the time required for their complete sedimentation or by the extent of their sedimentation after a given time. These alternatives form the basis for the separation of biological macromolecules and of cell organelles from tissue homogenates. The order of separation of the major cell components is generally whole cells and cell debris first, followed by nuclei, chloroplasts, mitochondria, lysosomes (or other microbodies), microsomes (fragments of smooth and rough endoplasmic reticulum) and ribosomes.

Considerable discrepancies exist between the theory and practice of centrifugation. Complex variables not accounted for in equation 2.10 and 2.11, such as the concentration of the suspension, nature of the medium, and the characteristics of the centrifuge, will affect the sedimentation properties of a mixed population of particles. Moreover, aspherical particles exhibit a modified relationship between the sedimentation rate and the particle size. Where there is an appreciable asymmetry, as in the case of rod-like molecules such as DNA and proteins like F-actin and myosin, the frictional coefficient (f) of the molecule can be increased by as much as ten times that of the frictional coefficient of a sphere (f_o). This results in particles sedimenting at a slower rate. Equation 2.10 can therefore be modified to give equation 2.12:

$$v = \frac{dr}{dt} = \frac{2}{9} \frac{r_p^2(\rho_p - \rho_m)\omega^2 r}{\eta(f/f_o)} \qquad 2.12$$

which takes into account the effect of varying size and shape on the sedimentation rate of a particle. The frictional ratio, f/f_o is approximately 1 for spherical molecules, larger values being observed for non-spherical molecules. Hence particles of a given mass, but different shape, sediment at different rates. This point is exploited in the study of conformation of molecules by analytical ultracentrifugation (Section 2.10.3).

Whilst it is convenient to consider the sedimentation of particles in a uniform centrifugal field, in practice this does not occur using preparative rotors. Due to the nature of rotor design (Fig. 2.1) the effective radial dimension of a given particle will change according to its position in the sample container and will vary between r_{min} and r_{max}. Since the centrifugal field generated is proportional to $\omega^2 r$, a particle will experience a greater field the further away it is from the axis of rotation. The operative centrifugal field, in a fixed angle rotor for example, can differ by a factor of up to two between the top and bottom of the centrifuge tube, thus the sedimentation rate of particles at the bottom of the tube will be twice that of identical particles near the top of the tube. As a result, particles will tend to move faster

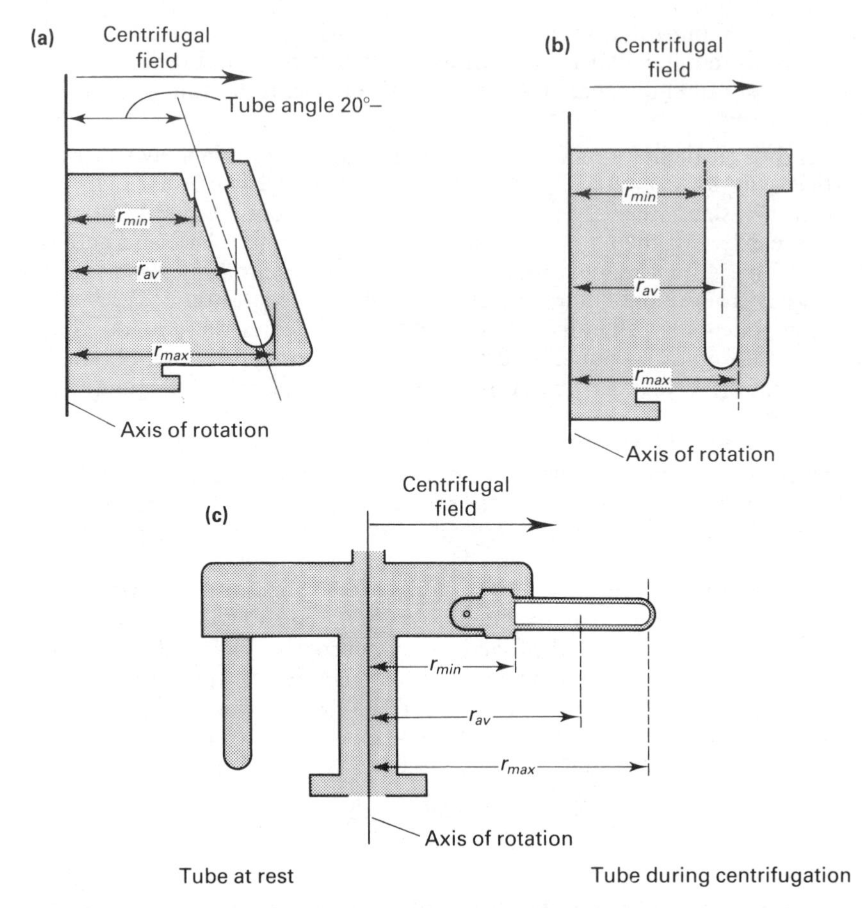

Fig. 2.1 Cross sectional diagrammatic representation of **(a)** fixed angle; **(b)** vertical tube; and **(c)** swinging bucket rotors showing the minimum, average and maximum radii from the axis of rotation.

as they sediment through a non-viscous medium. It is normal, therefore, to record the relative centrifugal field calculated from the average radius of rotation (r_{av}) of the column of liquid in the tube (i.e. the distance from the centre of rotation to the middle of the liquid column in the centrifuge tube). The average relative centrifugal field (RCF_{av}) is therefore the numerical average of the values exerted at r_{min} and r_{max}. If the sample container is only partially filled then, in the case of fixed angle and swinging bucket rotors, the minimum radius (r_{min}) is effectively increased and the particles will therefore start to sediment in a higher gravitational field and have a reduced path length to travel. Consequently sedimentation will be quicker. Centrifuge manuals normally provide details of the maximum permitted speed for a rotor, maximum relative centrifugal fields generated and graphs which enable the ready conversion of RCF to rev min^{-1} at r_{min}, r_{av}, and r_{max}.

The sedimentation rate or velocity (v) of a particle can also be expressed in terms of its sedimentation rate per unit of centrifugal field, commonly referred to as its *sedimentation coefficient*, s. From equation 2.12 it can be seen that, if the composition of the suspending medium is defined, then the sedimentation rate is proportional to $\psi^2 r$, the centrifugal field, and equation 2.12 simplifies to:

$$v = s\omega^2 r \qquad 2.13$$

or

$$s = \frac{v}{\omega^2 r} = \frac{dr}{dt} \cdot \frac{1}{\omega^2 r} \qquad 2.14$$

Since sedimentation rate studies may be performed using a wide variety of solvent–solute systems, the measured value of the sedimentation coefficient, which is affected by temperature, solution viscosity and density, is often corrected to a value that would be obtained in a medium with a density and viscosity of water at 20°C, and expressed as the *standard sedimentation coefficient* or $s_{20,w}$. For many macromolecules including nucleic acids and proteins the sedimentation coefficient usually decreases in value with increase in the concentration of solute, this effect becoming more severe with increase both in molecular weight and the degree of extension of the molecule. Hence $s_{20,w}$ is usually measured at several concentrations and extrapolated to infinite dilution to obtain a value of $s_{20,w}$ at zero concentration, the $s^o_{20,w}$. The sedimentation coefficients of most biological particles are very small, and for convenience its basic unit is taken as 10^{-13} seconds which is termed one *Svedberg unit* (S), in recognition of Svedberg's pioneering work in this type of analysis. Therefore, a ribosomal RNA molecule possessing a sedimentation coefficient of 5×10^{-13} seconds is said to have a value of 5S.

Examination of equation 2.12 and 2.14 shows that the sedimentation coefficient is influenced by such features as the shape, size and density of the particle, hence it is commonly used to characterise a particular molecule or structure. Generally, the larger the molecule or particle, the larger its Svedberg unit and hence the faster its sedimentation rate. Sedimentation coefficients for enzymes, peptide hormones and soluble proteins are 2 to 25S, nucleic acids 3 to 100S, ribosomes and polysomes 20 to 200S, viruses 40 to 1000S, lysosomes 4000S, membranes 100 to 100×10^3S, mitochondria 20×10^3S to 70×10^3S, and nuclei 4000×10^3S to $40\,000 \times 10^3$S.

2.3 Centrifuges and their use

Centrifuges may be classified into four major groups – the small bench centrifuges, large capacity refrigerated centrifuges, high speed refrigerated centrifuges and ultracentrifuges of two types, preparative and analytical.

2.3.1 Small bench centrifuges

These are the simplest and least expensive centrifuges and exist in many types of design. They are often used to collect small amounts of material that rapidly sediment (yeast cells, erythrocytes, coarse precipitates), and generally have a maximum speed of 4000 to 6000 rev min^{-1} with maximum relative centrifugal fields of 3000 to 7000*g*. Most operate at ambient temperature, the flow of air around the rotor controlling rotor temperature. Some of the latest designs, however, incorporate a refrigeration system to keep rotors cool thus preventing denaturation of proteins. Small *microfuges* are available providing virtual instant acceleration to maximum speeds of 8000 to 13 000 rev min^{-1} developing fields of approximately 10 000*g*. These centrifuges have proved extremely useful for sedimenting small volumes (250 mm^3 to 1.5 cm^3) of material very quickly (one or two minutes). Typical applications include the rapid sedimentation of blood samples, and of synaptosomes used to study the effect of drugs on the uptake of biogenic amines.

2.3.2 Large capacity refrigerated centrifuges

These have a maximum speed of 6000 rev min^{-1} and produce a maximum relative centrifugal field approaching 6500*g*. They have refrigerated rotor chambers and vary only in their maximum carrying capacity, all being capable of utilising a variety of interchangeable *swinging bucket* and *fixed angle rotors* enabling separation to be achieved in 10, 50 and 100 cm^3 tubes. Large total capacity (4 to 6 dm^3) centrifuges are also available which, in addition to accommodating smaller tubes, are also capable of holding bottles, each of 1.25 dm^3 capacity. In all these centrifuges, the rotors are usually mounted on a rigid suspension, hence it is extremely important that the centrifuge tubes and their contents are balanced accurately (to within 0.25 gram of each other). Rotors must never be loaded with an odd number of tubes and, where the rotor is only partially loaded, the tubes must be located diametrically opposite each other in order that the load is distributed evenly around the rotor axis. These instruments are most often used to compact or collect substances that sediment rapidly, for example, erythrocytes, coarse or bulky precipitates, yeast cells, nuclei and chloroplasts.

2.3.3 High speed refrigerated centrifuges

These instruments are available with maximum rotor speeds in the region of 25 000 rev min^{-1} generating a relative centrifugal field of about 60 000*g*. They generally have a total capacity of up to 1.5 dm^3, and a range of interchangeable fixed angle and swinging bucket rotors. These instruments are most often used to collect microorganisms, cellular debris, larger cellular organelles and ammonium sulphate protein precipitates. They cannot generate sufficient centrifugal force to effectively sediment viruses or smaller organelles such as ribosomes.

2.3.4 Continuous flow centrifuges

The *continuous flow centrifuge* is a relatively simple high-speed centrifuge. The rotor, through which particles suspended in medium flow continuously (usually 1 to 1.5 dm^3 min^{-1}), is, however, long, tubular, and non-interchangeable. As medium enters the rotating rotor, particles are sedimented against its wall and excess clarified medium overflows through an outlet port. The major application of this type of centrifuge is in the harvesting of bacteria or yeast cells from large volumes of culture medium (10 to 500 dm^3). Some high speed centrifuges can be adapted to function in the continuous flow mode by accepting a specially designed rotor (Section 2.4.4).

2.3.5 Preparative ultracentrifuges

Preparative ultracentrifuges are capable of spinning rotors to a maximum speed of 80 000 rev min^{-1} and can produce a relative centrifugal field of up to 600 000*g*. The rotor chamber is refrigerated, sealed and evacuated to minimise excessive rotor temperatures being generated by frictional resistance between the air and the spinning rotor. The temperature monitoring system is more sophisticated than in simpler instruments, employing an infrared temperature sensor which can continuously monitor rotor temperature and control the refrigeration system. An overspeed control system is also incorporated into these instruments to prevent operation of the rotor above its maximum rated speed and there are electronic circuits to detect rotor imbalance. In order to minimise vibration, caused by slight rotor imbalance, which may arise due to unequal loading of the centrifuge tubes, ultracentrifuges are fitted with a flexible drive shaft system. Centrifuge tubes and their contents, however, must still be accurately balanced to within 0.1 gram of each other. For safety reasons, rotor chambers of high speed and ultracentrifuges are always enclosed in heavy armour plating.

An air-driven table-top preparative ultracentrifuge, called an *airfuge*, is available capable of accelerating a 3.7 cm diameter rotor, accommodating 6 × 175 mm^3 tubes on a virtual friction-free cushion of air in a non-vacuated chamber, to 100 000 rev min^{-1} (160 000*g*) in approximately 30 seconds. The airfuge has found applications in biochemical and clinical research where only small volume samples are available requiring high centrifugal forces. Examples include steroid hormone receptor assays, macromolecule/ligand binding studies, separation of the major lipoprotein fractions from plasma and deproteinisation of physiological fluids for amino acid analysis.

2.3.6 Analytical ultracentrifuges

These instruments are capable of operating at speeds approaching 70 000 rev min^{-1} (500 000*g*) and consist (Fig. 2.2a) of a motor, a rotor contained in a protective armoured chamber which is refrigerated and evacuated, and an optical system which enables the sedimenting material to be observed so as to determine concentration distributions within the sample at any time during centrifugation.

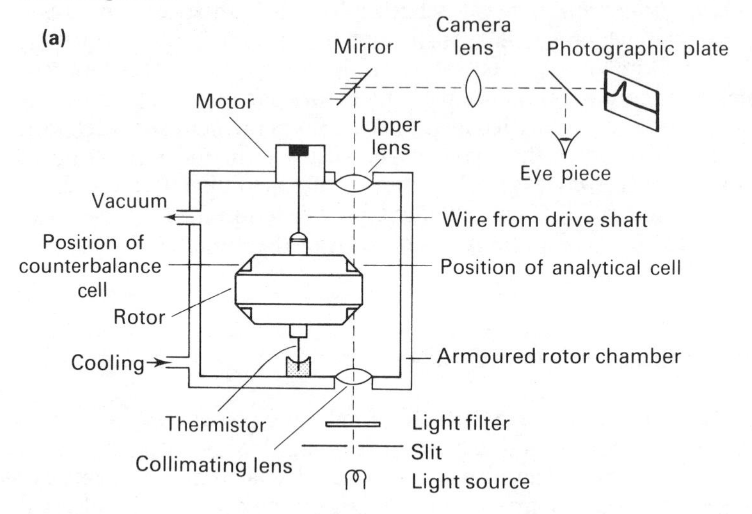

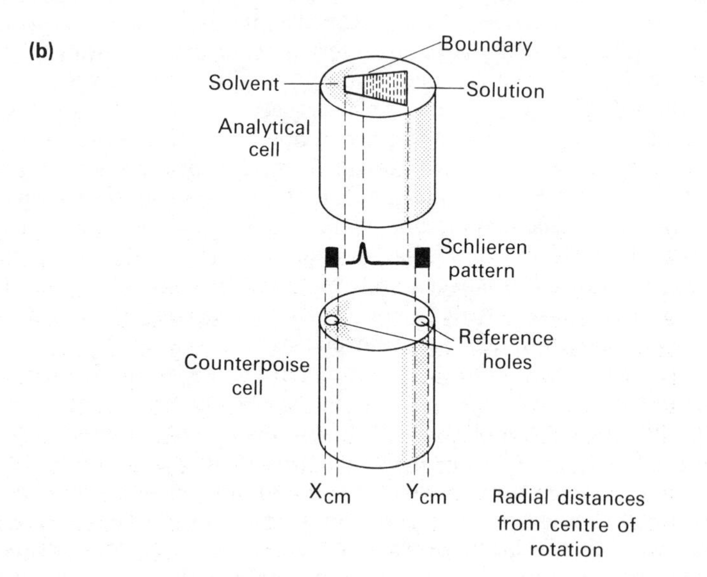

Fig. 2.2 Diagrammatic representation of **(a)** an analytical ultracentrifuge system and **(b)** an analytical and counterpoise cell.

The rotor is suspended on a wire coming from the drive shaft of a high speed motor which permits the rotor to find its own axis of rotation. The tip of the rotor contains a thermistor for measuring temperature. Housed in the rotor are two cells, the *analytical cell* and the *counterpoise cell* which counter-balances the analytical cell. Two holes (at calibrated distances from the centre of rotation) are drilled through the counterpoise cell (Fig. 2.2b) to facilitate the calibration of distances in the analytical cell. A commonly used analytical cell has a single sector centrepiece with a 4° sector shape to prevent convection and is capable of holding a column of liquid approximately 14 mm high when completely filled (usually 1 cm^3 capacity). The centrepiece is so designed that, when correctly aligned in the rotor, the walls will be parallel to the lines of centrifugal force, behaving on the same principle as the swinging bucket rotor (Section 2.4.2) to give almost ideal conditions for sedimentation. This ensures that there will be no accumulation of material against the wall of the analytical cell during centrifugation. Analytical cells have upper and lower plane windows of quartz or synthetic sapphire, the latter being used with interference optics, as they have less tendency to distort under a high gravitational field.

The rotor chamber contains an upper and lower lens, the former, together with a camera lens, focuses light onto a photographic plate whilst the latter collimates the light so that the sample cell is illuminated by parallel light. Monitoring of the sedimenting material may be achieved by utilising either an ultraviolet absorption system or differences in refractive index utilising a Schlieren optical system or a Rayleigh interferometric system. The Schlieren optical system makes use of the fact that, if light passes through a solution of uniform concentration it does not deviate but, on passing through a solution having different density zones, it is refracted at the boundary between these zones. The optical system records the change in refractive index of the solution which will vary as the concentration changes. In the case of sedimenting materials in an analytical cell, a boundary is formed between the solvent, which has been cleared of particles, and the remainder of the solution containing the sedimenting material. This behaves like a refraction lens resulting in the production of a peak on the photographic plate, which is used as the detector system. The concentration can be determined from the area of the peak. As sedimentation proceeds the boundary, and hence the peak, shifts, the rate at which the peak moves giving a measure of the rate at which the material is sedimenting (Fig. 2.3). After a period of sedimentation, the peak height diminishes and the width increases owing to diffusion of the sample. Its area, however, is unchanged. The Schlieren optical system plots the refractive index gradient against distance along the analytical cell which makes it useful for locating boundaries in sedimentation velocity measurements (Section 2.10.1). For some techniques (e.g. the sedimentation equilibrium method for molecular weight determinations (Section 2.10.1)), the Schlieren system is not sufficiently sensitive to detect small concentration differences. Use is therefore made of the more sensitive Rayleigh interference system which employs a double-sector cell, in which one sector contains the solvent and the other the solution. The optical system measures the difference

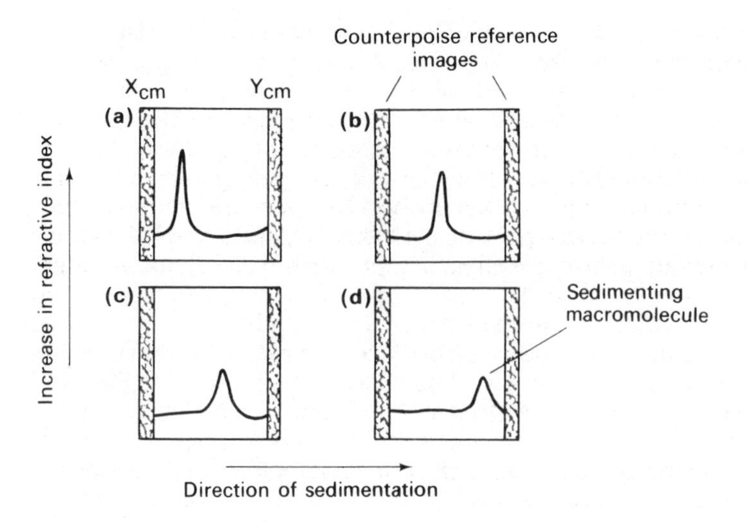

Fig. 2.3 Schematic diagrams of the stages (**(a)** to **(d)**) of sedimentation of a macromolecule using the technique of sedimentation velocity. Sedimentation patterns are obtained using the Schlieren optical system which measures the refractive index gradient at each point in the cell at varying time intervals.

in refractive index between the reference solvent and the solution by the displacement of interference fringes caused by slits placed above the analytical cell, each fringe tracing a curve of the refractive index gradient against distance in the cell. Since the position of the fringes is determined by solute concentration, it is possible to measure the concentration of solute at any point along the cell.

In advanced analytical ultracentrifuges, the photographic plate detector system is replaced by an accurate electronic scanning system which can plot the concentration of the sample at all points in the analytical cell at any particular time. Different wavelengths of light can be selected enabling the separate movement of single components in a mixture of substances to be monitored provided they absorb light at different wavelengths.

2.4 Design and care of preparative rotors

2.4.1 Materials used in rotor construction

The centrifugal force created by a spinning rotor generates load or stress on the rotor material. Since rotors used for low speed centrifugation experience a much lower degree of stress in comparison with high speed rotors they can be made of brass, steel or Perspex. The higher stress forces generated during high speed centrifugation necessitates the use of aluminium alloy or titanium alloy. Rotors made from titanium alloy have a greater strength to weight ratio and are therefore capable of withstanding nearly twice the centrifugal force of rotors made from aluminium alloy. They are also more resistant to chemical corrosion and do not suffer from metal fatigue. Aluminium alloy rotors, although less expensive, are far more susceptible to corrosion, being readily attacked by acids and alkalis, and high concentrations of salt solutions (e.g. NaCl and KBr used in lipoprotein fractionations, ammonium sulphate used in protein precipitation, and caesium or rubidium salts used in the preparation of density gradients). Aluminium rotors are therefore anodised to provide a protective coating to the underlying metal rotor surface.

2.4.2 Fixed angle and swinging bucket rotors

In *fixed angle rotors* (Fig. 2.1), particles move radially outward under the influence of the centrifugal field and have only a short distance to travel before reaching the outer wall of the centrifuge tube. On hitting the outer wall the particles slide down the tube wall to form a pellet at the bottom of the tube and hence give a rapid collection of sediment. However, strong convection currents caused by the dense particle layer against the outer wall, tend to have undesirable effects when attempts are made to separate particles of similar sedimentation characteristics. Thus, fixed angle rotors have proved valuable in the separation of particles whose sedimentation rates differ by a significant order of magnitude.

Convection currents are also produced in centrifuge tubes used in the *swinging bucket rotor* (Fig. 2.1), although to a lesser extent. This is because particles in a centrifugal field fan out radially from the centre of rotation rather than sedimenting in parallel lines. Particles again strike the wall of the tube and travel down the wall to the tube base. Control of convection and swirling effects has been achieved by use of a sector-shaped centrifuge tube (the *Strohmaier cell*) in the swinging bucket rotor, by slowly accelerating and decelerating the rotor, and by the use of density gradients.

2.4.3 Vertical tube rotors

Since sedimentation of particles occurs across the diameter of the tube, the *vertical tube rotor* (Fig. 2.1) presents the shortest possible pathlength for the particle. Sedimentation of particles can thus be achieved more quickly than using either the swinging bucket rotor or the fixed angle rotor and since r_{min} is greater, due to the tubes being at the rotors' edge, a larger minimum centrifugal field is generated. In this type of rotor, however, any pellet formed is deposited along the entire length of the outer wall of the centrifuge tube, which could be a disadvantage since it tends to fall back into the solution at the end of centrifugation.

2.4.4 Continuous flow rotors

Continuous flow rotors are designed for high speed separation of relatively small quantities of solid matter from large volumes of suspension and are particularly useful for the harvesting of cells and the large scale isolation of viruses. The suspension is continuously fed into the rotor during centrifugation. Throughput rates vary with the sample to be isolated, but can range from 100 $cm^3 min^{-1}$ to 1.4 $dm^3 min^{-1}$. These rotors can be used for pelleting samples or, by the use of a density gradient, to separate particles according to differences in their density.

2.4.5 Elutriator rotors

The *elutriator rotor* (Fig. 2.4a) is a type of continuous flow rotor which contains a single conical shaped separation chamber and a bypass chamber on the opposite side of the rotor which serves as a counterbalance. With the aid of a windowed centrifuge door and a synchronised stroboscopic lamp,

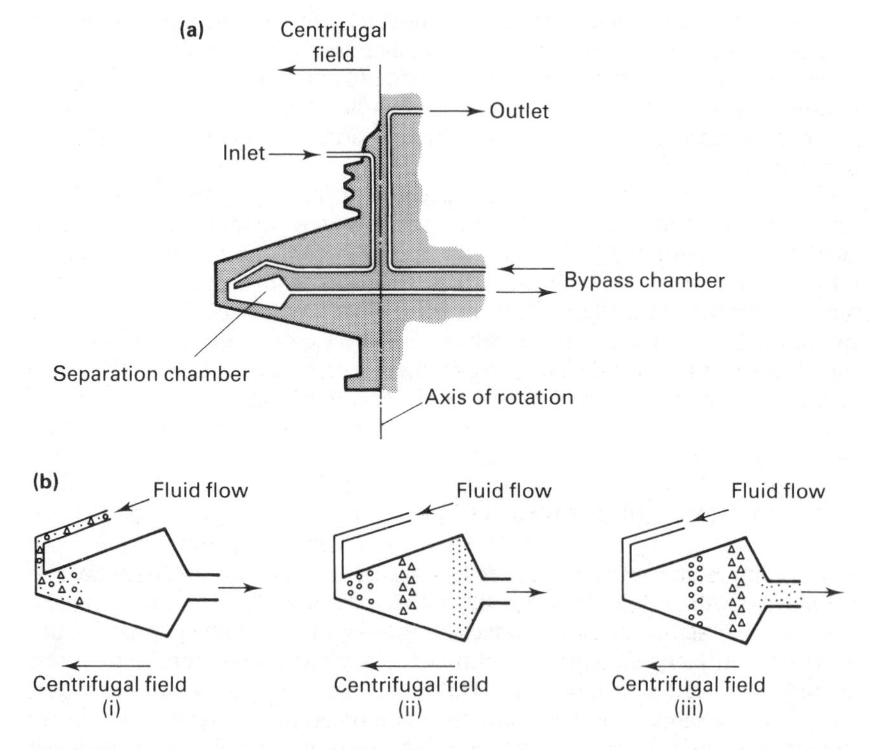

Fig. 2.4 Diagrammatic representations of **(a)** cross section through an elutriator rotor and **(b)** the separation of particles in the separation chamber of an elutriator rotor by centrifugal elutriation.

apertures in the rotor allow the rotor chamber contents to be observed during centrifugation. Elutriating fluid is pumped into the rotor chamber at its *peripheral edge* via a rotating seal assembly whilst the rotor is spinning at a preselected speed (usually between 1000 and 3000 rev min^{-1}). Since the separation chamber is conically shaped (Fig. 2.4b), a gradient of liquid flow velocity, which gradually decreases as the diameter of the chamber increases towards its *centripetal end* (i.e. towards the axis of rotation), will therefore exist in the chamber which opposes the applied centrifugal field. The tendency of particles of differing sedimentation rate to sediment in the centrifugal field (Fig. 2.4b(i)) is balanced against the controlled flow of liquid being pumped through the separation chamber towards its centripetal end. Particles then band in the separation chamber at a position where the opposing centrifugal force and fluid velocity acting on them are at equilibrium. The position of this equilibrium will depend upon the shape, density and particularly the size of the particle (Fig. 2.4b(ii)). Either by a stepwise decrease in rotor speed or a stepwise increase in liquid flow rate through the separation chamber, collection of the separated uniformly-sized particles can be made in order of successively increasing diameter by elutriation from the chamber (Fig. 2.4b(iii)).

Using the technique of *centrifugal elutriation* (Section 2.7.3) the elutriator rotor has been successfully used to separate different types of mononuclear leucocytes from human blood, the purification of Kupffer and endothelial cells from sinusoidal liver cells, fat-storing cells from rat liver, the bulk separation of rat brain cells and the fractionation of yeast cell populations.

2.4.6 Zonal rotors

Zonal rotors are designed to minimise the wall effects which are encountered in swinging bucket and fixed angle rotors, and to increase sample size. The capacity of zonal rotors range from 300 cm^3 to 2000 cm^3. Low speed zonal rotors, designed to operate near 5000 rev min^{-1} (5000*g*) are made of aluminium having a thick transparent Perspex top and bottom to permit direct examination of particle sedimentation during the course of centrifugation. High speed zonal rotors are made of aluminium or titanium alloy and can operate at speeds up to 60 000 rev min^{-1} (256 000*g*). A typical zonal rotor consists of either a large cylindrical container or a hollow bowl enclosed by a lid. The centre of the rotor has a core to which is attached a vane assembly dividing the rotor internally into four sector-shaped compartments. The vanes or septa have radial ducts to allow gradient to be pumped to the periphery of the rotor from the centre core. Swirling of the rotor contents is minimised by the vane assembly. The rotor core may be of two main types. The most commonly used *standard core* permits the loading and unloading of the rotor while it is spinning (*dynamic method*) whilst the second core type (*the reorienting gradient core*) is designed to allow the rotor to be loaded and unloaded with a reorienting gradient whilst it is at rest (*static method*).

In the dynamic mode of operation, loading of the standard core type rotor is achieved whilst the rotor is revolving at approximately 2000 rev min^{-1}. Unlike most gradients produced in a centrifuge, the lighter end of the preformed gradient is pumped into the rotor first through a fixed or a removable seal to emerge at the periphery and form a uniform layer held in a vertical orientation against the outer rotor wall by centrifugal force (Fig. 2.5a). The successive addition of denser gradient results in a continuous centripetal displacement of the lighter gradient towards the rotor core (Fig. 2.5b). When the gradient has been pumped into the rotor, a *fluid cushion*, as dense or denser than the heaviest end of the preformed gradient, is introduced into the rotor to fill it completely. The sample is then introduced by the fluid line leading to the rotor centre (Fig. 2.5c) from which it is subsequently displaced by the addition of an *overlay* of low density liquid (Fig. 2.5d), an equal volume of cushion being displaced from the periphery. After removal of the gradient lines to the rotor, the rotor is accelerated to the operating speed, for the required time interval, to give either *rate zonal* or *isopycnic zonal separation* (Fig. 2.5e). Recovery of the gradient and separated particles is then accomplished by decelerating the rotor to its original 2000 rev min^{-1} and the rotor contents displaced, lighter end first, by introducing additional cushion to the periphery of the rotor (Fig. 2.5f). A *modified rotor core* is available which adds versatility to the operation of the zonal rotor by allowing fractions to be recovered at the rotor's edge as well as its centre. Edge unloading is accomplished by the introduction of a buffer or distilled water at the rotor centre and collecting the fractions through the edge ports of the core. This modification has the advantage that it is more economic in the use of the displacing gradient.

In the static method of zonal centrifugation, using the reorienting (Reograd) gradient core, the sample solution is layered on top of a density gradient in the rotor whilst it is at rest (Fig. 2.6a). The rotor is then slowly accelerated to about 1000 rev min^{-1} to prevent mixing of the rotor contents, during which time the sample and gradient layers reorientate under centrifugal force (Fig. 2.6b). Near operating speeds the zones approach a vertical orientation, and at very high speeds, where the ratio between the centrifugal force and the acceleration due to gravity in a downward direction is very high, the zones become vertical (Fig. 2.6c). Particle separation occurs with the rotor at its operating speed (Fig. 2.6d). At the completion of the separation the rotor is decelerated from its operating speed to 1000 rev min^{-1} and then slowly decelerated to rest. This is to prevent mixing of the rotor contents when sample and gradient layers reorientate back to the horizontal position (Fig. 2.6e). With the rotor at rest, the contents can be displaced from it and recovered by either drawing the contents out of the bottom of the rotor or by displacing the gradient out through the top (Fig. 2.6f). Static loading and unloading of the rotor is most suitable for the isolation of long fragile particles, such as DNA strands, which would otherwise be damaged by the rotating seal assembly used in the dynamic method and has become an accepted method for the separation of lipoprotein fractions from large volumes of plasma or serum. The design of all of the zonal rotor cores enables

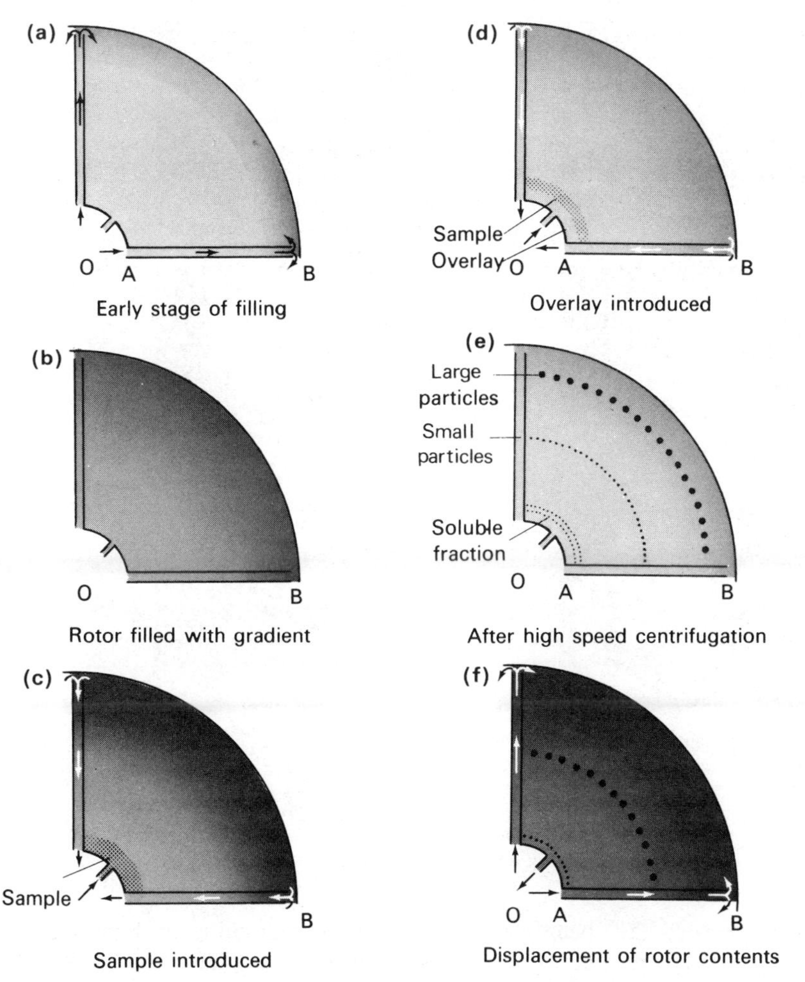

Fig. 2.5 Schematic diagram of stages (**(a)** to **(f)**) in the operation of a zonal rotor. (Reproduced with kind permission of Measuring and Scientific Equipment Ltd.)

the zones to be collected without any appreciable loss of resolution achieved during centrifugation. The static and dynamic methods give equally good resolutions.

To aid zone isolation, the gradient emerging from the rotor is passed through a suitable monitoring device, for example a photocell to determine protein content by its ultraviolet absorption at 280 nm (Section 3.2.2), or a suitable monitor to detect radioactivity (Section 9.2.3), and then collected into fractions to be analysed for concentration of gradient material (using a refractometer), for specific biological activity, or for some other appropriate property.

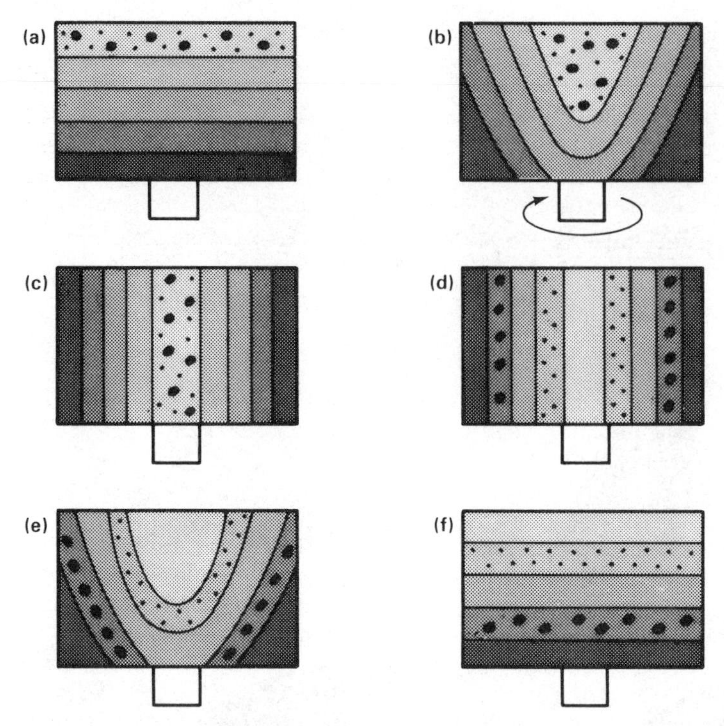

Fig. 2.6 Schematic diagram of a reorienting gradient (Reograd) rotor system. **(a)** Rotor is filled at rest with density gradient and sample layer; **(b)** rotor accelerated, layers reorienting under centrifugal force; **(c)** layers become vertical at sufficiently high rotor speed; **(d)** particles now separate through the gradient with the rotor at speed; **(e)** during rotor deceleration, layers containing separated particles reorientate; **(f)** at rest, rotor contents are displaced and various zones recovered. (• = small sized particles, ● = large sized particles.)

Zonal rotors have been used to remove contaminating proteins from a variety of preparations and for the separation and isolation of hormones, enzymes, macroglobulins, ribosomal subunits, viruses and subcellular organelles from animal or plant tissue homogenates.

2.4.7 Care of rotors

The protective coating on aluminium rotors is very thin (approximately 0.025 mm) and does not provide a high degree of protection against corrosion, thus rotors should always be handled with care. Scratches and strong alkaline detergents (e.g. Decon 90) can easily damage the protective coating leading to corrosion and eventually failure of the rotor. After use, therefore, rotors should be thoroughly washed with lukewarm water and, since moisture is a potential source of corrosion, dried and stored in a clean dry environment. Rotor outer surfaces only can be given a protective coat of lanolin or silicone polish. Swinging bucket rotors, however, should never be completely immersed in water since the bucket hanging mechanism is difficult to dry and can rust.

It is important to note that all rotors are designed to carry a maximum load at a maximum speed, which is based on the rotor tubes or bottles being filled with solution of 1.2 specific gravity. A reduction in the maximum rated speed (known as *derating* the rotor) is therefore required if the specific gravity of the solution exceeds this value. This reduction can be calculated from the equation:

$$N_s = \sqrt{\frac{1.2 \times N^2}{S}} \qquad 2.15$$

where N = maximum rotor speed using a solution of 1.2 specific gravity
N_s = new maximum rotor speed when using a solution of specific gravity S

Speed reductions are also required when using stainless steel caps and/or tubes and to prevent recrystallisation of high density salt solutions, which could occur when the salt concentration exceeds the solubility limit of the solution.

To prevent possible damage to the drive shaft of the centrifuge due to vibration caused by rotor imbalance, sample loads should be balanced within the limits specified by the manufacturer - each opposing pair of sample containers being balanced individually, and the total load balanced symmetrically in the rotor. Swinging bucket rotors should not be run with any buckets or caps removed or individual rotor buckets interchanged since they form an integral part of the balance of the rotor.

During acceleration and deceleration of the rotor, cyclic stretching and relaxing of the metal can cause metal fatigue leading to the eventual failure of the rotor. To avoid overstressing the rotor and ensure its continued safe operation an accurate record should be kept of its total usage, that is, the number of runs (at any speed up to its maximum rev min^{-1}) and the time of each run, so that the rotor can either be derated after a certain number of runs (e.g. 1000) or hours of centrifugation (e.g. 2500 hours) or replaced after a set period of time as specified by the manufacturer.

2.5 Sample containers

Centrifuge tubes and *bottles* are manufactured in a range of different sizes (100 mm^3 to 1 dm^3) from a wide variety of materials including glass, cellulose esters, polypropylene, polycarbonate, polyethylene, nylon and stainless steel.

The correct choice of sample container is important in order to achieve the desired degree of separation of particles from a sample mixture. It is important before commencing centrifugation to consult manufacturers' data sheets to determine the limitations of the container material. The type of container used will depend upon such factors as the nature and volume of the sample to be centrifuged, the type of rotor to be used, the available centrifuge, the centrifugal forces to be withstood, its chemical resistance to various solvents, upper and lower temperature limits and physical properties, i.e. whether the tube is transparent or opaque, and can be sliced or punctured for post-centrifugation analysis.

Centrifuge tubes and bottles should always be filled to the correct level and maximum allowable rotor speeds, using a particular container, observed. The need to cap bottles and tubes usually depends upon the speed at which the sample is to be centrifuged, the type of container and the nature of the sample. Thin-walled plastic tubes, used in fixed angle ultracentrifuge rotors, should always be completely filled and tightly capped in order to support the tube against the very high centrifugal forces generated. It is especially important that tubes are capped and sealed with special leakproof sealing caps when centrifuging material that may be a biohazard or radioactive. Capping and sealing of tubes used in vertical tube rotors is also important since they have to withstand a large upward hydrostatic force generated during centrifugation by the liquid in the tube.

2.6 Density gradient centrifugation

2.6.1 Formation and choice of density gradients

All *density gradient* methods involve a supporting column of liquid, the density of which increases towards the bottom of the centrifuge tube. The gradient stabilises the column of liquid in the centrifuge tube, preventing mixing of the separated particles due to convection currents, improves resolution of the separated components by eliminating, or largely alleviating, factors such as mechanical vibration and thermal gradients, which disturb the smooth migration of particles through the suspending medium, and permits the quantitative separation of several or all components in a mixture.

Density gradients may be produced by techniques which fall into two major groups, the *discontinuous* or *step gradient technique* and the *continuous density gradient technique*.

In the discontinuous technique, solutions of decreasing density are carefully layered over each other in the centrifuge tube, by means of a pipette. This discontinuous gradient can be used directly by layering the sample to be separated as a narrow zone on the top (lowest density) layer. The tube is then centrifuged under the appropriate experimental conditions. Alternatively, if the discontinuous gradient is allowed to stand, the layers slowly merge to produce a continuous linear gradient. The higher the viscosity, the longer it takes for the layers to merge. The merging of the layers may be accelerated either by gentle stirring with a piece of wire or by carefully tipping the tube.

The continuous density gradient technique, which is probably the most common of the two, requires the use of a special piece of apparatus known as a *gradient former*, many varieties of which are commercially available. It consists of two precision-bored cylindrical chambers of identical diameter which are interconnected at their base by a tube containing a control valve which allows the mixing of the contents of the two chambers to be regulated. One chamber (the mixing chamber), which is filled with a dense solution, contains a stirrer and possesses an outlet to the centrifuge tubes. The second chamber contains an equal amount (by weight) of a less dense solution. The hydrostatic pressures of the two liquid columns need to be equal otherwise liquid will flow through the connecting tube as soon as the control valve is opened. The dense liquid is then allowed to run through a filling pipe from the mixing chamber into the centrifuge tube and is immediately replaced in the mixing chamber by an equivalent amount of less dense solution via the control valve. This re-establishes hydrostatic equilibrium. As a result the concentration and density of the solution in the mixing chamber, which is maintained homogeneous by constant stirring, constantly and linearly decreases as the device empties. The concentration of the gradient in the centrifuge tube will therefore decrease in a linear manner as the tube is filled, during which time the filling pipe is gradually withdrawn. Non-linear gradients (i.e. either convex or concave in concentration as a function of volume) may be produced by choosing chambers with unequal volumes or different geometries. Alternatively, two mechanically driven syringes containing solutions of different densities may be used. In this case the shape of the gradient may be varied by altering the speed of pumping of one syringe with respect to that of the other.

Gradients of different shape (i.e. with a concentration profile along the tube) are designed for specific purposes and are important in achieving the desired separation and purification. Discontinuous gradients have therefore been found to be the most suitable for the separation of whole cells or subcellular organelles from plant or animal tissue homogenates. For most purposes linear gradients are used since the gradual change in density along the gradient has been found to yield a much higher resolution of components such as ribosomal subunits and certain viruses. However, for some applications better resolution is achieved using non-linear gradients. Thus concave gradients can be used to separate light particles by flotation (e.g. serum lipoproteins) whilst heavier particles in the mixture quickly sediment through the less dense upper regions of the gradient to be banded in the region of high viscosity near the bottom of the centrifuge tube. Large particles such as ribosomal subunits, polyribosomes and some viruses usually require steep gradients and long gradient columns to enhance separation. This may be achieved by using rotors with long slender tubes which provide the longer gradient columns and *linear-log gradients* in which the logarithm of the gradient column depth is a linear function of the logarithm of the sedimentation coefficient of the particle.

2.6.2 Sample application to the gradient

Before the sample is applied to the density gradient its optimum volume and concentration should be determined. The volume of the sample which can be applied to a centrifuge tube is a function of the cross-sectional area of the gradient exposed to the sample. Thus sample volumes in the range of 0.2 to 0.5 cm^3 may be added to tubes of 1.0 to 1.6 cm diameter and sample volumes of up to 1 cm^3 to tubes having a diameter of approximately 2.5 cm. Effective separation of particles would not be achieved with larger sample volumes due to insufficient radial distance in the centrifuge tube. Equally, if the sample concentration on the gradient is either too high or too low then the gradient may either become overloaded, resulting in a broadening of the separated zones and loss of resolution, or difficulty may be encountered in the identification of the separated bands. A reasonable guideline is that density gradients can usually support most samples if the ratio of sample concentration (% w/w) to starting gradient concentration (% w/w) is in the order of 1:10. Thus, a 5 to 30% (w/w) sucrose gradient would be capable of supporting a sample of 0.5% (w/w) concentration.

Application of the sample to the gradient is generally made using a syringe, or in the case of fragile samples like DNA, by pipette. The tip of the pipette is held approximately 2 to 3 mm above the gradient and inclined at 45° to the wall of the centrifuge tube, allowing the sample to run down the side of the tube onto the gradient.

2.6.3 Recovery and monitoring of gradients from centrifuge tubes

After particle separation has been achieved, it is necessary to remove the gradient solution in order to isolate the bands of separated material. Removal of gradients from centrifuge tubes can be achieved by a number of techniques. If the bands can be visually detected, recovery can be achieved using a hypodermic needle or syringe. A common method, however, is that of *displacement*. A dense liquid, for example 60 to 70% (w/w) sucrose solution, is pumped to the bottom of the centrifuge tube through a long needle. The gradient is displaced upwards and the fractions removed in sequence using a syringe or pipette, or by being channelled out through a special cap to which is attached a collection pipe either leading to a fraction collector or directly into a flow cell of an ultraviolet spectrophotometer.

Alternatively, the centrifuge tube may be punctured at its base using a fine hollow needle. As the drops of gradient pass from the tube through the needle they may be collected using a fraction collector and further analysed. Analysis of the contents of the displaced gradient can be achieved by ultraviolet spectrophotometry (Section 8.2), refractive index measurements, scintillation counting (Section 9.2.3), enzymic (Section 3.4), or chemical analysis.

2.6.4 Nature of gradient materials and their use

There is no ideal all-purpose gradient material, the choice of solute depending upon the nature of the particles to be fractionated. The gradient material should permit the desired type of separation, be stable in solution, inert towards biological materials, should not absorb light at wavelengths appropriate for spectrophotometric monitoring (visible or ultraviolet range), or otherwise interfere with assaying procedures, be sterilisable, non-toxic and non-flammable, have negligible osmotic pressure and cause minimum changes in ionic strength, pH and viscosity, be inexpensive and readily available in pure form and capable of forming a solution covering the density range needed for a particular application without overstressing the rotor.

Gradient-forming materials which provide the densities required for the separation of subcellular particles include salts of alkali metals (e.g. caesium and rubidium chloride), small neutral hydrophilic organic molecules (e.g. sucrose), hydrophilic macromolecules (e.g. proteins and polysaccharides), and a number of miscellaneous compounds more recently introduced and not included in the above group, such as colloidal silica (e.g. Percoll and Ludox) and nonionic iodinated aromatic compounds (e.g. Metrizamide, Nycodenz and Renograffin).

Sucrose solution, whilst suffering from the disadvantages of being very viscous at densities greater than 1.1 to 1.2 g cm^{-3} and exerting very high osmotic effects even at very low concentrations (i.e. at approximately 10% w/v concentration) has been found to be the most convenient gradient material for rate zonal separations. Ficoll (a copolymer of sucrose and epichlorhydrin) has been successfully used instead of sucrose for the separation of whole cells and subcellular organelles by rate zonal and isopycnic centrifugation, but whilst being relatively inert osmotically at low concentrations, both osmolarity and viscosity rise sharply at higher concentrations (i.e. above 20% w/v). Caesium and rubidium salts have been most frequently used for isopycnic separation (Section 2.7.2) of high density solutes such as nucleic acids. Some of the more commonly used gradient-forming materials and their applications are listed in Table 2.1.

2.7 Preparative centrifugation

2.7.1 Differential centrifugation

This method is based upon the differences in the sedimentation rate of particles of different size and density. As can be seen from equation 2.12, centrifugation will initially sediment the largest particles. For particles of the same mass but different density, the ones with the highest density (e.g. peroxisomes, ρ = 1.23 g cm^{-3} in sucrose solution) will sediment at a faster rate than the less dense particles (e.g. plasma membranes, ρ = 1.16 g cm^{-3} in sucrose solution). Particles having similar banding densities (i.e. most of the subcellular organelles, where ρ = 1.1 to 1.3 g cm^{-3} in sucrose solution) can usually be efficiently separated from one another by differential centrifugation or the rate zonal method (Section 2.7.2) provided there is about a ten fold difference in their sedimentation rate.

In *differential centrifugation*, the material to be separated (e.g. a tissue

Table 2.1 Commonly used gradient materials and their use.

Material	Ionic strength of solution	Maximum density of aqueous solution at 20 °C ($g\ cm^{-3}$)	Ultraviolet absorbance	Osmotic effect	Common uses
Caesium chloride	+ + +	1.91	+	+ + +	Banding of DNA, nucleoproteins, viruses; plasmid isolation
Caesium sulphate	+ + +	2.01	+	+ + +	Banding of DNA and RNA; purification of proteoglycans
Sodium bromide	+ + +	1.53	+	+ + +	Fractionation of lipoproteins
Sodium iodide	+ + +	1.90	+ + +	+ + +	Banding of DNA and RNA
Glycerol	–	1.26	+	+ + +	Banding of membrane fragments; protein separation
Sucrose	–	1.32	+	+ + +	Separation of subcellular particles, proteins, viruses, nucleic acids and membranes
Ficoll (Pharmacia)	–	1.17	+	+ (†)	Separation of whole cells, subcellular particles, nucleic acids
Dextran	–	1.13	+	+ (†)	Separation of whole cells; banding of microsomes
Bovine serum albumin	–	1.35	+ + +	+	Separation of whole cells
Percoll (Pharmacia)	–	1.30	+ + +	+	Separation of whole cells and subcellular particles
Metrizamide (Nyegaard)	–	1.46	+ + +	+ + (*)	Separation of whole cells, subcellular particles, nuclei, ribonucleoprotein particles, membranes
Nycodenz (Nyegaard)	–	1.42	+ + +	+ + (*)	Separation of whole cells, subcellular particles, nucleoproteins, membranes, viruses

+ + + High; + + Medium; + Low; – Non ionic
(†) Very low osmotic effect below 20% (w/v) concentration; increasing almost exponentially above 30% (w/v) concentration
(*) Osmotic effect increases almost linearly with concentration

homogenate) is centrifugally divided into a number of fractions by increasing (step-wise) the applied centrifugal field. The centrifugal field at each stage is chosen so that a particular type of material sediments, during the predetermined time of centrifugation, to give a *pellet* of particles sedimented through the solution, and a *supernatant* solution containing unsedimented material. Any type of particle originally present in the homogenate may be found in the pellet or the supernatant or both fractions depending upon the time and speed of centrifugation and the size and density of the particle. At the end of each stage, the pellet and supernatant are separated and the pellet washed several times by resuspension in the homogenisation medium followed by recentrifugation under the same conditions. This procedure minimises cross contamination, improves particle separation and eventually gives a fairly pure preparation of pellet fraction. To appreciate why, however, the pellet is never absolutely pure (homogeneous), it is necessary to consider the conditions prevailing in the centrifuge tube at the beginning of each stage.

Initially all particles of the homogenate are homogeneously distributed throughout the centrifuge tube (Fig. 2.7a). During centrifugation particles move down the centrifuge tube at their respective sedimentation rates (Fig. 2.7b to e) and start to form a pellet on the bottom of the tube. Ideally, centrifugation is continued long enough to pellet all the largest class of particles (Fig. 2.7c), the resulting supernatant then being centrifuged at a higher speed to separate medium sized particles and so on. However, since particles of varying sizes and densities were distributed homogeneously at the commencement of centrifugation, it is evident that the pellet will not be homogeneous, but contain a mixture of all the sedimented components, being enriched with the fastest (heaviest) sedimenting particles. In the time required for the complete sedimentation of heavier particles, some of the lighter and medium sized particles, originally suspended near the bottom of the tube, will also sediment and thus contaminate the fraction. Pure preparations of the pellet of the heaviest particle cannot therefore be obtained in one centrifugation step. It is only the most slowly sedimenting

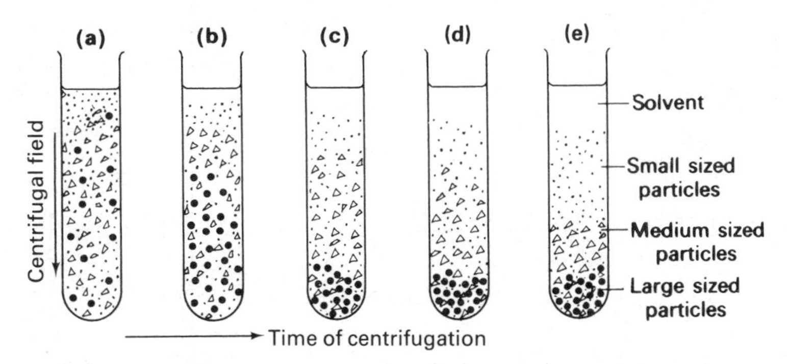

Fig. 2.7 Differential sedimentation of a particulate suspension in a centrifugal field. **(a)** Particles uniformly distributed throughout the centrifuge tube; **(b)** to **(e)** sedimentation of particles during centrifugation dependent upon their size and shape.

component of the mixture that remains in the supernatant liquid after all the larger particles have been sedimented which can be purified by a single centrifugation step but its yield is often very low.

The separation achieved by differential centrifugation may be improved by repeated (2 to 3 times) resuspension of the pellet in homogenisation medium

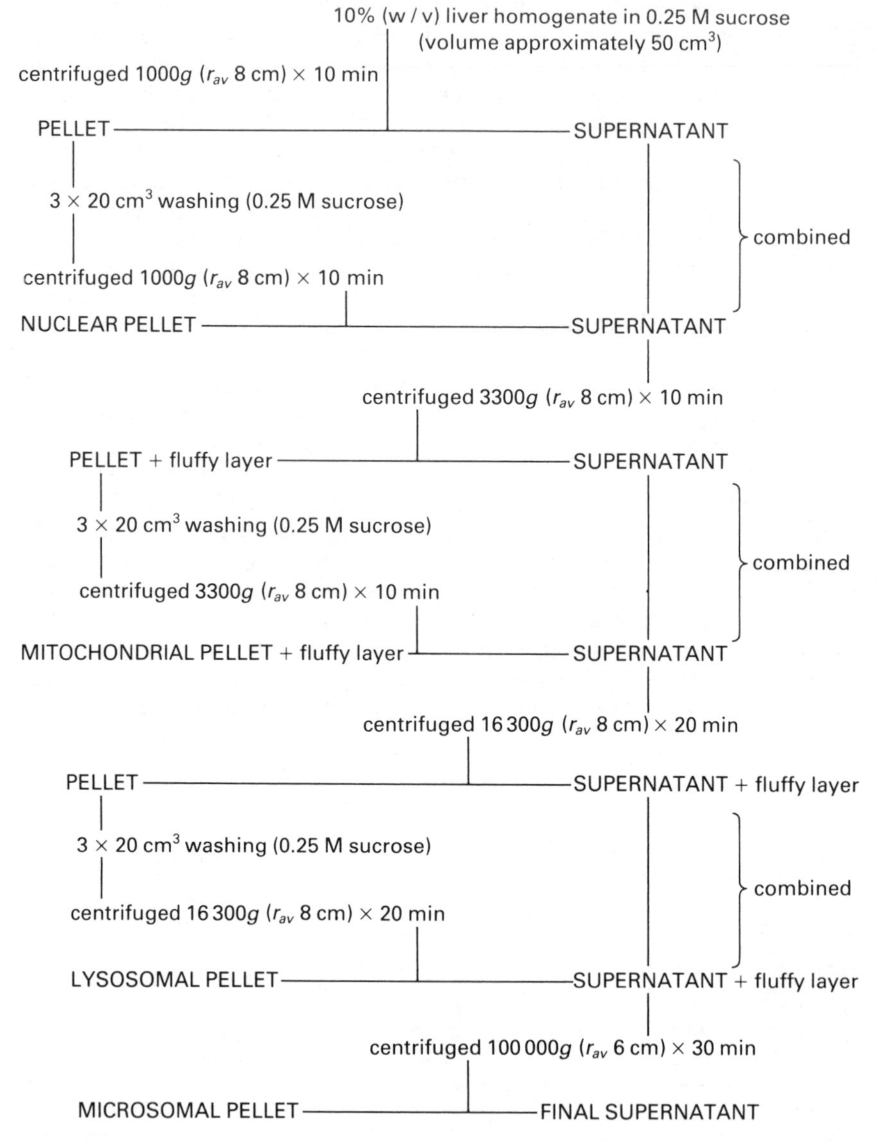

Fig. 2.8 Scheme for the fractionation of rat liver homogenate into the various subcellular fractions.

and recentrifugation under the same conditions as in the original pelleting, but this will inevitably reduce the yield obtained. Further centrifugation of the supernatant in gradually increasing centrifugal fields results in the sedimentation of the intermediate and finally the smallest and least dense particles. A scheme for the fractionation of rat liver homogenate into the various subcellular fractions is given in Fig. 2.8. In spite of its inherent limitations, differential centrifugation is probably the most commonly used method for the isolation of cell organelles from homogenised tissue.

2.7.2 Density gradient centrifugation

There are two methods of density gradient centrifugation, the *rate zonal technique* and the *isopycnic (equal density) technique*, and both can be used when the quantitative separation of all the components of a mixture of particles is required. They are also used for the determination of buoyant densities and for the estimation of sedimentation coefficients.

Particle separation by the rate zonal technique is based upon differences in size or sedimentation rates. The technique involves carefully layering a sample solution on top of a preformed liquid density gradient, the highest density of which does not exceed that of the densest particles to be separated. The sample is then centrifuged until the desired degree of separation is effected, i.e. for sufficient time for the particles to travel through the gradient to form discrete zones or bands (Fig. 2.9) which are spaced according to the relative velocities of the particles. Since the technique is time dependent, centrifugation must be terminated before any of the separated zones pellet at the bottom of the tube. The method has been used for the separation of enzymes, hormones, RNA-DNA hybrids, ribosomal subunits, subcellular organelles, for the analysis of size distribution of samples of polysomes and for lipoprotein fractionations.

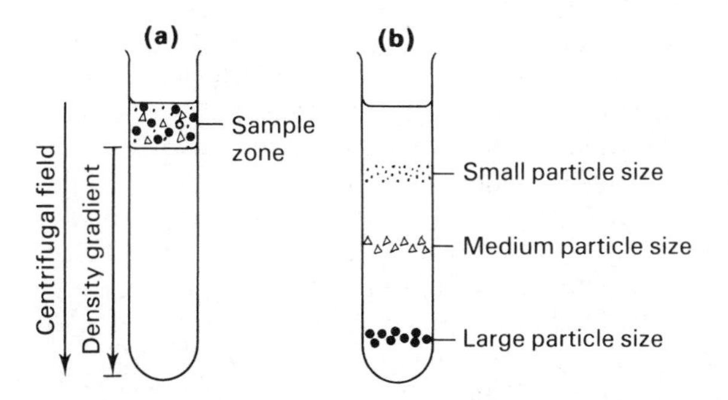

Fig. 2.9 Rate separation and isopycnic separation using a density gradient. **(a)** Mixture of particles layered on top of a preformed liquid density gradient prior to centrifugation; **(b)** centrifugation of particles. For rate separation, the required fraction does not reach its isopycnic position. For isopycnic separation, centrifugation is continued until the desired particles have reached their isopycnic position in the gradient.

Isopycnic centrifugation depends solely upon the buoyant density of the particle and not its shape or size and is independent of time. Hence soluble proteins, which have a very similar density (e.g. $\rho = 1.3$ g cm^{-3} in sucrose solution), cannot usually be separated by this method, whereas subcellular organelles (e.g. Golgi apparatus, $\rho = 1.11$ g cm^{-3}, mitochondria, $\rho = 1.19$ g cm^{-3} and peroxisomes, $\rho = 1.23$ g cm^{-3} in sucrose solution) can be effectively separated.

The sample is layered on top of a continuous density gradient which spans the whole range of the particle densities which are to be separated. The maximum density of the gradient, therefore, must always exceed the density of the most dense particle. During centrifugation, sedimentation of the particles occurs until the buoyant density of the particle and the density of the gradient are equal (i.e. where $\rho_p = \rho_m$ in equation 2.12). At this point no further sedimentation occurs, irrespective of how long centrifugation continues, because the particles are floating on a cushion of material that has a density greater than their own. Isopycnic centrifugation, in contrast to the rate zonal technique, is an equilibrium method, the particles banding to form zones each at their own characteristic buoyant density (Fig. 2.9). In cases where, perhaps, not all the components in a mixture of particles are required, a gradient range can be selected in which unwanted components of the mixture will sediment to the bottom of the centrifuge tube whilst the particles of interest sediment to their respective isopycnic positions. Such a technique involves a combination of both the rate zonal and isopycnic approaches.

As an alternative to layering the particle mixture to be separated onto a preformed gradient, the sample is initially mixed with the gradient medium to give a solution of uniform density, the gradient 'self-forming', by sedimentation equilibrium, during centrifugation. In this method (referred to as the *equilibrium isodensity method*), use is generally made of the salts of heavy metals (e.g. caesium or rubidium), sucrose, colloidal silica or Metrizamide. The sample (e.g. DNA) is mixed homogeneously with, for example, a concentrated solution of caesium chloride (Fig. 2.10a). Centrifugation of the concentrated caesium chloride solution results in the sedimentation of the CsCl molecules to form a concentration gradient and hence a density gradient. The sample molecules (DNA), which were initially uniformly distributed throughout the tube now either rise or sediment until they reach a region where the solution density is equal to their own buoyant density, i.e. their isopycnic position, where they will band to form zones (Fig. 2.10b). This technique suffers from the disadvantage that often very long centrifugation times (e.g. 36 to 48 hours) are required to establish equilibrium. However, it is commonly used in analytical centrifugation to determine the buoyant density of a particle, the base composition of double stranded DNA and to separate linear from circular forms of DNA. Many of the separations can be improved by increasing the density differences between the different forms of DNA by the incorporation of heavy isotopes (e.g. ^{15}N) during biosynthesis, a technique used by Meselson and Stahl to elucidate the mechanism of DNA replication in *Escherichia coli*, or by the binding of heavy metal ions or dyes such as ethidium bromide (Section 5.5.1).

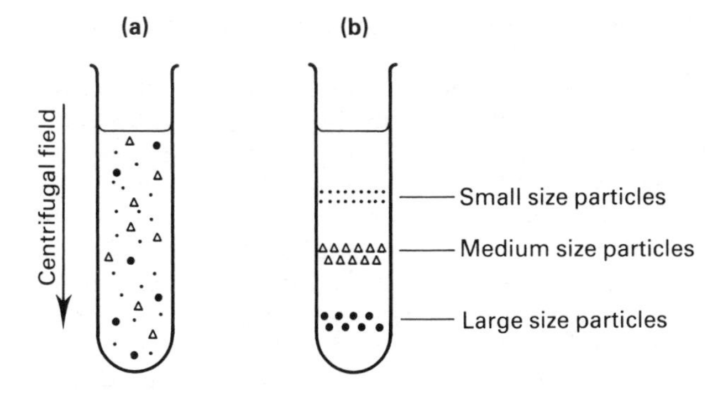

Fig. 2.10 Isopycnic centrifugation using the equilibrium isodensity method. **(a)** Particles distributed homogeneously throughout the tube prior to centrifugation; **(b)** during centrifugation the gradient is allowed to establish itself, sample particles redistribute and band in a series of zones at their respective isopycnic positions.

Isopycnic gradients have also been used to separate and purify viruses and analyse human plasma lipoproteins.

2.7.3 Centrifugal elutriation

In this technique the separation and purification of a large variety of cells from different tissues and species can be achieved by a gentle washing action using an elutriator rotor (Section 2.4.5 and Fig. 2.4). The technique is based upon differences in the equilibrium, set up in the separation chamber of the rotor, between the opposing centripetal liquid flow and applied centrifugal field being used to separate particles mainly on the basis of differences in their size. The technique does not employ a density gradient. Any medium in which cells sediment can be used, and since pelleting of the particles does not occur, fractionation of delicate cells or particles, between 5 and 50 μm diameter, can be achieved with minimum damage so that cells retain their viability. Separations can be achieved very quickly giving high cell concentrations and a very good recovery.

2.8 Selection, efficiency and application of preparative rotors

The selection of a suitable rotor for a particular separation will depend upon such considerations as the quantity and sample type, the time required for the separation and the centrifugation technique to be used.

In practice, the swinging bucket rotor is used principally for incomplete sedimentation of particles through a density gradient. It is relatively inefficient for differential centrifugation, being more effectively used for the rate zonal separation of particles. Fixed angle rotors have proved to be most effective when the rapid, total sedimentation of particles is required by differential centrifugation. Fixed angle rotors, whilst significantly reducing centrifugation time because of their reduced sedimentation pathlength are, however, seldom used for rate zonal centrifugation because undesirable wall effects can disrupt the sedimentation of the zones limiting particle separation. Vertical tube rotors, although also having a short sedimentation pathlength, are unsuitable for pelleting of particles by differential centrifugation. They can be used for the rate zonal separation of whole cells and subcellular organelles, but this does not achieve the quality of separation obtained using the swinging bucket rotor when separating macromolecules with a molecular weight of less than approximately 2×10^5 daltons. Swinging bucket, fixed angle and vertical tube rotors may be used for isopycnic centrifugation. However, fixed angle and vertical tube rotors are often used in preference because larger volumes can be handled and a better degree of resolution achieved in shorter times.

Zonal rotors do not differ significantly from conventional rotors in their ability to separate particles, but holding large volumes of gradient make them especially useful for large scale preparative separations and analytical separations where a number of tests may be required on the separated samples. The continuous flow rotor gives a poor separation with the rate zonal technique due to the very short sedimentation pathlength (approximately 1 cm) but it can be used for pelleting and isolation by isopycnic banding of particles which might otherwise be damaged by the pelleting technique.

The elutriator rotor provides a means for the rapid, high resolution, large capacity separation of whole cells and larger subcellular organelles without the need to use a gradient or to form a pellet. It is, however, of limited use in general teaching laboratories due to the requirement of expensive equipment, specialist accessories and an experienced user. The various types of rotors together with their applications are summarised in Table 2.2.

In order to determine the suitability of a preparative rotor for particle sedimentation, use can be made of the *k* and *k' factors* which provide a measure of efficiency of the rotor for the material that is under investigation. The *k* factor provides an estimate of the time (in hours) that will be required to sediment a particle of known sedimentation coefficient, $s_{20,w}$ (in Svedberg

Table 2.2 Preparative rotor types and their applications.

Rotor type	Centrifugation technique			
	Differential	Rate zonal	Isopycnic zonal	Elutriation
Swinging bucket	−	+ +	+	o
Fixed angle	+ + +	−	+ +	o
Vertical tube	−	+	+ + +	o
Zonal	−	+ + +	+ +	o
Continuous flow	+ + +	−	+ +	o
Elutriator	o	o	o	+ + +

Use: + + + Excellent; + + Good; + Reasonable; − Poor; o Not applicable.

units), at the maximum speed of the rotor. The k' factor is applied to rate zonal work in a density gradient and provides an estimate of the time required to move a band of particles of known sedimentation coefficient and density to the bottom of a centrifuge tube through a linear sucrose gradient (usually 5 to 20% w/v, at 5°C) at the maximum speed of the rotor, The k factor for the rotor can either be obtained from the manufacturer's data sheets or calculated from the equation

$$k = 2.53 \times 10^{11} \frac{[\ln(r_{max}) - \ln(r_{min})]}{(\text{rev min}^{-1})^2_{max}} \qquad 2.16$$

Therefore, if the k factor for the rotor and the sedimentation coefficient (in Svedberg units) of the particle are known, it is possible to estimate the time, in hours, to sediment a particle at the maximum speed of the rotor since

$$t = \frac{k}{s_{20,w}} \qquad 2.17$$

When the rotor is operated below its maximum speed, the k factor is increased to:

$$k_{revised} = k \frac{(\text{rev min}^{-1})^2_{max}}{(\text{rev min}^{-1})^2_{selected}} \qquad 2.18$$

The k' factor is calculated from the equation:

$$k' = 2.53 \times 10^{11} \frac{[I(Z_2) - I(Z_1)]}{(\text{rev min}^{-1})^2_{max}} \qquad 2.19$$

where Z_1 = minimum sucrose concentration (% w/w)
Z_2 = maximum sucrose concentration (% w/w)
I = time integral values obtained from tables (supplied by the instrument manufacturer) after determining Z_o from the equation:

$$Z_o = \frac{Z_1 r_{max} - Z_2 r_{min}}{r_{max} - r_{min}} \qquad 2.20$$

where Z_o = solute concentration corresponding to extrapolation of a linear gradient distribution to zero radius

and the time taken, in hours, for the particles to sediment through the gradient calculated from:

$$t = \frac{k'}{s_{20,w}} \qquad 2.21$$

If the time required to achieve a desired degree of separation of particles using a certain rotor (A) is known, then use of equation 2.22 makes it possible to estimate the sedimentation time, in hours, needed to reproduce these results using an alternative rotor (B) if rotor (A) is unavailable.

$$t_1 = t_2 \left[\frac{k_1}{k_2}\right] \tag{2.22}$$

where t_1 = sedimentation time using rotor B
t_2 = sedimentation time using rotor A
k_1 = k or k' factor for rotor B at its maximum speed
k_2 = k or k' factor for rotor A at its maximum speed

Since the k and k' factors are computed from equations (i.e. 2.16, 2.19 and 2.20) which use the r_{min} and r_{max} of a rotor, a new r_{min} has to be determined and hence a revised k or k' factor calculated should the sedimentation be performed using partially filled tubes.

2.9 Analysis of subcellular fractions

2.9.1 Assessment of homogeneity

It is only when an isolation technique leads to preparations of subcellular particles completely free from contamination by other particles that the properties of the preparations may be attributed to the particles themselves. The evaluation of *purity* is therefore essential. Microscopic examination has been used as a means of analysing the success of a particular homogenisation technique and degree of contamination of a fraction after centrifugation. Absence of visible contamination, however, is not conclusive proof of purity. Quantitative determination of purity has to be obtained by chemical analyses, for example protein, DNA, assay of enzyme activity, and possibly immunological properties. Organelles and molecules lacking assayable enzyme activites can be located either by their light absorption or by radioactive labelling techniques.

As a basis for the interpretation of patterns of enzyme distribution in tissue fractionation studies, two general postulates have been put forward. The first postulate presupposes that all members of a given subcellular population have the same enzyme composition. The second postulate assumes that each enzyme is entirely restricted to a single site within the cell. If valid, these postulates would enable enzymes to be used as markers for their respective organelles, for example cytochrome oxidase and monoamine oxidase as mitochondrial marker enzymes, acid hydrolases as lysosomal marker enzymes, catalase as a marker for peroxisomes and glucose-6-phosphatase as a marker enzyme for microsomal membranes. However, it has been demonstrated that some enzymes are located in more than one fraction, for example malate dehydrogenase, β-glucuronidase, NADPH cytochrome c reductase. Caution must therefore be used in the selection of an enzyme as a marker for a particular subcellular fraction. Further, the absence of marker enzymes cannot be taken as proof of the absence of a particular organelle. It is possible that enzymes released from their respective organelles may have been inhibited or inactivated in some manner during the fractionating process, hence it is normal practice to assay for at least two marker enzymes in each fraction.

2.9.2 Presentation of results

Enzyme activity and protein content are determined both in the whole homogenate and in each subcellular fraction isolated. The sum of the enzyme activity and protein content of the respective fractions should not differ appreciably from that in the initial homogenate and hence should represent the total recovery. Calculations are then made (Table 2.3) of enzyme activity and protein content in each fraction as a percentage of the total recovery. For this reason there is no need to convert absorbance readings into precise units of enzyme activity or mg protein.

The results obtained from tissue fractionation studies may also be represented graphically where enzyme distribution patterns, presented in the form of histograms, provide a visual appreciation of the results. In the histogram, each fraction is then presented separately on the ordinate scale, by its own relative specific activity which is a measure of the degree of purification achieved. On the abscissa, each fraction is represented cumulatively, from left to right, in the order in which it is isolated, by its percentage of total protein. The area of each rectangle is then equal to the percentage of the enzyme activity in that fraction.

2.10 Some applications of the analytical ultracentrifuge

The analytical ultracentrifuge has found many applications in biology, especially in the fields of protein and nucleic acid chemistry, yielding information from which the molecular weight, purity and shape of the material may be deduced.

2.10.1 Determination of molecular weight

Two main approaches are available using the analytical ultracentrifuge to determine molecular weights. These are *sedimentation velocity* and *sedimentation equilibrium*.

In the sedimentation velocity method the ultracentrifuge is operated at high speeds which cause the randomly distributed particles to migrate through the solvent radially outwards from the centre of rotation. A sharp boundary is formed between the portion of solvent that has been cleared of particles and the portion of the solvent still containing the sedimenting material. The movement of the boundary with time, which is a measure of the rate of sedimentation of the particles, is followed by one of the previously described methods and recorded photographically. The rate at which the

Table 2.3 Distribution pattern of liver lysosomal β-glycerophosphatase as established by differential centrifugation.

Enzyme assay

Fraction	Volume (cm^3)	Total dilutions	Absorbance (660 nm)	Enzyme activity in fraction (arbitrary units)	% recovered activity in fraction
Whole homogenate	120	1:30	0.50	1800	—
Nuclear	30	1:20	0.22	132	7.4
Mitochondrial	20	1:100	0.33	660	36.9
Lysosomal	16	1:100	0.34	544	30.4
Microsomal	20	1:25	0.44	220	12.3
Supernatant	290	1:20	0.04	232	13.0
				1788	100.0

Protein assay

Fraction	Volume (cm^3)	Total dilutions	Absorbance (540 nm)	Protein in fraction (arbitrary units)	% recovered protein in fraction
Whole homogenate	120	1:100	0.16	1920	—
Nuclear	30	1:80	0.13	312	16.4
Mitochondrial	20	1:200	0.13	520	27.4
Lysosomal	16	1:100	0.08	128	6.7
Microsomal	20	1:100	0.18	360	19.0
Supernatant	290	1:25	0.08	580	30.5
				1900	100.0

Relative specific activity calculations

$$\text{Relative specific activity} = \frac{\text{\% Enzyme activity in fraction}}{\text{\% Protein in fraction}}$$

Nuclear $\frac{7.4}{16.4} = 0.45$ Mitochondrial $\frac{36.9}{27.4} = 1.35$

Lysosomal $\frac{30.4}{6.7} = 4.53$ Microsomal $\frac{12.3}{19} = 0.65$

Supernatant $\frac{13}{30.5} = 0.43$

particle sediments is given by equation 2.14, from which the molecular weight of the particle may be determined using the *Svedberg equation*:

$$M = \frac{RTs}{D(1 - \bar{v}\rho)} \qquad 2.23$$

where M = anhydrous molecular weight of the molecule
D = diffusion coefficient of the molecule
$\bar{v}$ = partial specific volume of the molecule (volume increase when 1 gram of solute is added to an infinite volume of solution)
ρ = density of solvent at 20°C

and the measured values of s and D being corrected to standard conditions of zero concentration of solute in water at 20°C.

Sedimentation equilibrium methods, however, are the more versatile and in the majority of cases most accurate. They can be used to determine the molecular weights of molecules ranging from a few hundred to several million daltons. This versatility is due to the large range of centrifugal fields available to the ultracentrifuge. The centrifugal field, which varies with the square of the rotor speed, can cover a seven thousand-fold range at rotor speeds of 800 to 68 000 rev min^{-1} developed by the analytical ultracentrifuge.

In the equilibrium method, the ultracentrifuge is operated until a balance is established between sedimentation, under the influence of the centrifugal field, and diffusion of material in the opposite direction, *i.e.* until there is no net migration of solute throughout the length of the cell. Molecular weights can then be calculated from the concentration gradient of the solute which is set up, by using the equation:

$$M = \frac{2RT\ln(c_2/c_1)}{\omega^2(1 - \bar{v}\rho)(r_2^2 - r_1^2)} \qquad 2.24$$

where c_2 and c_1 = the concentrations of solute at distances r_2 and r_1 from the centre of rotation

The molecular weight of the molecule to be studied will dictate the rotor velocity to be used. In general, for *low speed sedimentation equilibrium* a ratio of approximately four to one between the ends of the solution column is desirable (i.e. $c_2/c_1 \cong 4$ in equation 2.24). Therefore it can be calculated, using equation 2.24, that for a molecule of molecular weight 50 000 daltons, a rotor speed of 10 000 rev min^{-1} should be selected. Molecules of higher molecular weights require correspondingly lower rotor speeds. However, it is difficult to perform a good low speed sedimentation equilibrium experiment for materials of molecular weight greater than 5×10^6 daltons due to the problem of rotor wobble at the low centrifuge speeds (e.g. 1000 rev min^{-1}) used.

A major disadvantage of low speed sedimentation equilibrium used to be the long time periods (several days to several weeks) necessary for equilibrium to be achieved. Modern techniques, however, employ analytical cells using short column depths of liquid, usually 1 to 3 mm. As the time taken to reach equilibrium varies with the square of the depth of solution, a great saving of time is possible. The technique of sedimentation equilibrium, unlike that for sedimentation velocity, does not require a knowledge of the diffusion coefficient (compare equations 2.23 and 2.24), making this method more convenient and widely used for the determination of the molecular weights of proteins.

High speed equilibrium or *meniscus depletion (Yphantis method)* methods use short column lengths of liquid (1 to 3 mm), and in principle a similar technique to the low speed method. However, the centrifuge is operated at such a high speed that the concentration of the solute at the meniscus becomes essentially zero. The concentrations throughout the cell are then proportional to the refractive index difference between the meniscus region and any point in the cell. The high speed method does, however, suffer from the disadvantage that for molecular weights below 10 000 daltons, speeds of rotation in excess of 65 000 rev min^{-1} would be required to ensure zero solute concentration at the meniscus. These excessive speeds can produce cell window distortion even when using sapphire windows. Nevertheless, this technique is extremely useful for molecular weight determination if the material is homogeneous.

2.10.2 Estimation of purity of macromolecules

The analytical ultracentrifuge has been used extensively in the investigation of the purity of DNA preparations, viruses and proteins. Sample purity is of course extremely important if an accurate estimation of the molecular weight of the molecule is required. The most widely used methods for the determination of the homogeneity of a preparation include the analysis of the sedimenting boundary using the technique of sedimentation velocity. Homogeneity is usually recognised by a single sharp sedimenting boundary. Impurities in the preparation are displayed as additional peaks, shoulders on the main peak, or asymmetry of the main peak.

2.10.3 Detection of conformational changes in macromolecules

Analytical ultracentrifugation has been successfully applied to the detection of conformational changes in macromolecules. DNA, for example, may exist as single or double strands each of which may be either linear or circular in nature. If exposed to a variety of agents, for example organic solvents or elevated temperature, the DNA molecules may undergo a number of conformational changes which may or may not be reversible. Changes in conformation may be ascertained by examining differences in the sedimentation velocity of the sample. The more compact the molecule, the lower would be its frictional resistance in the solvent. The more disorganised the molecule becomes, the greater is the frictional resistance, and sedimentation then occurs more slowly. Changes in conformation may therefore be detected by differences in sedimentation rates of the sample before and after treatment.

In the case of allosteric proteins (e.g. aspartate transcarbamylase) conformational changes may accompany combination of the protein with substrate and/or small ligands (activators or inhibitors). In addition, treatment of the protein with such reagents as urea and 4-chloromercuribenzoate may result in disaggregation of the protein into its subunits (protomers). All of these changes may readily be studied by analytical ultracentrifugation.

2.11 Safety aspects in the use of centrifuges

Centrifuges can be extremely dangerous instruments if not properly maintained or if incorrectly used. It is therefore essential that all centrifuge users read and understand the operating manual for the centrifuge.

The new British Standard (BS 4402) governing the manufacture of centrifuges has ensured that effective lid locks are fitted which prevent access to the rotor chamber whilst the rotor is still spinning. To prevent possible physical injury when filling and emptying zonal rotors, and in the operation of continuous flow rotors, care must be taken to ensure that the moving rotor is not touched and that long hair and loose clothing (e.g. ties) do not get caught in any rotating part. This is especially important when using older centrifuges where the lid can be opened before the rotor has stopped spinning.

To minimise the risk of rotor failure, which is one of the more serious hazards likely to arise, manufacturers' instructions regarding rotor use and care should always be followed (Section 2.4.7). It is important when centrifuging hazardous materials (e.g. pathogenic microorganisms, infectious viruses, carcinogenic, corrosive or toxic chemicals, radioactive materials), especially in low-speed non-refrigerated centrifuges in which rotor temperature is controlled by air flow through the rotor bowl, that the samples are kept in air-tight leak-proof containers. This is to prevent aerosol formation arising from accidental spillage of the sample, which would contaminate the rotor, centrifuge and possibly the whole laboratory.

2.12 Suggestions for further reading

Birnie, G.D. and Rickwood, D. (1978). *Centrifugal Separations in Molecular and Cell Biology*. Butterworth, London. (Gives in detail the theory and practice of modern centrifugation techniques in molecular and cell biology.)

Griffith, O.M (1983). *Techniques in Preparative, Zonal and Continuous Flow Ultracentrifugation*, 4th edition. Beckman Instruments Inc., Palo Alto. (Covering the current techniques of preparative and density gradient ultracentrifugation and their application.)

Rickwood, D. (Ed.)(1984). *Centrifugation*, 2nd edition. Published in Practical Approaches to Biochemistry Series, IRL Press Ltd, Oxford/Washington DC. (Covering the theory and practice of centrifugation, important criteria for optimising centrifugal separations and the protocols of illustrative experiments.)

3
Enzyme techniques

3.1 Introduction

The majority of cellular processes are catalysed by enzymes which have the ability to promote a specific reaction under the chemically mild conditions which prevail in living organisms. All enzymes are *globular proteins* whose catalytic properties rely on the three-dimensional arrangement of their polypeptide chain(s). The catalytic properties of an enzyme may also be dependent upon the presence of non-peptide *cofactors* or *coenzymes* which may be tightly bound to the polypeptide chain as a *prosthetic group.* Examples of cofactors include NAD^+, $NADP^+$, FMN, FAD, more complex organic structures such as haem groups (haemoproteins) and oligosaccharides (glycoproteins), and simple metal ions such as Mg^{2+}, Fe^{2+} and Zn^{2+}. An enzyme lacking its cofactor is termed an *apoenzyme* and the active enzyme, the *holoenzyme.*

Four levels of protein structure can be identified: primary, secondary, tertiary and quaternary. The *primary structure* of a protein is genetically determined and defines the amino acid sequence. Proteins composed only of amino acids are termed *simple proteins* whilst those which contain a covalently bound cofactor are termed *conjugated proteins. Secondary structure* defines the spatial organisation of a polypeptide chain due to hydrogen bonding and includes such structures as the α-helix and β-pleated sheet. Certain of the 20 different amino acids commonly found in proteins, for example proline and leucine, tend to disrupt secondary structure, so that the content of secondary structure in most proteins is limited to about 70%. *Tertiary structure* defines the remaining spatial arrangement of a polypeptide chain and is stabilised mainly by electrostatic and van der Waals forces and, in some peptides, by disulphide bridges. *Quaternary structure* is restricted to the *oligomeric* proteins which consist of two or more polypeptide chains that are held together by weak non-covalent forces and occasionally by disulphide bridges.

The individual polypeptide chains in an oligomeric protein are referred to as *subunits* and these may be identical or different. In both monomeric and oligomeric enzymes, the specific folding of the polypeptide chain(s) results in the juxtaposition of certain amino acid residues which constitute the *active or catalytic site.* Oligomeric enzymes may possess several such sites. Generally, an active site is located in a cleft which is lined with *hydrophobic* (non-polar) amino acid residues. Amino acid residues, different from those involved in the conversion of substrate to product, may be involved in binding the substrate to the active site.

Some oligomeric enzymes exist in multiple forms called *isoenzymes* or *isozymes*. Their existence relies on the presence of two genes which give similar but not identical subunits. One of the best known examples of isoenzymes is lactate dehydrogenase which reversibly interconverts pyruvate and lactate. It is a tetramer and exists in five forms (LDH1–5) corresponding to the five permutations of arranging the two types of subunits into a tetramer. Each isoenzyme promotes the same reaction but has different kinetic constants, thermal stability and electrophoretic mobility. The tissue distribution of isoenzymes within an organism is frequently different. Thus in man, LDH1 is the dominant isoenzyme in heart muscle, but LDH5 is the most abundant form in liver. These differences are exploited in diagnostic enzymology to identify specific organ damage and thereby to aid a clinical diagnosis and prognosis.

The process of protein *denaturation* results in the loss of enzyme activity, decreased aqueous solubility and increased susceptibility to proteolytic degradation. It can be brought about by heat and by treatment with reagents such as acids and alkalis, detergents, organic solvents and heavy metal cations such as mercury and lead. It is associated with the loss of organised (tertiary) three-dimensional structure and the exposure to the aqueous environment of numerous hydrophobic groups normally located within the folded structure. Experimental procedures using isolated enzymes therefore need to incorporate adequate precautions to minimise the possibility of denaturation. These include a medium with adequate buffer capacity, working in the temperature range 0 to 40°C and the use of clean glassware. In contrast, as a preliminary step in some analytical techniques such as the assay of substrates or products by chromatography, it is advantageous to deproteinise the sample before carrying out the analysis. This is commonly achieved by heating or by treatment with *protein precipitants* such as perchloric acid, trichloroacetic acid, sulphosalicylic acid, tungstic acid, zinc hydroxide, ammonium sulphate and phenol/chloroform.

By international convention, enzymes are classified into one of six classes on the basis of the chemical reaction they catalyse. Each class is sub-divided according to the nature of the chemical group, coenzymes and other groups involved in the reaction. In accordance with the Enzyme Commission rules, each enzyme can be assigned a unique four figure code and an unambiguous systematic name based upon the reaction catalysed. The six classes are:

Group 1: *oxidoreductases* which transfer hydrogen atoms, oxygen atoms or electrons from one substrate to another;
Group 2: *transferases* which transfer chemical groups between substrates;
Group 3: *hydrolases* which catalyse hydrolytic reactions;
Group 4: *lyases* which cleave substrates by reactions other than hydrolysis;
Group 5: *isomerases* which interconvert isomers by intramolecular rearrangements;

Group 6: *ligases* (*synthetases*) which catalyse covalent bond formation with the concomitant breakdown of a nucleoside triphosphate.

The comprehensive study of an enzyme involves the investigation of its *molecular structure* (primary, secondary, tertiary, quaternary, cofactors), *protein properties* (isoelectric point, electrophoretic mobility, pH and temperature stability, spectroscopic properties), *enzymic properties* (specificity, reversibility, kinetics), *thermodynamics* (activation energy, free energies and entropies), *active site* (number, molecular nature, mechanism) and *biological properties* (cellular location, isoenzymes, metabolic relevance of the reaction promoted). To undertake such a study, the enzyme has to be isolated in pure form and studied *in vitro*. Enzyme purification is often a difficult process especially in those cases where the enzyme is located within a membrane structure. The study of purified enzymes is fundamental to biochemistry since it generates data which allow the biochemist to understand and exploit the *in vivo* cellular situation. This exploitation extends to the design of selective inhibitors which can be used as biocides, to the industrial use of enzymes to promote specific chemical conversions and to the use of enzymes in biosensors which are used to assay specific substrates.

3.2 Enzyme units and enzyme purification

3.2.1 Enzyme units

The amount of enzyme present in a particular preparation is conventionally expressed not in terms of units of mass or moles but in terms of units based upon the rate of the reaction that the enzyme promotes. The *international unit* (U) is defined as that amount of enzyme which will convert 1 micromole of substrate to product in 1 minute under defined conditions (generally 25° or 30°C and the optimum pH). The *SI unit* of activity is defined as the amount of enzyme which will convert 1 mole of substrate to product in 1 second. It has units of katal (kat) such that 1 kat $= 6 \times 10^7$ U and 1U $= 1.7 \times 10^{-8}$ kat. For some enzymes, especially those where the substrate is a macromolecule of unknown molecular weight (amylase, pepsin, RNase, DNase), it is not possible to define either of these units. In such cases arbitrary units are used generally based upon some observable change in chemical or physical property of the substrate.

The purity of enzyme preparation is expressed by its *specific activity* which relates its total catalytic activity to the total amount of protein present in the preparation:

$$\text{specific activity} = \frac{\text{total units of enzyme}}{\text{total amount of protein}}$$

The measurement of units of enzyme relies on an appreciation of certain basic kinetic concepts and upon the availability of a suitable analytical procedure. These will be discussed in Sections 3.3 and 3.4.

3.2.2 Protein estimation

The protein content of a preparation can be determined in several ways some of which are not absolute but require the use of a standard. This is most commonly bovine serum albumin because of its low cost and general availability in pure form.

Ultraviolet Absorption The aromatic amino acid residues (tyrosine, phenylalanine, tryptophan) in a protein absorb light in the region of 280 nm. The intensity of this absorption is taken as a direct measure of protein content. Since the number of such residues varies between proteins the method is only a relative one though it is sensitive down to about 10 $\mu g\ cm^{-3}$. Nucleic acids also absorb at 280 nm but, by measuring the absorption at 280 and 260 nm, it is possible to correct for the nucleic acid present in a particular biological sample.

Lowry (Folin-Ciocalteau) Method The phenolic group of tyrosine residues in a protein will produce a blue/purple colour, with maximum absorption in the region of 660 nm, with Folin and Ciocalteau reagent which consists of sodium tungstate, molybdate and phosphate. The method is sensitive down to about 20 $\mu g\ cm^{-3}$ and is probably the most widely used protein assay in spite of the facts that it is only a relative method, is subject to interference from Tris, zwitterionic buffers such as PIPES and HEPES and EDTA and that the incubation time is critical for a reproducible assay. The basis of the assay is thought to be the production of the cuprous ion which reduces the Folin-Ciocalteau reagent. Recently, a number of alternative reagents to detect the cuprous ion have been introduced and claimed to offer a more convenient and reproducible assay. One such reagent is bicinchoninic acid.

Biuret Method Biuret reagent consists of alkaline copper sulphate solution containing sodium potassium tartrate. The cupric ions form a co-ordination complex with four -NH groups present in peptide bonds giving an absorption maximum at 540 nm. Since the method is based on peptide bonds, it is an absolute one and is very reproducible. Its main disadvantage is its lack of sensitivity, being unsuitable for the assay of proteins at concentrations much less than 1 $mg\ cm^{-3}$.

Coomassie Brilliant Blue Dye Method Coomassie Brilliant Blue is one of a number of dyes which complex with proteins to give an absorption maximum in the region of 595 nm. The practical advantages of the method are that the reagent is simple to prepare, the colour develops rapidly and is stable. Although it is sensitive down to 20 $\mu g\ cm^{-3}$ it is only a relative method as the amount of dye binding appears to vary with the content of basic amino acid residues in the protein. This makes difficult the choice of a standard. In addition, many proteins will not dissolve properly in the acidic reaction medium.

Kjeldahl Analysis This is a general chemical method for determining the nitrogen content of any compound. The sample is digested by boiling with concentrated sulphuric acid in the presence of sodium sulphate (to raise the boiling point) and a copper and/or selenium catalyst. The digestion converts all the organic nitrogen to ammonia which is trapped as ammonium sulphate. Completion of the digestion stage is generally recognised by the formation of a clear solution. The ammonia is released by the addition of excess sodium hydroxide and removed by steam distillation in a Markham still. It is collected in boric acid and titrated with standard hydrochloric acid using methyl red/methylene blue as indicator. It is possible to carry out the analysis automatically in an autokjeldahl apparatus. Alternatively, a selective ammonium ion electrode (Section 10.3) may be used to determine directly the ammonium ion content of the digest. Whilst Kjeldahl analysis is a precise and reproducible method for the determination of nitrogen, the determination of the protein content of the original sample is complicated by the variation of the nitrogen content of individual proteins and by the presence of nitrogen in contaminants such as DNA. In practice the nitrogen content of proteins is generally assumed to be 16% by weight.

Turbidimetric Methods Strong organic acids such as trichloroacetic and sulphosalicylic cause proteins to precipitate. The amount of precipitate formed is measured by its light scattering intensity (*turbidimetry*) (Section 8.2.3). The method requires precalibration and relies upon the conditions being controlled so that precipitation is complete but such that flocculation does not occur.

3.2.3 Enzyme purification

Cytoplasmic enzymes and those found in the organelles of eukaryotes may be released by the disruption techniques described in Section 1.8.2. Membrane-bound enzymes present a greater problem for isolation and normally cannot be released by simple cell disruption procedures. Enzymes which are simply bound to the surface of the membrane (*extrinsic proteins*) can generally be dissociated by treatment with 1M NaCl or by freezing and thawing. Enzymes embedded in the membrane (*intrinsic proteins*) are best released by treatment with either a detergent or organic solvent. Sodium dodecylsulphate (SDS) will disrupt the protein-lipid complex and allow the protein to be extracted. Organic solvents have the same effect with n-butanol being most commonly used because of its miscibility with water. Since organic solvents will denature proteins, the extraction has to be carried out at 0 to $-5\,^{\circ}C$. Whatever extraction procedure is adopted care must be taken to minimise denaturation and proteolytic degradation by contaminating enzymes. Thiol compounds such as mercaptoethanol, glutathione or dithiothreitol (*Cleland's reagent*) are commonly added to prevent the oxidation of sulphydryl groups and compounds such as *N*-tosyl-L-phenylalanylchloromethyl ketone to inhibit proteolytic enzymes.

If an enzyme is to be isolated in order to investigate its biological properties then stringent precautions must be taken to preserve the native structure. In contrast, if the enzyme is to be subject to degradative structural investigations, then denaturation considerations are not so critical, indeed in some cases denaturation is desirable since the denatured protein is more susceptible to chemical and enzymic hydrolysis. Proteins differ in their sensitivity to denaturation and this is exploited in enzyme purification. The solubility of native proteins is influenced by pH (solubility being a minimum at the isoelectric point), by the addition of salts such as ammonium sulphate and by the presence of small amounts of organic solvents at low temperatures. Small differences in these properties of enzymes are exploited in fractionation and purification procedures.

There are numerous fractionation procedures available which enable the concentration of a particular enzyme in a complex, heterogeneous mixture to be relatively cheaply, quickly and simply increased at the expense of contaminating proteins. It is important to appreciate, however, that there is no universal order in which these procedures should be employed. An ideal sequence of steps for one enzyme may be unsuccessful for another and could even lead to its denaturation. This is a direct consequence of small but vital differences in protein structure and hence stability. In practice, therefore, the development of a successful protocol involves a considerable amount of trial and error and the use of a number of small-scale pilot experiments to assess the potential of each proposed stage. This is turn relies upon the availability of a sensitive and specific assay for the enzyme being purified and for total protein.

Each fractionation divides the total protein in the mixture into a series of fractions (6 to 8 in pilot experiments, usually 2 or 3 in large-scale preparations) each of which is then assayed for total protein and for enzyme activity. A successful fractionation is recognised by a fraction with a high *specific activity* and hence *fold purification*, and high *yield* where:

$$\text{yield} = \frac{\text{units of enzyme in fraction}}{\text{units of enzyme in original preparation}}$$

$$\text{fold purification} = \frac{\text{specific activity of fraction}}{\text{original specific activity}}$$

In practice, few fractionation procedures have the desired selectivity and a balance has to be made between a large increase in specific activity and purification on the one hand and low yield on the other.

In the preliminary stages of enzyme fractionation and purification, two procedures are commonly included. The first is designed to remove contaminating nucleic acid by the addition of the strongly basic protein protamine. This forms an insoluble electrostatic complex with the nucleic acid which can then be removed by centrifugation. The second is a *heat denaturation fractionation* which exploits differences in the heat sensitivity of proteins. The temperature at which the enzyme being purified is denaturated is determined by a small scale experiment. Once this temperature is known, it is possible to remove more thermo-labile contaminating proteins by heating the mixture to a temperature 5 to 10°C below this critical temperature for a period of 15 to 30 minutes. The denatured, unwanted protein is then removed by centrifugation. The presence of the substrate, product or a competitive inhibitor often stabilises the enzyme and allows an even higher heat denaturation temperature to be employed.

Salt fractionations are carried out by the stepwise addition of a suitable salt. In practice, ammonium sulphate is most commonly used since it is highly water soluble, can be obtained in a high degree of purity, is cheap and has no deleterious effect on the structure of proteins. After each salt addition, care must be taken to ensure its complete dissolution and the production of a homogeneous solution. The precipitated protein is centrifuged off, redissolved in fresh buffer and assayed for total protein and enzyme activity. All stages are generally carried out at 0 to 10°C to minimise denaturation. A given protein is normally precipitated (*salted-out*) over a small range of ammonium sulphate concentration. This reflects the fact that protein–protein aggregation suddenly becomes predominant over protein–water and protein–salt interactions. The results of a successful ammonium sulphate fractionation are shown in Fig. 3.1.

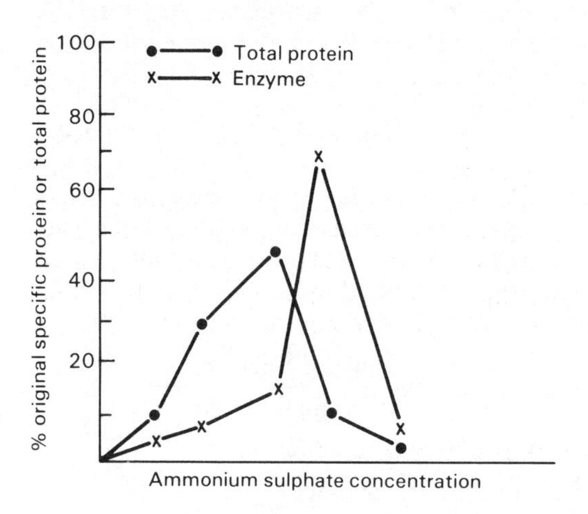

Fig. 3.1 Ammonium sulphate fractionation of an enzyme preparation.

Organic solvent fractionations are based upon differences in the solubility of proteins in aqueous solutions of such organic solvents as ethanol, acetone and butanol. The organic solvent lowers the dielectric constant of the medium thereby increasing the attraction between charged protein molecules and decreasing their interaction with water. The protein solubility therefore decreases. Since the aqueous organic solvents may simultaneously cause denaturation, this form of fractionation must be carried out at between −10 and −20°C and each fraction must be carefully redissolved in fresh buffer before it is allowed to reach room temperature. Nevertheless, even such precautions cannot avoid some denaturation so that unlike salt fractionation, the technique does result in some loss of enzyme. The experimental procedure for organic solvent fractionation is similar to that for salt fractionation.

Fractional gel adsorption relies upon the fact that under certain conditions of pH and ionic concentration, proteins are selectively adsorbed by such materials as calcium phosphate gel (hydroxyapatite) and alumina (γ) gel. Incremental amounts of the gel are added to the heterogeneous protein solution. After each addition, the system is allowed to reach equilibrium and the gel removed by centrifugation. The adsorbed protein in each fraction is released from the gel either by altering the pH and/or the ionic concentration or by adding another compound which has a greater affinity than the protein for the gel. Fractional gel adsorption can also be carried out as a form of column chromatography. Whereas fractionation by heat denaturation may only be included once in a given fractionation protocol, it is frequently advantageous to alternate the other forms of fractionation employing one or more of them several times such that the protein-enriched fraction from one technique becomes the starting material for the next.

The degree of purification achieved by these fractionation procedures is limited and more specific chromatographic and electrophoretic techniques have to be applied to achieve complete purification. Analytical methods particularly suited to the preparation of a pure enzyme are ion-exchange chromatography (Section 6.5), exclusion chromatography (Section 6.6) affinity chromatography (Section 6.7), block and polyacrylamide gel electrophoresis (Section 7.6), isotachophoresis (Section 7.8) and isoelectric focusing (Section 7.7). So-called fast protein liquid chromatography (Section 6.8.3) is a particularly efficient procedure. The purity of the final preparation is best checked by ultracentrifugation (Section 2.10), by electrophoresis or by an immunologically based method (Chapter 4). The molecular weight of the protein may be determined by ultracentrifugation (Section 2.10.1), exclusion chromatography (Section 6.6) or SDS-gel electrophoresis (Section 7.7.6).

3.3 Steady-state enzyme kinetics

3.3.1 Initial rates

When an enzyme is mixed with an excess of substrate there is an initial short period of time (a few hundred microseconds) during which intermediates leading to the formation of the product gradually build-up (see Fig. 3.9). This so-called *pre-steady-state* requires special techniques for study and these are discussed in Section 3.6. After this initial state, reaction rates and the concentration of intermediates change relatively slowly with time and so-called *steady-state kinetics* exist. Measurement of the progress of the reaction during this phase gives the relationships shown in Fig. 3.2. Tangents drawn through the origin to the substrate and product concentration versus time curves allows the *initial rate*, v_0, to be calculated. For simplicity, initial rates are sometimes determined on the basis of a single measurement of the amount of substrate consumed or product produced rather than by the tangent method. This approach is valid only over the short period of time when the reaction is proceeding effectively at a constant rate. This linear

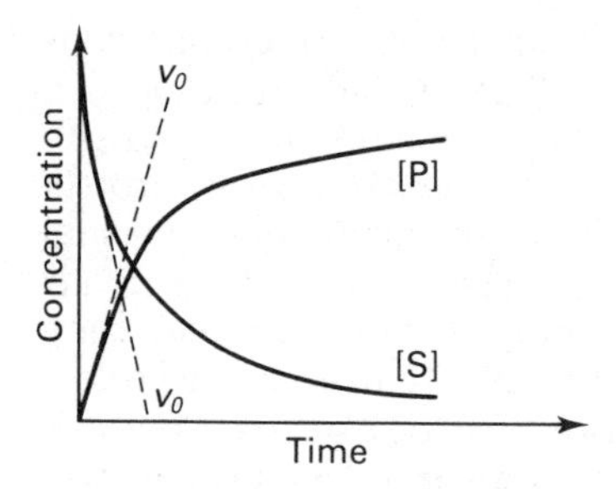

Fig. 3.2 Calculation of initial rate (v_o) from the time dependent change in the concentration of substrate [S] and product [P] of an enzyme catalysed reaction.

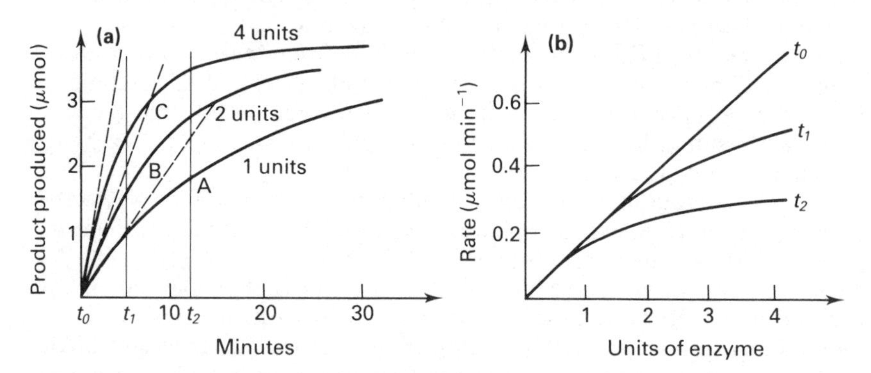

Fig. 3.3 Importance of measuring initial rate in the assay of an enzyme. **(a)** Time dependent variation in the concentration of product on the presence of 1, 2 and 4 units of enzyme; **(b)** variation of reaction rate with enzyme concentration using true initial rate (t_o) and two fixed times (t_1 and t_2).

section comprises at the most the first 10% of the total possible change and clearly the error is smaller the quicker the rate is measured. In such cases, the initial rate is proportional either to the reciprocal of the time to produce a fixed change (*fixed change assays*) or to the amount of substrate reacted in a given time (*fixed time assays*). The potential problems with fixed-time assays is illustrated in Fig. 3.3 which represents the effect of enzyme concentration on the progress of the reaction in the presence of a constant initial substrate concentration (Fig. 3.3a). Measurement of the rate of the reaction at time t_0 (by the tangent method to give the true initial rate) or two fixed times, t_1 and t_2, gives the relationship between initial rate and enzyme concentration shown in Fig. 3.3b. Only the tangent method gives the correct linear relationship. Since the determination of initial rate means that the changes in the concentration of substrate or product are relatively small, it is inherently more accurate to measure the increase in product concentration since the relative increase in its concentration is significantly larger than the corresponding decrease in substrate concentration.

Measurement of the initial rate of an enzyme catalysed reaction is fundamental to a complete understanding of the mechanism by which the enzyme works as well as to the estimation of the activity of an enzyme in a biological sample. Its numerical value is influenced by many factors including substrate and enzyme concentration, pH, temperature and the presence of activators or inhibitors. Each of these variables will now be discussed.

3.3.2 Variation of initial rate with substrate concentration

For the majority of enzymes, the initial rate of reaction varies hyperbolically with substrate concentration (Fig. 3.4a). At low substrate concentrations, approximately *first order kinetics* are observed, but at high substrate concentrations *saturation* (*zero order*) *kinetics* exist and the initial rate is independent of substrate concentration. The mathematical relationship between initial rate and substrate concentration, is expressed by the *Michaelis-Menten equation*:

$$v_0 = \frac{V_{max}\,[S]}{K_m + [S]} \qquad 3.1$$

where $[S]$ = substrate concentration
V_{max} = limiting value of v_0
K_m = Michaelis constant

It can be seen from equation 3.1, that when $v_0 = \frac{1}{2} V_{max}$, $K_m = [S]$. Thus K_m is numerically equal to the substrate concentration at which the initial rate

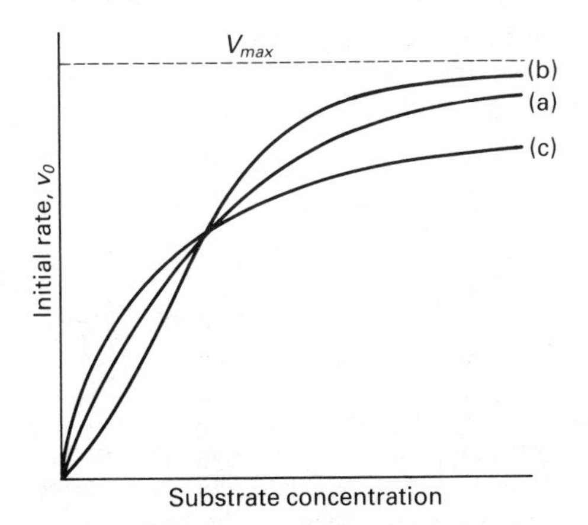

Fig. 3.4 The effect of substrate concentration on the initial rate of an enzyme catalysed reaction: **(a)** when substrate binding is non-cooperative; **(b)** when substrate binding is positively cooperative; and **(c)** when substrate binding is negatively cooperative. (Note that negative homotropic cooperativity gives a curve which is neither a rectangular hyperbola nor sigmoidal.)

is one-half the maximum rate. K_m, which therefore has units of molarity, is independent of enzyme concentration and is a characteristic of the system. It has values usually in the range 10^{-2} to 10^{-5}M and is important since its value, in conjunction with the value of V_{max} and the relative amounts of enzymes, influences the importance of competing metabolic pathways in a cell. In interpreting K_m and V_{max} values in these terms, however, it must be appreciated that the *in vitro* conditions used to determine their value may not reflect the *in vivo* situation. In practice K_m values are important since they enable the concentration of substrate required to saturate all the active sites of an enzyme to be calculated. When $[S] \geqslant K_m$, equation 3.1 reduces to $v_0 \sim V_{max}$, but a simple calculation reveals that when $[S] = 10K_m$, v_0 is only 90% V_{max} and that when $[S] = 100K_m$, $v_0 = 99\%\ V_{max}$. Appreciation of this relationship is vital in enzyme assays. V_{max} is related to the *turnover number* of the enzyme, which expresses the number of moles of substrate converted to product by one mole of enzyme in one second. Values range from 1 to 10^6.

A number of oligomeric enzymes do not display simple Michaelis-Menten kinetics, but give a *sigmoidal* relationship between initial rate and substrate concentration (Fig. 3.4b). Such a curve is indicative of an *allosteric enzyme* which is one whose molecular conformation changes as a result of binding substrate, a process referred to as a *homotropic effect*. This conformational change may result in either increased (*positive cooperativity*) or decreased (*negative cooperativity*) activity towards further substrate molecules. Changes in activity towards the substrate may also be induced by molecules other than the substrate. Compounds which induce such change are referred to as *heterotropic effectors*. They are commonly key metabolic intermediates such as ATP, ADP, AMP and Pi, which bind to an *allosteric* (regulatory) *site* so that the structure of the active site is modified. Heterotropic activators increase the reactivity of the enzyme, making the curve less sigmoidal and moving it to the left whilst heterotropic inhibitors cause a decrease in activity, making the curve more sigmoidal and moving it to the right. The operation of cooperative effects may be confirmed by a *Hill plot*:

$$\log\left(\frac{v_0}{V_{max} - v_0}\right) = h\log[S] + \log K \qquad 3.2$$

where h = Hill constant or coefficient
K = an overall binding constant related to the individual binding constants for n sites by the expression $K = (K_{a_1} \cdot K_{a_2} \cdot \ldots \cdot K_{a_n})^{1/n}$.

The Hill constant, which is equal to the slope of the graph, is a measure of the cooperativity such that if $h = 1$ binding is non-cooperative and normal Michaelis-Menten kinetics exist, if $h > 1$ binding is positively cooperative and if $h < 1$ binding is negatively cooperative. At very low substrate concentrations which are insufficient to fill more than one site and at high concentrations at which most of the binding sites are occupied, the slopes of Hill plots tend to a value of 1. The Hill coefficient is therefore taken from the linear central portion of the plot. One of the problems with Hill plots is the difficulty of estimating V_{max} accurately. This is best obtained from a Lineweaver-Burk (equation 3.7) or Eadie-Hofstee (equation 3.9) plot. The Michaelis constant K_m is not used with allosteric enzymes. Instead, the term $S_{0.5}$, which is the substrate concentration required to produce 50% saturation of the enzyme, is used. It is important to appreciate that sigmoidal kinetics do not confirm the operation of allosteric effects since sigmoidicity may be the consequence of the enzyme preparation containing more than one enzyme capable of acting on the substrate. The presence of more than one enzyme is easy to establish since there will be a discrepancy between the substrate consumed and the expected product produced. It is equally important to appreciate that not all enzymes subject to allosteric control display sigmoidal kinetics. Some monomeric enzymes, for example wheat hexokinase, have been shown to be subject to such control but to display simple Michaelis-Menten kinetics.

Enzyme catalysed reactions proceed via the formation of an *enzyme-substrate complex* (ES) in which the substrate (S) is non-covalently bound to the active site of the enzyme (E). The formation of this complex is rapid and reversible and is characterised by the dissociation constant, K_s, of the complex:

$$E + S \underset{k_{-1}}{\overset{k_{+1}}{\rightleftharpoons}} ES \qquad 3.3$$

$$K_s = \frac{[E][S]}{[ES]} = \frac{k_{-1}}{k_{+1}}$$

where k_{+1} and k_{-1} are first order rate constants.

The conversion of the bound substrate to product is a slower and rate determining step. In the simplest situation where the product is formed in a single step and at a rate such that the equilibrium concentration of ES is maintained, it can be shown that the observed K_m is equal to K_s, *i.e.* the Michaelis-Menten equation becomes:

$$v_0 = \frac{V_{max}\,[S]}{K_s + [S]} \qquad 3.4$$

However, in the majority of cases the conversion of ES to EP is such that, whilst the concentration of ES is essentially constant, it is not the equilibrium concentration. In these cases, *Briggs-Haldane kinetics* prevail and it can be shown that:

$$K_m = \frac{k_{+2} + k_{-1}}{k_{+1}} = K_s + \frac{k_{+2}}{k_{+1}} \qquad 3.5$$

where k_{+2} is the first order rate constant for the conversion of ES to EP. Thus, in these cases, K_m, is numerically larger than K_s. Detailed studies using the approaches outlined in Section 3.6 have revealed that the conversion of ES to EP frequently proceeds through a number of intermediates, some of which involve covalent bond formation. An example is the action of chymotrypsin on some amides in which an acylated form of the enzyme (EA) is an additional intermediate and two products, P_1 and P_2, are produced sequentially:

$$E + S \rightleftharpoons ES \xrightarrow{k_{+2}} EA \xrightarrow{k_{+3}} E + P_2$$
$$\downarrow$$
$$P_1$$

In such circumstances it can be shown that:

$$K_m = K_s \left(\frac{k_{+3}}{k_{+2} + k_{+3}} \right) \qquad 3.6$$

so that K_m is numerically smaller than K_s. It is obvious therefore that care must be taken in the interpretation of the significance of K_m relative to K_s. Only when the complete reaction mechanism is known can the relationship between K_m and K_s be fully appreciated.

Whilst the Michaelis-Menten equation can be used to calculate K_m and V_{max}, its use is subject to error due to the difficulty of experimentally measuring initial rates at high substrate concentrations. Linear transformations of the Michaelis-Menten equation are therefore preferred. The most popular of these is the *Lineweaver-Burk equation*:

$$\frac{1}{v_0} = \frac{K_m}{V_{max}} \frac{1}{[S]} + \frac{1}{V_{max}} \qquad 3.7$$

A plot of $1/v_0$ against $1/[S]$ (Fig. 3.5) gives a straight line of slope K_m/V_{max} with an intercept on the $1/v_0$ axis of $1/V_{max}$ and an intercept on the $1/[S]$ axis of $-1/K_m$. This double reciprocal plot identifies substrate inhibition (Section 3.3.6) by an upward curve at high substrate concentrations (low $1/[S]$ values) (Fig. 3.7b). Positive cooperativity is characterised by a concave upwards curve and negative cooperativity by a concave downwards curve (Fig. 3.7c). Unless very careful thought is given to the planning of the series of substrate concentrations, the Lineweaver-Burk equation tends to give an unequal distribution of points and greatest emphasis to the points at low substrate concentrations. Alternative plots (Fig. 3.5) are based on the *Hanes equation*:

$$\frac{[S]}{v_0} = \frac{K_m}{V_{max}} + \frac{[S]}{V_{max}} \qquad 3.8$$

and the *Eadie-Hofstee equations*:

$$v_0 = V_{max} - K_m \frac{v_0}{[S]} \qquad 3.9$$

Bisubstrate reactions (Fig. 3.6) such as those catalysed by the transferases, kinases and dehydrogenases, in which two substrates S_1 and S_2 are converted to two products P_1 and P_2 (*two substrate-two product, bi-bi, reactions*), are inherently more complicated than monosubstrate reactions. They may be *sequential* in which case both substrates bind to give a *ternary complex* before the products are formed. Sequential reactions may be *compulsory ordered*, in

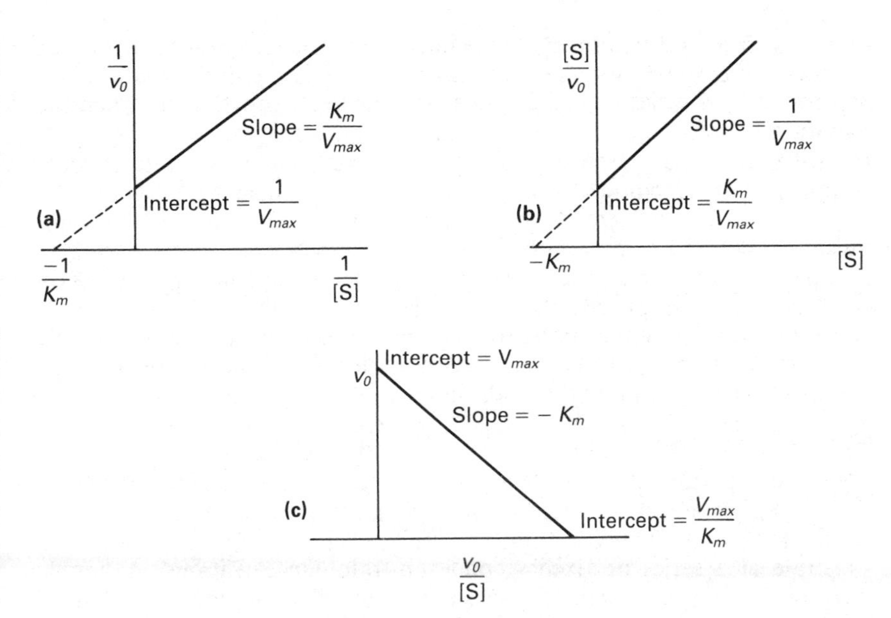

Fig. 3.5 Linear plots for the calculation of K_m and V_{max}. **(a)** Lineweaver-Burk; **(b)** Hanes; and **(c)** Eadie-Hofstee.

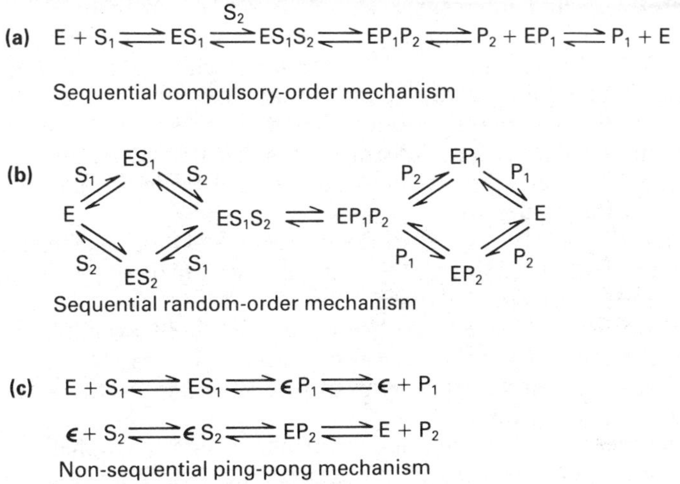

(a) $E + S_1 \rightleftharpoons ES_1 \overset{S_2}{\rightleftharpoons} ES_1S_2 \rightleftharpoons EP_1P_2 \rightleftharpoons P_2 + EP_1 \rightleftharpoons P_1 + E$

Sequential compulsory-order mechanism

(b) $E \overset{S_1}{\rightleftharpoons} ES_1 \overset{S_2}{\rightleftharpoons} ES_1S_2$; $E \overset{S_2}{\rightleftharpoons} ES_2 \overset{S_1}{\rightleftharpoons} ES_1S_2$

$ES_1S_2 \rightleftharpoons EP_1P_2$

$EP_1P_2 \overset{P_2}{\rightleftharpoons} EP_1 \overset{P_1}{\rightleftharpoons} E$; $EP_1P_2 \overset{P_1}{\rightleftharpoons} EP_2 \overset{P_2}{\rightleftharpoons} E$

Sequential random-order mechanism

(c) $E + S_1 \rightleftharpoons ES_1 \rightleftharpoons \epsilon P_1 \rightleftharpoons \epsilon + P_1$

$\epsilon + S_2 \rightleftharpoons \epsilon S_2 \rightleftharpoons EP_2 \rightleftharpoons E + P_2$

Non-sequential ping-pong mechanism

Fig. 3.6 Possible reaction mechanisms for bisubstrate reactions.

which case the two substrates bind in a definite sequence, or *random ordered* in which case either substrate can bind first. In both cases there are separate binding sites for each substrate. Alternatively the reaction may be *non-sequential* in which case one product is released before the second substrate is bound. One example of this type of mechanism is a *ping–pong* reaction which proceeds via a modified form of the enzyme (ϵ) which may take the form of an acylated enzyme. A *ping-pong bi-bi* mechanism is indicated but not confirmed by a series of parallel lines in double reciprocal plots when the variation of initial rate with increasing concentrations of one substrate is investigated in the presence of a series of fixed second substrate concentrations. Double reciprocal plots give a progressively smaller intercept on the 1/[S] axis as the concentration of second substrate is increased. A compulsory order and a random order ternary complex mechanism both give non-parallel double reciprocal plots with a progressively smaller intercept on the $1/v_0$ axis as the concentration of fixed second substrate is increased.

For all bisubstrate reactions, by holding the concentration of one of the two substrates constant and studying the influence of varying the concentration of the second substrate on the initial rate, the V_{max} value and a value of K_m for each substrate may be obtained. In these bisubstrate reactions, V_{max} is defined as the maximum initial rate when both substrates are saturating and the K_m for a particular substrate as the concentration of that substrate which gives ½ V_{max} when the other substrate is saturating. To determine these K_m values, the initial velocity is studied as a function of the concentration of one substrate at a series of fixed second substrate concentrations. A double reciprocal plot is then made for each second substrate concentration giving a series of straight lines called *primary plots* and a *secondary plot* made of the $1/v_0$ intercepts of the primary Lineweaver–Burk plots against the reciprocal of the second (fixed) substrate. This gives a straight line slope K_m (for the *second* substrate)/V_{max} and intercept $1/V_{max}$. The study is then repeated reversing the roles of the two substrates. The principle of secondary plots is illustrated in Fig. 3.8. The elucidation of the reaction mechanism associated with a particular bisubstrate reaction generally involves the study of the variation of the initial rate with the concentration of one substrate at a series of fixed concentrations of the second substrate in the absence and presence of the two reaction products and the application of a series of rules formulated by Cleland. Two of these rules are:

(*i*) that the intercept on the $1/v_0$ axis of double reciprocal plots is affected only by an inhibitor which binds reversibly to an enzyme form other than that to which the variable substrate binds; and

(*ii*) that the slope of double reciprocal plots is affected by an inhibitor which binds to the same enzyme form as the variable substrate or to an enzyme form which is connected by a series of reversible steps to that with which the variable substrate binds.

The consequence of the first rule is that if characteristic competitive inhibition (Section 3.3.6) behaviour is observed the inhibitor and the substrate whose concentration is being varied bind at the same site. The consequence of rule two is that if characteristic uncompetitive inhibition (Section 3.3.6) is observed there must be no reversible link between the inhibitor and the substrate whose concentration is being varied. Studies applying these rules have, for example, revealed that histamine-*N*-methyltransferase operates via a compulsory order mechanism whilst phosphoglycerate mutase operates via a ping-pong mechanism involving a phosphorylenzyme intermediate.

3.3.3 Variation of initial rate with enzyme concentration

It can be shown that:

$$v_0 = \frac{k_{+2}[\mathrm{E}][\mathrm{S}]}{K_m + [\mathrm{S}]} \text{ and hence that } v_0 = \frac{k_{+2}[\mathrm{E}]}{\frac{K_m}{[\mathrm{S}]} + 1} \qquad 3.10$$

Thus when the substrate concentration is very large, the initial rate is directly proportional to the enzyme concentration. This is the basis of the determination of enzyme activity in a particular biological sample (Section 3.4). The importance of the correct measurement of initial rate is illustrated by Fig. 3.3.

3.3.4 Variation of initial rate with temperature

The rate of an enzyme reaction varies with temperature according to the Arrhenius equation:

$$\text{rate} = Ae^{-E/RT} \qquad 3.11$$

where A = constant
E = activation energy (in joules per mole)
R = gas constant (8.2 J mol^{-1} K^{-1})
T = absolute temperature

The equation explains the sensitivity of enzyme reactions to temperature since the relationship between reaction rate and temperature is exponential. The rate of most enzyme reactions approximately doubles for every 10°C rise in temperature (*Q_{10} value*). At a temperature characteristic of the enzyme and generally in the region 40° to 70°C, the enzyme is denatured and enzyme activity is lost. The activity displayed in this 40° to 70°C temperature range partly depends upon the equilibration time before the reaction is commenced. Thus the so-called *optimum temperature*, at which the enzyme appears to have maximum activity, is time-dependent and for this reason is not the one normally chosen to study enzyme activity. Enzyme assays are routinely carried out at 30° or 37°C (Section 3.4.1).

3.3.5 Variation of initial rate with pH

The state of ionisation of amino acid residues in the active site of an enzyme is pH dependent. Since catalytic activity relies on a specific state of ionisation of these residues enzyme activity is pH dependent. As a consequence the pH enzyme activity profile is either bell-shaped (two important amino acid residues in active site), giving a narrow *pH-optimum*, or it has a plateau (one important amino acid residue in active site). In either case an enzyme is generally studied at a pH at which its activity is maximal. By studying the variation of $\log K_m$ and $\log V_{max}$ with pH, it is possible to identify the pK_a values of key amino acid residues involved in the catalytic process. Table 1.1 (p. 8) lists the ionisable groups found in proteins.

3.3.6 Influence of inhibitors on initial rate

Enzyme inhibitors (I) specifically combine with an enzyme to reduce its ability to convert substrate to product. *Irreversible inhibitors*, such as the organophosphorus and mercury compounds, cyanide, carbon monoxide and hydrogen sulphide, combine by covalent forces and the extent of their inhibition is dependent upon their reaction rate constant (and hence time) and upon the amount of inhibitor present. The effect of irreversible inhibitors, which cannot be overcome by simple physical techniques such as dialysis, is to reduce the amount of enzyme available for reaction. The inhibition involves reaction with a functional group such as hydroxyl or sulphydryl or with a metal atom in a prosthetic group in the active site or a distinct allosteric site. *Reversible inhibitors* combine non-covalently with the enzyme so that they can be readily removed by dialysis. *Competitive* reversible inhibitors combine at the same site as the substrate and must therefore be structurally related to the substrate. An example is the inhibition of succinate dehydrogenase by malonate. A *non-competitive* reversible inhibitor combines at a distinct site from the substrate but in such a way as to produce a so called *dead-end complex* irrespective of whether or not the substrate is bound. *Un-competitive* reversible inhibitors can only bind to the ES complex and not to the free enzyme so that binding may be at a site created by the binding of the substrate to the active site (*i.e.* a conformational change occurs on substrate binding) or it may be to the bound substrate molecule. The resulting ternary complex, ESI, is also a dead-end complex. All types of reversible inhibitors are characterised by their dissociation constant K_i, called the *inhibitor constant*, which may relate to the dissociation of EI (K_{EI}) or ESI (K_{ESI}).

For competitive inhibition the following two equations can be written:

$$\mathrm{E} + \mathrm{S} \rightleftharpoons \mathrm{ES} \longrightarrow \mathrm{E} + \mathrm{P}$$
$$\mathrm{E} + \mathrm{I} \rightleftharpoons \mathrm{EI}$$

Since the binding of both substrate and inhibitor involves the same site, the effect of a competitive reversible inhibitor can be overcome by increasing the substrate concentration. The result is that V_{max} is unaltered but the concentration of substrate required to achieve it is increased so that when

$$v_0 = \tfrac{1}{2}V_{max}, \qquad [\mathrm{S}] = K_m\left(1 + \frac{[\mathrm{I}]}{K_i}\right)$$

where [I] = is the concentration of inhibitor

It can be seen from this that K_i is equal to the concentration of inhibitor which apparently doubles the value of K_m. With this type of inhibition K_i is equal to K_{EI} whilst K_{ESI} is infinite since no ESI is formed.

In the presence of a competitive inhibitor the Lineweaver-Burk equation (3.7) becomes:

$$\frac{1}{v_0} = \frac{K_m}{V_{max}} \frac{1}{[S]} \left(1 + \frac{[I]}{K_i}\right) + \frac{1}{V_{max}} \quad 3.12$$

allowing the diagnosis of competitive inhibition (Fig. 3.7a). The numerical value of K_i can be calculated from a Lineweaver-Burk plot for the uninhibited and inhibited reactions. In practice, however, a more accurate value is obtained from a secondary plot. The reaction is carried out in the presence of a series of fixed inhibitor concentrations and a Lineweaver-Burk plot for each inhibitor concentration constructed. Secondary plots of the slope of the primary plot against the inhibitor concentration or of the apparent K_m, K_m', (which is equal to K_m $(1 + [I]/K_i)$ and which can be calculated from the reciprocal of the negative intercept on the $1/[S]$ axis) against inhibitor concentration will both have intercepts on the inhibitor concentration axis of $-K_i$. Sometimes it is possible for two molecules of inhibitor to bind at the active site. In these cases although all the primary double reciprocal plots are linear, the secondary plot is parabolic. This is referred to as *parabolic competitive inhibition* to distinguish it from normal *linear competitive inhibition*.

For non-competitive inhibition the inhibitor may also bind to ES:

$$ES + I \rightleftharpoons ESI$$

Since this inhibition involves a site distinct from the active site, the inhibition cannot be overcome by increasing the substrate concentration. The consequence is that V_{max} is reduced but not K_m since the inhibitor and substrate do not affect the binding of each other. With this type of inhibition K_{EI} and K_{ESI} are identical and K_i is numerically equal to both of them.

The Lineweaver-Burk equation (3.7) therefore becomes:

$$\frac{1}{v_0} = \left(\frac{K_m}{V_{max}} \frac{1}{[S]} + \frac{1}{V_{max}}\right)\left(1 + \frac{[I]}{K_i}\right) \quad 3.13$$

Once non-competitive inhibition has been diagnosed (Fig. 3.7a) the K_i value is best obtained from a secondary plot of either the slope of the primary

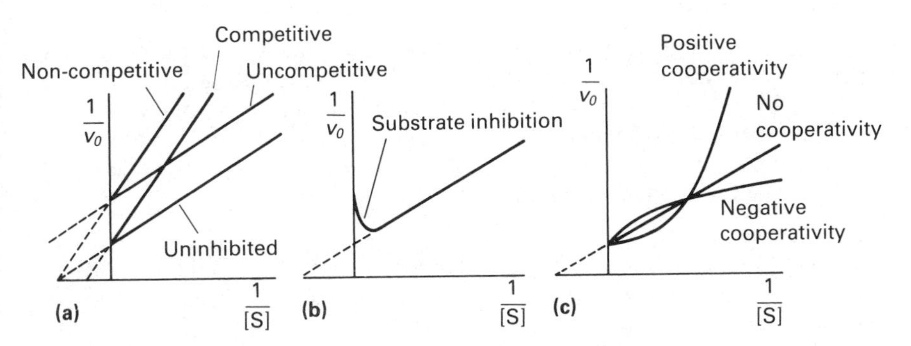

Fig. 3.7 Lineweaver-Burk plots showing **(a)** the effects of three types of reversible inhibitor; **(b)** substrate inhibition; and **(c)** homotropic cooperativity.

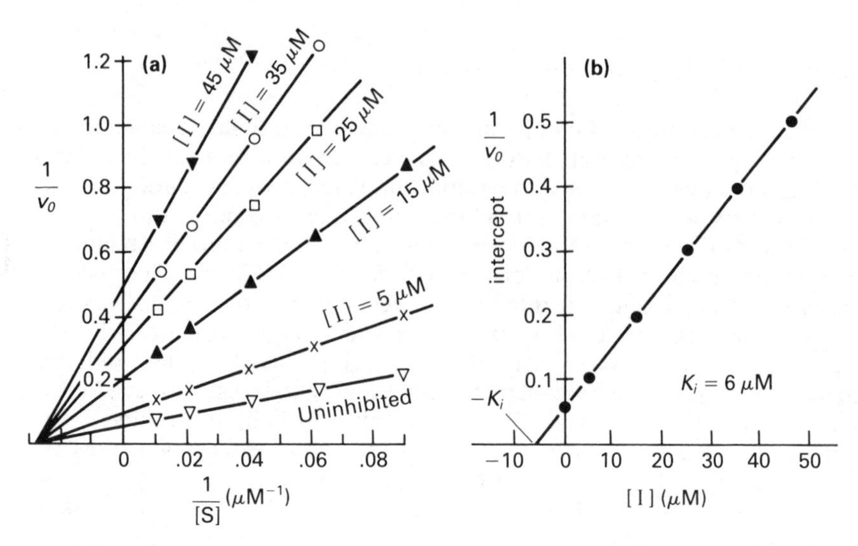

Fig. 3.8 **(a)** Primary Lineweaver-Burk plots showing the effect of a simple linear non-competitive inhibitor at a series of concentrations and **(b)** the corresponding secondary plot which enables the inhibitor constant, K_i, to be calculated.

plot or $1/V'_{max}$ (which is equal to the intercept on the $1/v_0$ axis) against inhibitor concentration. Both secondary plots will have an intercept of $-K_i$ on the inhibitor concentration axis (Fig. 3.8).

For uncompetitive inhibition the inhibitor *only* binds to ES:

$$\mathrm{E + S \rightleftharpoons ES \longrightarrow E + P}$$
$$\mathrm{ES \underset{-I}{\overset{+I}{\rightleftharpoons}} ESI}$$

As with non-competitive inhibition, the effect cannot be overcome by increasing the substrate concentration, but in this case both K_m and V_{max} are reduced by a factor of $(1 + [\mathrm{I}]/K_i)$. An inhibitor concentration equal to K_i will therefore halve the value of both K_m and V_{max}. With this type of inhibitor, K_{EI} is infinite since the inhibitor cannot bind to the free enzyme so K_i is equal to K_{ESI}. The Lineweaver-Burk equation (3.7) therefore becomes:

$$\frac{1}{v_0} = \frac{K_m}{V_{max}}\frac{1}{[\mathrm{S}]} + \frac{1}{V_{max}}\left(1 + \frac{[\mathrm{I}]}{K_i}\right) \qquad 3.14$$

The value of K_i is best obtained from a secondary plot of either $1/V'_{max}$ or $1/K'_m$ (which is equal to the intercept on the $1/[\mathrm{S}]$ axis) against inhibitor concentration. Both secondary plots will have an intercept of $-K_i$ on the inhibitor concentration axis.

The classification of the type of observed inhibition is not always straightforward or unambiguous. Thus it is possible to observe competitive inhibition in cases where the inhibitor binds to a site other than the active site but in such a way that a conformational change follows the binding of the substrate thereby preventing binding of the inhibitor. Equally, in some cases of bisubstrate reactions, an inhibitor may competitively inhibit the binding of one substrate and non-competitively inhibit the binding of the second (Section 3.3.2). It is also possible for an ESI complex to have some catalytic activity or for K_{EI} and K_{ESI} to be neither equal nor infinite in which case so called *mixed inhibition kinetics* are obtained. Mixed inhibition is characterised by a linear Lineweaver-Burk plot which does not fit any of the patterns shown in Fig. 3.7a. The uninhibited and inhibited plots may intersect either above or below the 1/[S] axis. The associated K_i can be obtained from a secondary plot of either the slope of the primary plot or of $1/V'_{max}$ for the primary plots against inhibitor concentration. In both cases the intercept on the inhibitor concentration axis is $-K_i$. Non-competitive inhibition may be regarded as a special case of mixed inhibition.

A number of enzymes at high substrate concentration display *substrate inhibition* characterised by a decrease in initial rate with increased substrate concentration. The graphical diagnosis of this situation is shown in Fig. 3.7b. It is explicable in terms of the substrate acting as an uncompetitive inhibitor and forming a dead-end complex.

The study of enzyme inhibitors helps our understanding of the mechanisms by which enzymes work. Inhibitors are also widely used in the study of metabolic pathways to help in the identification of intermediates. The whole area of selective toxicity, including the use of antibiotics and insecticides, is based on the exploitation of species differences in susceptibility to enzyme inhibitors.

3.3.7 Cellular control of enzyme activity

The *in vivo* control of the activity of enzymes is vital for the efficient regulation of cellular metabolism. Not all of the factors so far discussed which influence enzyme activity *in vitro* operate within a given organism. Coarse (long-term) control is achieved at the level of protein synthesis by induction or inhibition of enzyme synthesis. Fine (short-term) control is obtained in a number of ways. Fluctuations in the concentration of substrate or reversible inhibitor can influence activity either by a direct effect on the concentration of ES complex or by an allosteric effect. Studies have also identified the importance of *reversible covalent modification*. This may involve adenylation or phosphorylation of the enzyme, commonly triggered by cyclic AMP produced by adenylate cyclase and involving active and inactive protein kinases. Examples of such control are found in glycogenolysis and glyconeogenesis. It has also been shown that the activity of some enzymes is dependent upon the cellular concentration of calcium. The calcium is bound to specific binding proteins such as calmodulin and troponin C. Examples of enzymes subject to such control include actomyosin, phosphorylase kinase and cyclic nucleotide diesterase. Another mechanism important for the regulation of some enzymes is *irreversible limited proteolysis*. Enzymes subject to such control are synthesised in an inactive form called *proenzymes* or *zymogens*. These are activated when required by the proteolytic cleavage of peptide fragments by enzymes such as trypsin. Examples include chymotrypsinogen, the precursor of chymotrypsin, and thrombinogen, the precursor of thrombin.

3.4 Enzyme assay techniques

3.4.1 General considerations

The determination of the activity of an enzyme is based upon the rate of utilisation of substrate or formation of product under controlled conditions. Most assays are carried out at 30°C, but some are performed at 37°C because of the physiological significance of the temperature. Adequate buffering capacity must be used and care should be taken to ensure that all apparatus is scrupulously clean. Analytical methods may be classified as either *continuous* (*kinetic*) or *discontinuous* (*fixed-time*). Continuous methods monitor some property change (e.g. absorbance or gas volume) in the reaction mixture whereas discontinuous methods require samples to be withdrawn from the reaction mixture and analysed by some convenient technique. The inherent greater accuracy of continuous methods commends them whenever they are available.

Irrespective of the principle of the analytical method, enzyme assays require the use of excess substrate (zero order kinetics) (at least equal to 10 K_m) and an appropriate control. The control is in all respects the same as the test assay but lacking either enzyme or substrate. Changes in the experimental parameter in the control lacking the test enzyme will give an assessment of the extent of the non-enzymic reaction whilst changes in the control lacking added substrate evaluates any background reaction in the enzyme preparation. All reaction mixtures are incubated at the experimental temperature for at least two minutes before the reaction is started either by the addition of pre-equilibrated enzyme or substrate. It is worthwhile assaying the enzyme using different volumes of the test solution to confirm linearity between initial rate and enzyme concentration thereby confirming the absence of activators or inhibitors in the preparation.

3.4.2 Visible and ultraviolet spectrophotometric methods

Many substrates and products absorb light in the visible or ultraviolet region and provided the substrate and product do not absorb at the same wavelength and that the Beer-Lambert law (Section 8.1) holds, the change in absorbance can be used as the basis of the assays. Studies are best carried out in a double-beam recording instrument with a temperature-controlled cell housing.

A large number of assays are based on the interconversion of $NAD(P)^+$ and NAD(P)H. Both the oxidised and reduced forms of these two nucleotides absorb at 260 nm but only the reduced form at 340 nm (the molar extinction coefficient $\epsilon = 6.3 \times 10^3 \ dm^3 \ mol^{-1} \ cm^{-1}$). Enzymes which do not directly involve this interconversion may be assayed by it by using the concept of *coupled reactions*. In such reactions, the enzyme to be assayed is linked to one utilising the NAD^+/NADH system by means of common intermediates. The principle is illustrated by the assay of phosphofructokinase (PFK). It can be linked via aldolase to the glyceraldehyde-3-phosphate dehydrogenase (G3PDH) reaction:

$$\text{fructose-6-phosphate} \xrightarrow[Mg^{2+}]{\text{PFK}} \text{fructose-1-6-bisphosphate}$$

(ATP → ADP)

$$\text{fructose-1-6-bisphosphate} \xrightarrow{\text{aldolase}} \text{glyceraldehyde-3-phosphate} + \text{dihydroxyacetone phosphate}$$

$$\text{glyceraldehyde-3-phosphate} \xrightarrow{\text{Pi}} \text{1-3-bisphosphoglycerate}$$

(NAD^+ → NADH)

The assay mixture would therefore contain fructose-6-phosphate, ATP, Mg^{2+}, aldolase, G3PDH, NAD^+ and Pi all in excess so that the rate of NADH production, and hence increase in absorption at 340 nm, would be determined solely by the concentration of PFK in the known volume of preparation added to the assay mixture. In principle there is no limit to the number of reactions that can be coupled in this way provided the enzyme under investigation is present in limiting amounts. The number of units of enzyme in the test preparation is calculated as follows:

$$\underset{(\text{kat cm}^{-3})}{\text{enzyme units}} = \frac{\Delta E_{340}}{6.3 \times 10^3} \, \frac{a}{1000} \, \frac{1000}{x} \qquad 3.15$$

where ΔE_{340} = control-corrected change in absorption at 340 nm per second

a = total volume of reaction mixture (generally about 3 cm^3) in a cuvette of 1 cm light path

x = volume of test preparation included in the reaction mixture

A general form of this equation is applicable to all spectrophotometric enzyme assays. In some cases the stoichiometry of the reaction (the number of molecules of compound undergoing the observed change in absorbance) is not unity in which case a correction for the stoichiometry has to be introduced. The general equation is therefore:

$$\text{enzyme units} = \frac{\Delta E \, a}{\epsilon \, d \, n} \qquad 3.16$$

where ϵ = molar extinction coefficient of the chromophor
n = stoichiometry
d = light path in cuvette (cm)
ΔE = change in absorbance at experimental wavelength

By dividing equations 3.15 and 3.16 by Cp, the total concentration of protein in the enzyme preparation, the specific activity of the preparation can be calculated.

The scope of visible spectrophotometric enzyme assays can be extended by the use of artificial substrates and by the production of coloured derivatives of the substrate or product. Many enzymes, especially the hydrolases, will act on synthetic analogues of their natural substrate to release a coloured product such as *p*-nitrophenol and phenolphthalein. An example is the assay of α-glucosidase (maltase):

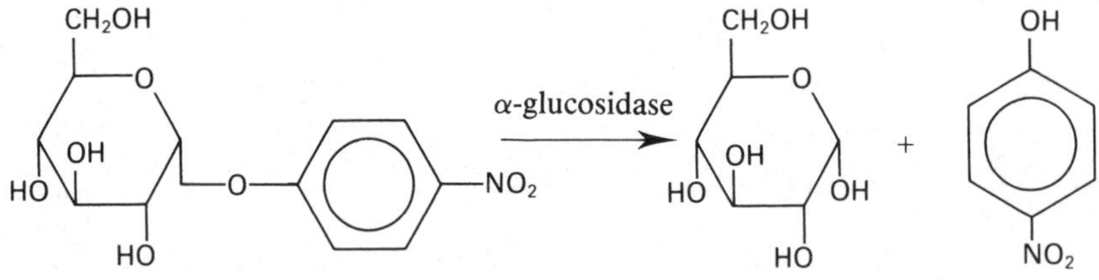

CH_2OH … NO_2 —(α-glucosidase)→ CH_2OH … + OH … NO_2

p-nitrophenol-α-D-glucopyranoside → D-glucose + *p*-nitrophenol (yellow)

An extension of this approach is the use of synthetic dyes for the study of the oxidoreductases. The oxidised and reduced forms of these dyes are different colours. Examples are the tetrazolium dyes, methylene blue, 2-6-dichlorophenol indophenol and methyl and benzyl viologen. Their use, which is discussed fully in Section 10.4.3, depends upon them having an appropriate oxidation-reduction potential relative to that for the substrate.

Substrates or products containing certain functional groups can be converted to a coloured derivative. An example is the orange dinitrophenylhydrazone derivatives of aldehydes and ketones. Thus an assay of isocitrate lyase is based on this reaction:

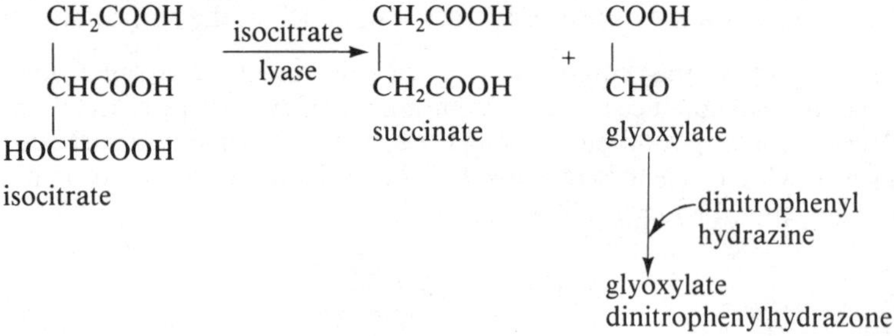

isocitrate		succinate		glyoxylate
CH_2COOH	isocitrate lyase →	CH_2COOH	+	COOH
$CHCOOH$		CH_2COOH		CHO
$HOCHCOOH$				

glyoxylate —(dinitrophenyl hydrazine)→ glyoxylate dinitrophenylhydrazone

Either samples of the reaction mixture are withdrawn periodically and reacted with dinitrophenylhydrazine or, for a fixed time assay, the whole reaction mixture is reacted with the reagent at a pre-established time. In some cases it is possible to convert the product of an enzyme reaction to a coloured derivative *in situ* without interfering with the enzyme reaction itself. An example is glucose oxidase whose product, hydrogen peroxide, can be used to oxidise *o*-dianisidine, incorporated in the assay mixture, to a yellow product.

3.4.3 Spectrofluorimetric methods

Although potentially very sensitive, fluorimetric enzyme assays have the practical limitations that trace impurities in the enzyme preparation can quench the emitted radiation (Section 8.5). Additionally many fluorescent compounds are unstable, especially in the presence of ultraviolet light. Nevertheless, the technique is widely applied to enzyme assays. NAD(P)H is fluorescent so that many enzymes can be assayed by coupling to an appropriate reaction as before (Section 3.4.2). Equally, synthetic substrates are available which release fluorescent products. Examples include the use of 4-methylumbelliferyl-β-D-glucuronide for the assay of β-glucuronidase (see Section 8.5.3).

3.4.4 Luminescence methods

The increasing popularity of bioluminescent reactions (Section 8.6), in which the intensity of emitted light is used to study enzyme reactions, is due to the high sensitivity of the technique. One of their problems is occasional lack of reproducibility. Firefly luciferase catalyses the oxidation of luciferin in an ATP-dependent reaction:

$$\text{luciferin} + \text{ATP} + O_2 \xrightarrow{\text{luciferase}} \text{oxyluciferin} + \text{AMP} + \text{PPi} + CO_2 + \text{LIGHT}$$

The reaction can be used to assay ATP and appropriate enzymes via coupled reactions (Section 8.6.3). The corresponding bacterial luciferase uses reduced FMN to oxidise long chain alphatic aldehydes. The resulting FMN can be coupled to NAD(P)H thus permitting the assay of many enzymes e.g. malate dehydrogenase:

$$\text{malate} + \text{NAD}^+ \xrightarrow[\text{dehydrogenase}]{\text{malate}} \text{oxaloacelate} + \text{NADH} + \text{H}^+$$

$$\text{NADH} + \text{H}^+ + \text{FMN} \xrightarrow{\text{oxidoreductase}} \text{FMNH}_2 + \text{NAD}^+$$

$$\text{FMNH}_2 + \text{RCHO} + O_2 \xrightarrow{\text{luciferase}} \text{FMN} + \text{RCOOH} + H_2O + \text{LIGHT}$$

The luminol/aminophthalic acid system involving microperoxidase can form the basis of the assay of many enzymes including acetylcholinesterase (ACE) involved in synaptic transmission:

$$\text{acetylcholine} \xrightarrow{\text{ACE}} \text{acetate} + \text{choline}$$

$$\text{choline} + O_2 + H_2O \xrightarrow{\text{choline oxidase}} \text{betaine} + 2H_2O_2$$

$$H_2O_2 + \text{luminol} \xrightarrow{\text{microperoxidase}} \text{aminophthalic acid} + N_2 + \text{LIGHT}$$

3.4.5 Radioisotope methods

Although potentially a very sensitive method, the use of radioisotope techniques (Section 9.5.2) for enzyme assays is restricted to applications where it is possible to separate, easily, the radiolabelled forms of substrate and product. In those cases where one of the products is a gas, this presents no problem. Thus the assay of glutamate decarboxylase could be based on the evolution of $^{14}CO_2$:

$$\underset{\text{glutamate}}{HOOCCH_2CH_2CH(NH_2)^{14}COOH} \xrightarrow[\text{decarboxylase}]{\text{glutamate}} {}^{14}CO_2 + \underset{\gamma\text{-aminobutyrate}}{HOOCCH_2CH_2CH_2NH_2}$$

The $^{14}CO_2$ evolved could be trapped in alkali and hence the rate of $^{14}CO_2$ evolution measured. In other cases, the substrate and product may be separated by solvent extraction. Thus in the assay of monoamine oxidase (MAO), samples of the assay mixture could be acidified (thereby converting the monoamine to its salt), the labelled aldehyde removed by ether extraction and the extract added to a scintillation cocktail for counting the radioactivity.

$$\underset{\text{monoamine}}{R^{14}CH_2NH_2} + O_2 + H_2O \xrightarrow{\text{MAO}} \underset{\text{aldehyde}}{R^{14}CHO} + H_2O_2 + NH_3$$

3.4.6 Manometric methods

Enzyme reactions resulting in the net evolution or uptake of a gas such as oxygen and carbon dioxide can be monitored by the Warburg manometer or the Gilson respirometer (Section 1.10). Both types of instrument can readily measure small changes in gas volume provided the temperature is adequately controlled and that corrections are applied for the solubility of the gas in the reaction mixture. The versatility of the method can be extended to examples where one gas is evolved and another taken up, by chemically removing the evolved gas. In the case of CO_2 this would be by absorption in sodium hydroxide solution and in the case of O_2 by absorption in pyrogallol or chromous chloride solution. Enzymes that can be assayed manometrically include glutamate decarboxylase (above), catalase, malic enzyme and monoamine oxidase (above):

$$2H_2O_2 \xrightarrow{\text{catalase}} 2H_2O + O_2$$

$$\underset{\text{pyruvate}}{CH_3COCOOH} + CO_2 + NADPH + H^+ \xrightarrow[\text{enzyme}]{\text{malic}} \underset{\text{malate}}{HOOCCH(OH)CH_2COOH} + NADP^+$$

3.4.7 Ion-selective and oxygen electrodes methods

The development of ion-selective electrodes (Section 10.3), such as those for the ammonium ion, and the oxygen electrode, has afforded attractive methods for enzyme assays. The methods are very sensitive, reproducible and can use very small volumes of reaction mixture. The glass electrode can be used to monitor pH changes due to the enzymatic release of an acid or alkali, for example by proteases in a poorly buffered medium. Since such release would alter the pH of the reaction mixture and hence enzyme activity, a *pH-stat* can be used to maintain a pre-selected pH and to record the amount of acid or alkali which it automatically adds to maintain the selected pH. The rate of reaction is therefore expressed in terms of the rate of addition of acid or alkali to the reaction mixture.

3.4.8 Immunochemical methods

Polyclonal or monoclonal antibodies (Section 4.2) raised to a particular enzyme can be used as the basis for a highly specific assay for the enzyme. Such systems can distinguish between isoenzymes which, in the context of clinical measurements of enzyme activities, is of considerable diagnostic value. A monoclonal assay is available for prostatic acid phosphatase which is one of the best means of diagnosing carcinoma of the prostate.

3.4.9 Microcalorimetric methods

Most biological reactions are accompanied by a minute heat (enthalpy) change which gives rise to a temperature change of the order of 10^{-2} to 10^{-4} °C. Measurement of such small changes is possible using *thermistors* which are temperature-sensitive metal oxides. The technique, which requires stringent insulation of the reaction vessel, may be improved by coupling the primary reaction to a secondary one which generates a larger heat evolution. Thus reactions releasing protons may be carried out in Tris buffer which has a large enthalpy change on protonation:

$$\text{glucose} + \text{ATP} \xrightarrow{\text{hexokinase}} \text{glucose-6-phosphate} + \text{ADP} + \text{H}^+ \quad \Delta H^0 = -28 \text{ kJ mol}^{-1}$$

$$\text{Tris} + \text{H}^+ \longrightarrow \text{TrisH}^+ \quad \Delta H^0 = -47 \text{ kJ mol}^{-1}$$

3.4.10 Automated enzyme analysis

Spectrophotometric methods are the most popular methods for enzyme assays and form the basis of many commercially available *reaction rate analysers*. These are instruments dedicated to the measurement of enzyme or substrate. Many are automated and based on fixed change or fixed time assays. *Discrete analysers* mix the enzyme and substrate by means of automatic pipettes according to a pre-programmed instruction. The reaction mixture is passed to the detector at a pre-determined time. *Continuous flow analysers* pump the substrate continuously through a flow tube and periodically introduce a sample of the enzyme to be analysed into the line, each sample being separated from the next by a bubble. The reactants are mixed in a narrow mixing coil and pumped to the detector. The flow rate ensures that the reactants reach the detector at the correct time. An interesting variant is the *fast (centrifugal) analyser*. In it the enzyme and substrate are placed in separate wells located near the centre of a horizontal centrifuge plate (rotor) which can accommodate up to 30 separate samples. When the plate is rotated the reactants are forced centrifugally into a cuvette at the edge of the plate thus initiating the reaction. The change in absorbance at an appropriate wavelength in each cuvette is continuously recorded enabling a good estimation to be made of the initial rate.

3.5 Substrate assay techniques

Enzyme-based assays are very convenient methods for the estimation of the amount of substrates present in a biological sample. In theory, the principle of using excess enzyme and relating the substrate concentration to the observed initial rate could be used but in practice this is not popular because of the difficulty of measuring a rapidly decreasing reaction rate at low substrate concentrations. The procedures actually employed overcome this problem in a variety of ways but basically they are all variants of the so-called *end-point technique*. In this approach all of the substrate is converted to product and the total change in parameter (e.g. ultraviolet absorption) recorded. This change is then used to compute the amount of substrate originally present. The technique relies on the use of sufficient enzyme to ensure that the reaction goes to completion in a reasonable time. For reversible reactions it is necessary to adjust the position of equilibrium so that the reaction is effectively complete. This can be achieved by such means as adjusting the pH away from the optimum for the enzyme, in the case of bisubstrate reactions using a high concentration of the second substrate or in the case of reactions using $NAD(P)^+$ using an analogue such as APAD (acetylpyridine adenine dinucleotide) which has a more favourable oxidation-reduction potential. Coupled reactions are commonly used in substrate assays. In all cases the substrate concentration should be very much smaller than the K_m and the amount of enzyme used should be adjusted so that the reaction is complete in 2 to 10 minutes. It can be shown that if the amount of enzyme is such that $V_{max}/K_m \sim 1$, then the reaction will be 99% complete in about 5 minutes. The detection limits for such assays are determined by the molar extinction coefficient of the compound being monitored. In the case of NAD(P)H, they are of the order of 10^{-2} to 10^{-3} μmol cm^{-3} in a 1 cm cuvette for ultraviolet spectrophotometric assays.

The sensitivity limits for substrate assays can be dramatically improved by the technique of *enzymatic cycling*. The method, which is particularly valuable in those cases where the substrate is present in very low concentrations, involves the regeneration of the substrate by means of a coupled reaction. The product accumulated in a given period of time (30 to 60 minutes) is then measured. A precalibration is necessary using known amounts of the test substrate with all the other components in the assay present in excess. The resulting 10^4 to 10^5 fold increase in sensitivity lowers the detection limits for visible and ultraviolet spectrophotometry to 10^{-6} to 10^{-8} μmol cm^{-3}. The method is commonly used for the assay of $NAD(P)^+$ and ATP/ADP. In the latter case, pyruvate kinase and phosphoenolpyruvate are used to regenerate ATP. In the case of NAD^+ or $NADP^+$ glutamate dehydrogenase may be used. For example in the assay of $NADP^+$, glucose-6-phosphate dehydrogenase (G6PDH) and glutamate dehydrogenase (GDH) may be coupled:

G6PDH
glucose-6-phosphate → 6-phosphogluconate
$NADP^+$ NADPH
L glutamate ← oxoglutarate + NH_4^+
GDH

3.6 Pre-steady-state enzyme kinetics

3.6.1 Rapid mixing methods

The experimental techniques so far discussed for steady-state kinetics allow the determination of K_m and V_{max} values, but special techniques must be employed for the determination of the rate constants of the individual steps in the conversion of substrate to product since the intermediates are transient. Figure 3.9 shows the progress curves in the initial stages of the conversion of substrate to product via ES. The induction period, t, is related to k_{+1}, k_{-1}, and k_{+2}.

In the *continuous flow method* solutions of the enzyme and substrate are introduced from syringes into a small mixing chamber (typically 100 mm^3

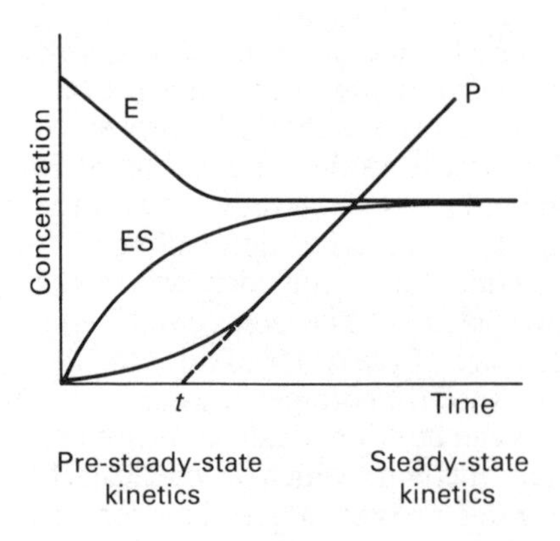

Fig. 3.9 Initial stage progress curve for the reaction E + S ⇌ ES → E + P when [S] ≫ [E].

capacity) and then pumped at a pre-selected speed through a narrow tube to which is attached a light source and a photomultiplier or similar detector. Flow through the tube is fast (typically 10 m s^{-1}) so that it is turbulent, thus ensuring homogeneity of the solution. The precise reaction time, after mixing, at the point of observation can be calculated from the known flow rates of the two solutions. By varying these rates, the reaction time at the observation point can be varied, thus allowing the extent of reaction to be studied as a function of time. From this data, rate constants can be calculated. The technique, which requires relatively large volumes of reactants, is limited only by the inherent time to get the reactants mixed.

The *stopped flow method* is a development of the continuous flow method in that, shortly after the reactants emerge from the mixing chamber, the flow is stopped. At this point the recorder is triggered and a continuous recording made of the change in experimental parameter (e.g. absorbance). The attraction of the method is its conservation of reactants. In both methods, the problem of studying the first few hundred microseconds of the reaction can be partially solved by altering the pH or temperature to slow down the reaction or by using alternative substrates with slower turnover times. The versatility of both the continuous flow and stopped flow techniques is increased by the use of synthetic substrates which either generate a chromophoric leaving group or which give a chromophoric acyl or phosphoryl intermediate. An example is the use of *p*-nitrophenylacetate to study chymotrypsin.

A variant of these rapid flow techniques is the *quenching method*. In this technique, the reactants from the mixing chamber enter a second chamber where they are mixed with a quenching agent such as trichloroacetic acid which stops the reaction. The reaction products are then analysed by an appropriate analytical technique. By quenching after a range of reaction times, the progress of the reaction can be followed.

3.6.2 Relaxation methods

The great limitation of the rapid mixing methods is the dead time during which the enzyme and substrate are mixed. In the relaxation method, an equilibrium mixture of the reactants is preformed and the position of equilibrium altered by a change in reaction conditions. The most common procedure for achieving this is the *temperature jump technique* in which the temperature is rapidly raised by 5° to 10°C by the discharge of a capacitor or infra-red laser. The rate at which the system adjusts to its new equilibrium (relaxation time τ) is inversely related to the first order rate constants involved in the reaction sequence. The rate of return to equilibrium is studied by spectrophotometric techniques. It is often advantageous to use more than one such technique, e.g. ultraviolet absorbance and fluorimetry, since they may yield complementary information. Careful analysis of the total number of relaxation times gives an indication of the number of intermediates involved in the overall process as well as the value of the related rate constants.

Whilst relaxation techniques are best for studying the fastest processes, rapid flow techniques are frequently best for studying the processes involved in the catalytic steps. Rapid kinetic techniques have revealed that enzyme and substrate generally associate very rapidly with first order rate constants in the range 10^6 to $10^8\ M^{-1}\ s^{-1}$ and dissociate more slowly with rate constants in the range 10 to $10^4\ s^{-1}$. The association process is generally slower than predicted by simple collision theory and indicates the need for specific orientation of the substrate and enzyme and perhaps conformation changes and the involvement of solvation processes.

3.6.3 Detection of intermediates

Although pre-steady-state kinetic measurements allow the calculation of rate constants they do not necessarily directly identify the intermediate(s) predicted by the proposed reaction mechanism. For this the intermediate(s) must be isolated and characterised. One way of achieving this is by the use of *affinity labels*. These are irreversible inhibitors which structurally resemble the substrate but which form a covalent bond with an amino acid residue in the active site. The resulting enzyme-inhibitor complex is then isolated and studied by conventional analytical techniques. Examples of affinity labels include organophosphorus compounds such as diisopropylfluorophosphate (DFP) which irreversibly bind to acetylcholinesterase and related esterases via a serine residue, and iodoacetone phosphate which binds to triose-phosphate isomerase. A variant of the technique is the use of *photoaffinity labels* which initially bind reversibly to the enzyme but which, on photolysis, are converted to reactive intermediates which bind irreversibly. Examples are diazo compounds which give carbenes and azides which give nitrenes:

$$\underset{\text{diazo compound}}{RCOCH_2N_2} \xrightarrow{\text{LIGHT}} \underset{\text{carbene}}{RCO\ddot{C}H_2} + N_2$$

$$\underset{\text{azide}}{RN_3} \xrightarrow{\text{LIGHT}} \underset{\text{nitrene}}{RN:} + N_2$$

Affinity labels of course identify specific amino acid residues in the active site in addition to providing information about the chemical nature of the intermediates. It is possible to expand this approach to the identification of amino acid residues in an active site by using reagents which do not resemble the substrate but which do react with specific amino acid residues. The success of the method relies upon the fact that certain amino acid residues in the active site are activated making them more reactive than similar residues elsewhere in the protein. Examples of these reagents include the reaction of *N*-tosyl-L-phenylalanylchloromethyl ketone (TPCK) with histidine residues and iodoacetate and iodoacetamide with cysteine residues.

3.7 Protein-ligand binding studies

3.7.1 General principles

The affinity of an enzyme for its substrate is expressed by the dissociation constant K_s of the ES complex. K_s values cannot be directly derived from steady state kinetics although under certain circumstances the Michaelis constant K_m, which is readily measured, may numerically approximate to K_s (Section 3.3.2). In contrast, the dissociation constant for reversible inhibitor–enzyme complexes, K_i, can be obtained directly from steady state studies (Section 3.3.6) without any assumption being made about the relative values of K_m, and K_s.

The approach to the direct measurement of K_s is similar to that for the study of other protein–ligand binding studies such as the binding of hormones to plasma membrane receptor proteins and the binding of drugs and hormones to plasma proteins such as albumin, α_1 acid glycoprotein and steroid hormone binding globulins. All such reversible binding can be represented by the general equation:

$$\underset{\text{protein}}{P} + \underset{\text{ligand}}{nL} \underset{K_s}{\overset{K_a}{\rightleftharpoons}} \underset{\text{complex}}{PLn}$$

where n = number of identical but non-cooperative ligand binding sites
K_a = protein–ligand binding (affinity) constant
K_s = protein–ligand dissociation constant = $1/K_a$

Applying the Law of Mass Action to the equilibrium and developing the expression for K_a algebraically, gives the *Scatchard equation*:

$$\frac{r}{[L]} = nK_a - rK_a \qquad 3.17$$

where r = number of moles of ligand bound to each mole of protein

i.e. $r = \dfrac{[PLn]}{[P]}$

[P] = molar concentration of protein

[L] = molar concentration of free (unbound) ligand

To determine the numerical value of n and K_a the equilibrium binding of ligand to protein is studied as a function of ligand concentration at a fixed protein concentration [P]. In cases of enzyme and substrate, the experimental conditions must be such that product formation is not possible. A plot is then made of $r/[L]$ against r (Fig. 3.10). In some situations the Scatchard plot (Fig. 3.10) is biphasic indicating either that binding is cooperative or that there are two sets of binding sites with different affinities. In this latter case the isolation of the two sets of binding constants is difficult. Although the Scatchard equation is widely used, it is worth noting that, like the Eadie-Hofstee plot, it is inherently unsatisfactory as one of the variables, r, is incorporated into both axes. A double reciprocal plot of $1/r$ against $1/[L]$ (equation 3.18), analogous to the Lineweaver-Burk plot, lacks this criticism but like the Lineweaver-Burk equation gives an uneven distribution of points

$$\frac{1}{r} = \frac{1}{n} + \frac{1}{n[L]K_a} \qquad 3.18$$

If the numerical value of [P] is not known, for example in a sample of plasma or serum, a plot is constructed of the concentration of bound ligand/concentration of unbound ligand against the concentration of bound ligand i.e. [PLn]/[L] against [PLn]. It will have a slope of $-K_a$, intercept on the x-axis

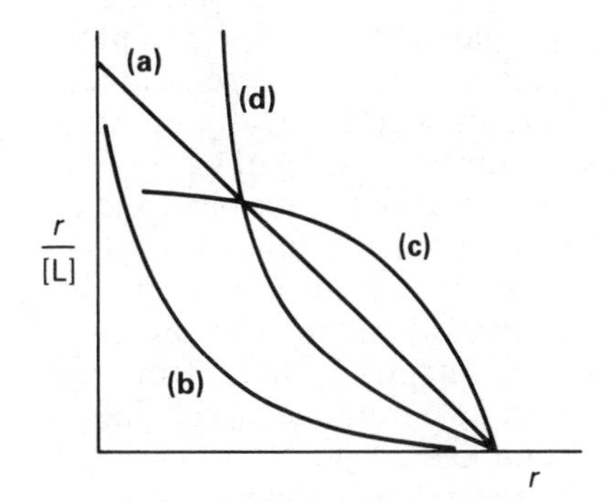

Fig. 3.10 Scatchard plot for **(a)** a single set of sites with no cooperativity; **(b)** two sets of sites with no cooperativity; **(c)** a single set of sites with positive cooperativity; and **(d)** a single set of sites with negative cooperativity.

of $n\,[P]K_a$ and on the y-axis of $n\,[P]$. The term $n\,[P]$ is frequently referred to as the *binding capacity*.

The binding of some peptide hormones to their membrane receptor proteins displays cooperativity. In such cases the Scatchard plot is concave upwards for negative cooperativity and concave downwards for positive cooperativity but in both cases the intercept on the x-axis is equal to n (Fig. 3.10). If cooperative ligand binding is suspected, it should be confirmed by a Hill plot (equation 3.2) which, in its non-kinetic form, is:

$$\log\left(\frac{\overline{Y}}{1-\overline{Y}}\right) = h\log[L] + \log K \qquad 3.19$$

where $\overline{Y}$ = fraction of binding sites filled

For a protein with multiple binding sites which function independently $h = 1$ and $n = 1, 2, 3, 4 \ldots$, whilst for a protein with multiple sites which are interdependent $n = 1, 2, 3, 4 \ldots$ and $1 < h < 1$.

Numerous techniques including ultracentrifugation (Section 2.10), exclusion chromatography (Section 6.6), circular dichroism (Section 8.4) and nuclear magnetic resonance spectrometry (Section 8.9) have been used to study ligand binding. In the case of spectrophotometric techniques, information regarding the functional groups in the protein involved in binding the ligand can be obtained in addition to data permitting the calculation of dissociation constants. The principles of these techniques are discussed in appropriate chapters in this book and only details of the two simplest techniques, equilibrium dialysis and ultrafiltration will be described here.

3.7.2 Equilibrium dialysis

The protein and ligand, each in a buffer of the same pH, are placed in opposite halves of a dialysis cell. The cell, which generally has a total internal volume of about 3 cm^3, is constructed of transparent plastic such as Perspex and unscrews into two halves which are separated by a cellulose acetate or nitrate semipermeable membrane mounted on an inert mesh support. Many commercial variants of the cell are available, some consisting of banks of up to six cells. The temperature of the cell is thermostatically controlled and the cell is slowly rotated to help the system reach equilibrium. The ligand molecules, which are small and diffusable, readily cross the membrane until their unbound concentration is the same on both sides. The protein is confined to one half of the cell. At equilibrium, samples are taken from each half of the cell and analysed for ligand. The sample from the protein half of the cell will give the sum of the bound and unbound ligand concentrations whilst that in the other half will simply give the unbound ligand concentration which will be the same as the unbound ligand concentration in the other half of the cell containing the protein. From a knowledge of the protein concentration, the values of r can be calculated and a Scatchard plot constructed. Most commonly the ligand is partly present as a ^{3}H or ^{14}C-labelled form so that the assay can simply be performed by scintillation counting. For reliable results the binding of both the ligand and protein to the membrane must be minimal and the total ion concentration in each half of the cell equalised to minimise any possibility of a charge inequality on either side of the membrane affecting the distribution of ligand (*Donnan effect*). The limitations of the technique are the relatively long period of time it takes to establish equilibrium and the fact that it cannot be applied to cases where the ligand is a macromolecule, for example in the study of the binding of tRNA to aminoacyl-tRNA synthetase.

3.7.3 Ultrafiltration

The protein and ligand in a buffered solution are contained in a thermostatically-controlled cell (generally 1 to 3 cm^3 capacity) containing a semi-permeable membrane on an inert mesh support at its base. Since no diffusion across the membrane is required to establish equilibrium, attainment of equilibrium is rapid (20 minutes). A small sample (100 mm^3) is then forced across the membrane into a collection cup either by applying a gas pressure to the mixture side or more simply by placing the cell in a low-speed centrifuge and centrifuging at about 3000*g* for a few minutes. By analysing the ultrafiltrate (representative of the unbound ligand concentration) and the reaction mixture (bound plus unbound ligand), the influence of ligand concentration on the extent of binding can readily be studied. The speed of the method is its attraction, but binding of the reactants to the membrane must be checked and the volume of the ultrafiltrate kept to a minimum to minimise any possibility that the sampling procedure displaces the position of equilibrium. As with equilibrium dialysis the method cannot be used to study macromolecular ligands. The binding of such ligands is best studied by ultracentrifugation or exclusion chromatography.

3.8 Immobilised enzymes

Although kinetic studies of purified enzymes have revealed detailed information about the mechanisms by which enzymes work, it is doubtful whether such information is entirely relevant to the *in vivo* situation. For example, some intracellular enzymes have been shown by histochemical studies to be present in high concentrations whilst others have been shown to be membrane-bound rather than being free in dilute solution. Studies of immobilised enzymes have shown that they display kinetic behaviour subtly different from that of free enzymes. The reasons for this are complex but involve the perturbation of the three-dimensional structure of the enzyme by the immobilising matrix and changes in the microenvironment of the active site of the immobilised enzyme. Kinetic studies of immobilised enzymes are therefore essential for our complete understanding of their *in vivo* action, especially their regulation. Immobilised enzymes, however, are also important because of their considerable analytical and industrial potential. Industrially, solutions of enzymes are commonly used to carry out chemical transformations especially in the synthesis of antibiotics and steroids and are used increasingly in preference to the techniques of conventional organic chemistry. Enzymes offer the advantages of both chemical and stereo-specificity and accordingly give purer products. However, they are frequently unstable and difficult to recover from the reaction mixture. In principle, however, the ability to immobilise enzymes, pack them into columns and re-use them many times is commercially most attractive.

Enzymes can be immobilised in many ways, and these can be divided into two types. In *physical immobilisation* no covalent bonds are formed between the enzyme and the supporting matrix. Adsorption on animal charcoal or alumina was initially used to achieve immobilisation but current approaches include ionic adsorption onto ion-exchangers especially those of the Sephadex (Section 6.6) type and adsorption onto controlled-pore glass. Immobilisation by adsorption has the advantages of simplicity, general applicability, high yield and ability to recharge when the catalytic activity of the immobilised enzyme decreases below an unacceptable level. A limitation of the technique is the need to strictly control the working conditions to prevent desorption. Enzyme entrapment in liposomes (artificially produced concentric spheres of phospholipid bilayers) and in water-insoluble polymers such as polyacrylamide and agarose is also a simple and generally applicable method but suffers from poor flow properties, inefficiency and progressive leaching of the enzyme.

Chemical immobilisation results in at least one covalent bond being formed between the enzyme and the matrix. The chemical procedures used to produce immobilisation are similar to those used in affinity chromatography (Section 6.7.3). Attachment must not involve amino acid residues in the active site of the enzyme. The matrix may be polysaccharide, polymers such as nylon or inorganic carriers such as glass and titanium dioxide.

Immobilised enzymes are finding an increasing number of analytical applications especially in clinical situations where they offer the potential for fast, sensitive and accurate determinations of analytes like blood glucose and urea. The most significant development is the combination of immobilised enzymes, with their high specificity, with electroanalytical chemistry with its inherent sensitivity. The so-called *enzyme electrode* offers the opportunity for accurate analysis without any sample preparation. The principles involved are discussed more fully in Section 10.6.

3.9 Suggestions for further reading

Bergmeyer, H.U. (Ed.) (1978). *Principles of Enzymatic Analysis*. Verlag-Chemie. (Authoritative accounts of enzyme assays.)

Cleland, W.W. (1970). Steady state kinetics. In *The Enzymes*, 3rd edition (Boyer, P.D., Ed.) pp. 1–65. Academic Press, New York. (Excellent review of kinetics and mechanisms.)

Colowick, S.P. and Kaplan, N.O. (Eds) *Methods in Enzymology*. Academic Press, New York. (Contributions by leading authorities on all aspects of the subject. Particularly recommended: vols 48, 49, 61, and 91 (structure), 22 and 104C (purification), 63, 64 and 87 (kinetics and mechanism), 46 (affinity labelling) and 44 (immobilised enzymes).)

Dixon, M., Webb, E.C., Thorne, C.J.R. and Tipton, K.F. (1979). *Enzymes*, 3rd edition. Longman, London. (Comprehensive account of enzymology).

Fersht, A. (1985). *Enzyme Structure and Mechanism,* 2nd edition. Freeman, Reading. (General text particularly good on enzyme mechanisms.)

Halford, S.E. (1974). Rapid reaction techniques. In *Companion to Biochemistry*, (Bull, A.T., Lagnado, J.R. Thomas, J.D. and Tipton, K.F., Eds). Longman, London. (A good account of this specialist technique.)

Palmer, T. (1985). *Understanding Enzymes*, 2nd edition. Ellis Horwood, Chichester. (A very readable undergraduate text on all aspects of enzymology.)

Trevan, M. (1980). *Immobilised Enzymes*. Wiley, Chichester. (An excellent account of the formation, kinetic properties and industrial uses of immobilised enzymes.)

4

Immunochemical techniques

4.1 General principles

4.1.1 Introduction

Immunology is the study of the immune responses, which are the processes by which an animal defends itself against invasion by foreign organisms. Immune responses can be divided into two general types: either (a) the *humoral* (antibody-mediated) *response* or (b) *the cell-mediated response*. Both types of immune response involve cells of the lympho-reticular system. Immunochemical techniques employ the antibodies involved in humoral immunity.

When an *antigen*, which may be a foreign organism or compound, enters the tissues of an animal, lymphocytes are stimulated to divide and differentiate which, in the case of the humoral response, results in the eventual production of *plasma cells*. Plasma cells secrete specific *antibody* molecules containing specific binding sites capable of binding tightly but non-covalently to the original antigen thereby causing its precipitation, neutralisation or death, via phagocytosis or complement-mediated cell lysis, depending on whether the antigen is a macromolecule, toxin or micro-organism. Antibodies belong to a group of proteins known as the *immunoglobulins* (Ig) which may be divided into five classes – IgG, IgM, IgA, IgD and IgE – some properties of which are listed in Table 4.1.

Immunochemical techniques utilise mainly IgG, which constitutes 80% of the serum immunoglobulin. All immunoglobulins have structures based on four polypeptide chains consisting of two identical heavy chains and two identical light chains. The immunoglobulin classes listed in Table 4.1 are defined by the nature of their heavy chains: γ, μ, α, δ, or ϵ. Only two types of light chains are known, κ and λ, and these are common to all classes of immunoglobulins. The proteolytic enzyme papain cleaves IgG antibody molecules into three fragments with similar molecular weights: two identical Fab or antibody binding fragments and one Fc or crystallisable fragment; the latter is capable of binding and activating the *complement system*. This information is summarised in Fig. 4.1, which also indicates that papain actually cleaves the IgG molecules in the so-called hinge region. This hinge region allows limited flexibility in the spatial positioning of the Fab and Fc

Table 4.1 Physico-chemical and biological properties of immunoglobulin classes.

Immunoglobulin class	IgG	IgM	IgA	IgD	IgE
Outdated synonyms	γG, 7SG	β_2M, 19SG	γA, β_2A	γD	γE
Heavy chain symbol	γ	μ	α	δ	ϵ
Heavy chain (molecular weight	50 000	70 000	55 000	65 000	65 000
Molecular composition	$\gamma_2\kappa_2$ $\gamma_2\lambda_2$	$(\mu_2\kappa_2)_5$ $(\mu_2\lambda_2)_5$	$\alpha_2\lambda_2, \alpha_2\kappa_2$ $(\alpha_2\lambda_2)_2$Sc* $(\alpha_2\kappa_2)_2$Sc*	$\delta_2\kappa_2$ $\delta_2\lambda_2$	$\epsilon_2\kappa_2$ $\epsilon_2\lambda_2$
Physical state	Monomer	Pentamer	Monomer, dimer	Monomer	Monomer
Svedberg value	7S	19S	7–11S*	7S	8S
Molecular weight (daltons)	150 000	900 000	160 000–380 000*	180 000	200 000
Valency	2	5–10	2, 4*	2	2
Electrophoretic mobility	mid	fast	slow	fast	fast
Carbohydrate content	3	12	7.5–10	12	12
Mercaptoethanol sensitivity	–	+	–	–	–
Heat stability (56°C, 4 h)	+	+	+	?	–
Complement fixation	+	++	–	–	–
Normal serum levels (mg cm^{-3})	8–16	0.6–2	1–4	0.001	0.0003
Half life (days)	23	5	6	3	2

*Dimer in external secretions carries the secretory component, Sc

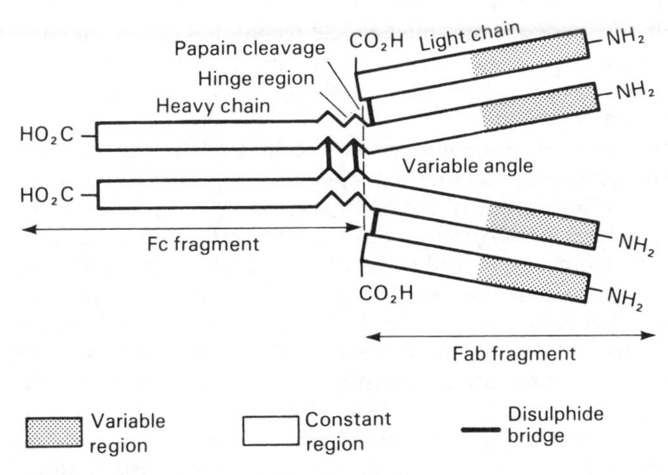

Fig. 4.1 Structure of the IgG molecule.

fragments in the intact IgG molecule. It should be noted that different IgG subclasses contain 2 to 4 inter-chain disulphide bridges in the hinge region, and that both the heavy and light chains of all immunoglobulins contain a number of intra-chain disulphide links, resulting in a compact globular structure.

Many of the 100 to 110 amino acid residues at the *N*-terminal end of both the light and heavy chains vary in different IgG molecules, defining this as the *variable region*. The other amino acid residues of the light and heavy chains are virtually constant within an IgG subclass (*constant region*). Variations in the composition of a few of the amino acid residues (the *hypervariable residues*) in the *N*-terminal portion (variable region) of each chain of an immunoglobulin result in the unique topography of the antigen binding site on each immunoglobulin and account for the uniqueness of each specific antibody. IgG, which is a monomer containing only two identical light and two identical heavy chains, has an antigen valency of two, i.e. it has two identical antigen binding sites. The variable regions of one light and one heavy chain interact to form a globular domain (*variable domain*). The constant region of one light chain and the appropriate constant region of one heavy chain form a globular *constant domain*, and the remaining constant region of the heavy chains form two further constant domains. Thus each Fc and Fab fragment contains two globular domains and a single IgG molecule contains two variable domains and four constant domains with limited flexibility between each compact domain.

The specificity of the immune reaction is well known to all those who find themselves immune to a disease caused by one microorganism but fall ill to infection with a different microorganism. It is even more apparent to those renal transplant patients who eventually reject a kidney from another person who apparently had an almost identical antigenic make-up to the donor, such as that of a very close relative. The analytical biochemist, however, is more interested in the *in vitro* use of immunochemical techniques rather than the detailed *in vivo* interactions involved in immunology. It should be stressed that it is not necessary to be a student of immunology to use and understand most immunochemical techniques.

The major attraction of immunochemical techniques is the extreme specificity conferred on them by biological recognition at the molecular level, even in the presence of high levels of contaminating molecules; for example at the submolecular level *monospecific antisera* can differentiate between macromolecules containing the + and − forms of a single amino acid. The multivalency of most antigens and antibodies enables them to interact to form a precipitate. Labelling antigens and antibodies with radioisotopes enables the biochemist to combine the specificity of biological activity with the sensitivity of liquid scintillation counting techniques (Section 9.2.3). Thus immunochemical techniques extend the ability of the biochemist to detect and quantify specific molecular structures, even in the presence of closely related material in, for example, sera, microorganism culture filtrates, tissue extracts, fractions from gradient centrifugation or the effluent of chromatography columns.

4.1.2 Definitions

Immunology is a subject which tends to have a vocabulary of its own and it is advisable to begin with a series of definitions of important terms used throughout this chapter.

Adjuvant A substance that increases the biosynthesis of antibody in response to antigens.

Affinity The intrinsic binding power of a single antibody combining site with a single antigen binding site, which may be expressed in terms of a single binding constant.

Antibody A protein, with the molecular properties of an immunoglobulin, capable of specific combination with the antigen that caused its production in a susceptible animal.

Antibody valency The number of antigen binding sites on an antibody molecule.

Antigen Any foreign substance that elicits an immune response (e.g. the production of specific antibody molecules) when introduced into the tissues of a susceptible animal and which is capable of combining with the specific antibody molecules produced. Antigens are usually of high molecular weight and are commonly either proteins or polysaccharides.

Antigen combining site or paratope The site on the antibody that combines specifically with its corresponding antigenic determinant. It usually involves only a few amino acid residues lining a shallow cleft in the antibody molecule but which are not usually covalently linked directly to each other.

Antigen valency The number of antibody binding sites on the antigen.

Antigenic determinant or epitope A small site on an antigen to which a complementary antibody molecule may be specifically bound through its combining site. This is usually 1 to 6 monosaccharides or amino acid residues on the surface of the antigen, not necessarily covalently linked to each other.

Antigenicity The potential of an antigen to stimulate an immune response in a particular host.

Antiserum A serum containing antibodies against a specific antigen or antigen mixture; for example anti-ovalbumin serum, or anti-sheep erythrocyte serum.

Autoantibody An antibody to one of the constituent molecules of the individual producing the antibody e.g. anti-DNA or antithyroid antibodies. Such antibodies do not arise naturally in healthy individuals and are indicative usually of a pathological condition.

Avidity The net combining power of an antibody molecule with its antigen.

B-lymphocyte A cell which on exposure to a particular epitope is stimulated to divide and differentiate to form a clone of plasma cells synthesising an immunoglobulin capable of combining specifically with that epitope.

Clone A family of cells of genetically identical constitution derived asexually from a single cell by repeated division.

Complement A group of nine serum proteins essential for antibody mediated immune haemolysis.

Epitope See antigenic determinant.

Equivalence The ratio of antigen and antibody giving complete precipitation of both molecules, leaving neither unbound in solution.

Hapten A substance that can combine with a specific antibody but which lacks antigenicity, i.e. it cannot initiate an immune response unless bound to an antigenic carrier molecule such as bovine serum albumin. Haptens are usually compounds with a molecular weight less than 1000, such as simple sugars, amino acids, small oligopeptides, phospholipids, triglycerides, drugs.

HAT medium A growth medium containing hypoxanthine, aminopterin and thymine, used for selecting HPGRT deficient cell lines.

HPGRT The enzyme hypoxanthine-guanine phosphoribosyl transferase involved in nucleic acid synthesis.

Hybridoma The name given to cell lines created by fusing B lymphocytes with myeloma cells. Hybridomas produce monoclonal antibodies.

Immunoglobulin A member of the largest family of proteins produced by animals. See antibody.

Lymphocyte A cell which on exposure to a specific antigenic determinant is capable of triggering an immune response.

Macrophage A general name for a large number of morphologically dissimilar, relatively long lived, phagocytic cells, which may be sessile or motile, and which play an important role in the immune response by processing antigens and presenting them to B and T lymphocytes.

Monoclonal antibody Immunoglobulin derived from a single clone and therefore homogeneous.

Myeloma or ***plasmacytoma*** A neoplasm of a B lymphocyte or plasma cell.

Plasma cell The terminal cell of the differentiation of a B lymphocyte on exposure to its specific epitope.

Neoplasm A cancer.

Polyclonal antiserum An antiserum, possibly to a single antigen, containing a number of antibodies to that antigen from different plasma cell clones. (All antisera are in fact polyclonal).

T-lymphocyte A cell matured in the thymus which on exposure to a particular epitope may regulate the immune response to antigens possessing that epitope.

Valency See antigen valency and antibody valency.

4.1.3 The precipitin reaction

The basic principle involved in most immunochemical techniques is that a specific antigen will combine with its specific antibody to give an antigen–antibody complex as illustrated in Fig. 4.4a. Since antigens and antibodies are both multivalent, the antigen–antibody complex is usually insoluble and may be seen with the naked eye.

4.1.4 Preparative uses of the antigen–antibody interaction

Most immunochemical techniques are used in an analytical capacity. However, a specific antiserum can be used in a preparative mode to isolate the appropriate antigen from a heterogeneous solution by mixing and centrifugation to remove the antigen–antibody precipitate. If the protein–antibody complex is not completely insoluble, it might be precipitated by the addition of a second antibody capable of precipitating the first antibody or by addition of Protein A from staphylococci which has the property of crosslinking IgG molecules. An alternative procedure, if the aim of the experiment is to isolate and purify a particular protein, is to use the specific antiserum to prepare an *immunoadsorbent* which may then be used to isolate the protein, either by affinity chromatography (Section 6.7) or as a batch procedure. Advantages of immunoadsorbent procedures include: the production of the antigen in purer form (no other serum components present); far less chance of proteolysis by serum proteases; and more rapid separation of antigen and antibody in the final stages of purification. Advantages of precipitation methods are that less antiserum is required and higher yields of antigen are often obtained.

The hardest part of using antisera to purify protein antigens is the final dissociation of the antigen–antibody complex. The dissociation conditions must be commensurate with the minimum denaturation of the desired protein, since dissociation usually requires exposure of the complex to either a very low pH or to high concentrations of guanidinium chloride or urea. The correct conditions must be identified by trial and error. It is usual to try to dissociate the complex using the least destructive system from a series such as: 10% dioxan pH 7.2; 25% ethylene glycol pH 6.5; 3M potassium thiocyanate pH 6.4; 0.2M glycine/HCl pH 2.8; 1M propanoic acid pH 2.4. One advantage of the affinity column chromatography method is that the eluted antigen can be collected into another buffer to reduce the exposure time to the denaturing pH (e.g. 0.3M borate buffer, pH 8.2).

4.2 Production of antibodies

4.2.1 Production of antisera (polyclonal antibodies)

Most of the antibodies used in immunochemistry are raised or induced by injection of a solution or suspension of the appropriate antigen into a rabbit. After a suitable period of time, 5 to 50 cm^3 of blood is obtained from the immunised rabbit by making an incision in its posterior marginal ear vein. The blood is allowed to clot at 37°C for one hour. The clot is detached from the sides of its container to allow it to retract, and left at 4°C to contract and exude 2 to 25 cm^3 of serum. The serum is separated from the clot and free cells by centrifugation. Proteases and complement are inactivated by incubating the serum at 56°C for 45 minutes and the serum is usually stored in small aliquots at −20°C. A control serum is usually obtained from the same rabbit prior to immunisation. Sheep, goats and horses are used for large scale antiserum production.

Inoculating a rabbit with a single injection of a saline solution of a strongly antigenic compound results in the production of specific antibodies, detectable in its serum after about 10 days, which reach a maximum after 15 to 20 days, and which then decline over a period of weeks. This is the *primary humoral immune response* and results entirely in the production of IgM antibodies. A secondary response may be induced by a further injection of the same antigen at any period after the primary response. The *secondary response* is more rapid and increased antibody levels may be detected after three days with the maximum level of antibody occurring only 10 days after the second injection. A similar level of IgM antibody to that in the primary response is found but, in addition, 3 to 10 times that amount of IgG antibody may also be detected. Further injections of the same antigen, at say fortnightly intervals, results in a hyperimmunised rabbit whose serum contains vastly increased levels of specific IgG antibody (1 to 5 mg specific IgG cm^{-3}).

When using weakly antigenic compounds two approaches may be used to obtain antisera of reasonable *titre* i.e. antisera containing readily detectable levels of specific antibodies. Firstly, the period of time that the immune system is exposed to the antigen may be extended, either by repeated inoculation or by establishing depots of antigen within the rabbit which slowly release the antigen over a period of weeks. The latter may be achieved by intramuscular, subcutaneous or intradermal inoculation of particulate or precipitated antigen. Secondly, the antigenticity of the compound may be enhanced by the use of adjuvants. Aluminium potassium sulphate is a simple adjuvent used to precipitate soluble protein antigens such as tetanus or diphtheria toxin prior to their inoculation. This results in the slow release of the antigen, which mimics a prolonged series of injections and which causes the antigen to be more readily trapped or phagocytosed by the *macrophages* involved in the immune response.

The most commonly used adjuvant is *Freund's complete adjuvant* which contains a pharmaceutical grade white mineral oil such as liquid paraffin, an emulsifier such as mannide mono-oleate, and heat-killed *Mycobacterium tuberculosis*. Often 1 cm^3 of an emulsion of equal parts of Freund's complete adjuvant and the antigen solution is divided into three aliquots and inoculated into three sites within the rabbit. For example instead of using a single large injection, one aliquot of the emulsion might be inoculated subcutaneously and the other two inoculated into the hind leg muscles, in order to extend the number of lymph nodes that the antigen in emulsion droplets might reach via the lymphatic system. Antigen entering the blood stream in emulsion droplets may activate lymphocytes in the spleen and bone marrow. A depot of antigen in emulsion tends to persist at each inoculation site and the killed tubercle bacilli stimulate invasion of the site by reticulo-endothelial cells. This results in the formation of a *granuloma* which further enhances the overall immune response to the antigen. When raising antisera for use in immunochemical studies, the final inoculations before bleeding usually contain the antigen in an emulsion with *Freund's incomplete adjuvant* which lacks the tubercle bacilli and hence tends to stimulate only the humoral immune response.

Antibodies are readily raised to particulate antigens such as erythrocytes or killed bacteria in isotonic saline following their intravenous inoculation into an appropriate mammal. However, before haptens such as drugs and non-peptide hormones can be used to raise antisera they must be coupled to antigenic structures such as proteins, polysaccharides or sheep erythrocytes by methods similar to those used to prepare affinity chromatography material (Section 6.7). Addition of excess carrier macromolecule or erythrocytes to the antiserum precipitates the carrier-antibody complexes. After centrifugation, the antiserum, now mono-specific for the hapten, may be decanted off.

It is most important to understand that even a small hapten of relative molecular mass as low as 200 can induce a mammal to produce up to six different immunoglobulins, each differing in amino acid composition and capable of combining with different parts of the surface of the hapten. Therefore it is not surprising that large antigens such as proteins induce the production of a multitude of different immunoglobulins, each reacting with a slightly different epitope on the surface of the antigen. This in turn can lead to slight cross reactions between an antiserum and antigens sharing only a minimum common structural feature. The different immunoglobulins are produced as a result of the interaction between the epitopes in the antigen surface and different clones of B lymphocytes each carrying a specific receptor for that epitope. Hence the name *polyclonal antiserum*. The relative amounts of each different immunoglobulin in such a polyclonal serum depends on the route of inoculation of antigen, the species and strain inoculated, as well as the adjuvants used and frequency of inoculation. Thus it is not surprising that the standardisation of so-called antigen specific polyclonal antisera is a very difficult task because most animals contain many different clones of B lymphocytes each capable of reacting with a different epitope on the antigen, which may or may not be induced to produce IgG. In practice it is impossible to obtain a homogeneous population of a single molecular species of immunoglobulin molecules (i.e. a *monoclonal antibody*) from polyclonal antiserum despite intensive efforts by immunochemists for many decades. However, modern cell biological techniques, including fusion of B lymphocytes with immortal cell cultures, have facilitated the production of monoclonal antibodies.

4.2.2 Production of monoclonal antibodies

The IgG fraction of a normal healthy individual's serum has a large range of electrophoretic mobilities, reflecting the 10^5 to 10^6 different IgG molecules present. However, the IgG fraction of the serum of unhealthy individuals suffering from myelomatous disease is much simpler and is dominated by a single type of IgG molecule. Most of the circulating lymphocytes of such individuals are *myeloma cells*, that is they are a single clone of neoplastic B lymphocytes. Myeloma cells, unlike normal B lymphocytes, continue to divide rapidly and secrete a single species of immunoglobulin molecule in the absence of any stimulating antigen. Whereas normal B lymphocytes do not survive for more than a few hours *in vitro*, myelomas like other neoplastic cells are virtually immortal in cell culture.

Normal mamalian cells have two different ways of synthesising DNA: a *de novo* pathway in which nucleic acids are built up from purine and pyrimidine bases, deoxyribose and inorganic phosphate; and a salvage pathway that converts similar nucleotides into the correct nucleic acid. The *de novo* pathway is completely inhibited by the metabolic poison aminopterin. Cells containing the enzyme hypoxanthine-guanine phosphoribosyl transferase (HGPRT) can utilize the salvage pathway and survive in the presence of aminopterin. However, as a result of the extra demand placed on the salvage pathway, extra hypoxanthine and thymidine must be added to the so-called HAT medium (*H*ypoxanthine, *A*minopterin, *T*hymidine). Thus the HAT medium will selectively allow cells containing HGPRT to grow, whilst preventing the growth of HGPRT$^-$ cells. Mouse myeloma cells have been selected that lack HGPRT (HGPRT$^-$ clones) and hence cannot survive in HAT medium. Furthermore, HGPRT$^-$ myeloma cells have been selected that secrete little or even no unwanted immunoglobulin.

The currently standard type of methodology used to produce monoclonal antibodies is illustrated in Fig. 4.2a. The spleen from a mouse immunised with the appropriate antigen is used as a source of a suspension of sensitised B lymphocytes. Sensitised B cells and HGPRT$^-$ myeloma cells grown *in vitro* are fused in the presence of 30 to 50% (w/v) polyethylene glycol for only a few minutes since polyethylene glycol is toxic to cells. The fusion mixture is seeded into culture vessels containing HAT medium. Lymphocytes and fused lymphocytes soon die as they cannot be cultured *in vitro*. Myeloma cells and fused myeloma cells are poisoned by the aminopterin in the HAT Medium, since they lack HGPRT. Only hybridomas, that is fusion products of B lymphocytes and HGPRT$^-$ myeloma cells can survive because the myeloma cell confers the ability to grow *in vitro* and the B lymphocytes contribute the HGPRT essential for growth in the HAT medium. After 10 to 14 days hybridomas will be the only surviving cells. However, most of these will be producing either unwanted immunoglobulins or no immunoglobulin. Unfortunately, non-immunoglobulin producing hybridomas have a tendancy to outgrow the immunoglobulin-producing hybridomas so that it is essential over the next 7 to 14 days to continually test individual cultures for required immunoglobulin production and to reject those that are non-productive. Also cultures that are making useful immunoglobulin should be subcultured and aliquots stored at −196°C (Section 1.6.6) in case contamination with microorganisms destroys months of work.

Finally single clones are obtained in one of two ways. Cultures may be diluted to limiting dilution and distributed so that each well of a *microtitre plate* contains only one hybridoma cell. In order for such clones to grow, it is usually necessary to supplement the medium regularly with feeder layers of normal cells such as macrophages. When the clones have grown sufficiently they may be tested for specific immunoglobulin production using sophisticated immunochemical techniques such as radioimmunoassay or enzyme-linked immunoassays (Sections 4.5 and 4.6 respectively). Alternatively hybrids may be diluted and seeded into sloppy nutrient agar so that individual clones may be observed growing and picked out with a Pasteur pipette. These clones can then be grown up in microtitre plates and tested for specific immunoglobulin production. Aliquots of each useful clone are further

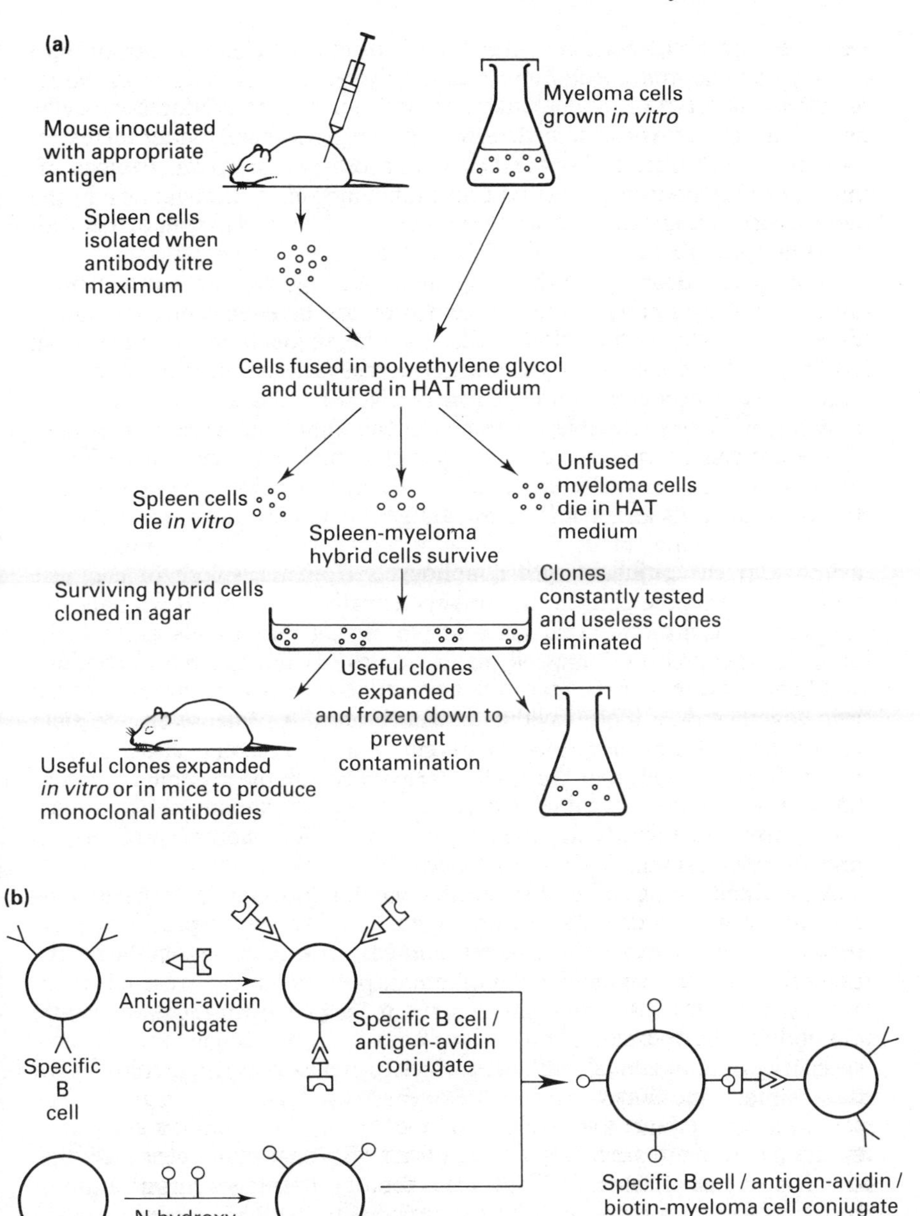

Fig. 4.2 Monoclonal antibody production. **(a)** General scheme for the procedure most commonly used; **(b)** scheme for ensuring a high proportion of specific hybridomas by antigen specific adhesion of the B cells of interest to the myeloma cells.

characterised and frozen at −196°C. It is usual to reclone at least once to ensure that the immunoglobulin produced is a true monoclonal antibody. Useable quantities of monoclonal antibody can then be produced either by large scale culture of cells, *in vitro*, which is expensive and labour intensive, or more easily by growing the hybridoma cells as an *ascites (single cell) tumour* in the peritoneal cavity of a suitable mouse. A disadvantage of the latter procedure is that the monoclonal antibody becomes contaminated with the other proteins found in ascitic fluid, but this is rarely important.

A major disadvantage of the usual methods of monoclonal antibody production is that they rely on the random fusion of a heterogeneous population of B lymphocytes with myeloma cells, resulting in much time being spent in growing and testing hybrids which secrete a vast variety of immunoglobulins. Alternative procedures (illustrated in Fig. 4.2b) endeavour to ensure that only B lymphocytes capable of producing immunoglobulins complementary to the antigen of interest lie close enough to myeloma cells for fusion to occur. This may be engineered by chemically coupling biotin to the surface of the myeloma cells and avidin to the antigen of interest (avidin is a protein which specifically binds biotin with high affinity). On mixing the avidin–antigen conjugate with lymphocytes from the spleen of a mouse immunised with the antigen, it binds specifically with those B lymphocytes that carry receptors for that antigen. On mixing this suspension with the biotin-conjugated myeloma cell suspension the avidin and biotin combine and hence ensure that the myeloma and antigen specific B lymphocytes are held together by a biotin–avidin–antigen bridge. An intense electric field across the bulk cell suspension (4 kV cm^{-3} for five seconds at 30°C) then causes fusion of cells that are in close proximity; i.e. the myeloma cells and those B lymphocytes capable of making the appropriate immunoglobulins. The appropriate hybrids may then be selected in HAT medium and cloned as previously described.

A major advantage of working with monoclonal antibodies is the extreme confidence with which claims can be made about such work. However, because of the intensive time, labour and cost of producing, and hence the expense of buying, monoclonal antibodies, polyclonal sera are used whenever there is no distinct advantage to be gained by using monoclonal antibodies. Furthermore, there are occasions on which the extreme specificity of monoclonal antibodies which recognise only single epitopes is a disadvantage and polyclonal sera are of more use.

Despite the labour and expense of their production, monoclonal antibodies have already been prepared against a vast range of antigens including serum proteins, enzymes, cell surface receptors, hormones, drugs, tumour specific antigens, viruses, and differentiation antigens. The latter are proving invaluable in the sub-classification of cells. The extreme specificity of monoclonal antibodies means that in general a clonal product will identify one target molecule. Thus monoclonal antibodies raise the exciting possibility of identifying and purifying previously unknown, or at least uncharacterised molecules, such as the T lymphocyte receptor for antigen. This is likely to become the area in which monoclonal antibodies have their biggest impact on research.

4.3 The precipitin reaction in free solution

4.3.1 Principles

If increasing amounts of an antigen (e.g. human serum albumin) are mixed with aliquots of suitable antibody solution (e.g. rabbit antihuman serum albumin) a precipitate forms in some of the tubes. After equilibration, the antigen-antibody precipitate may be isolated from the supernatant by centrifugation and may be estimated with a suitable protein assay (Section 3.2.2). The presence of excess antigen or antibody in each supernatant sample may be checked by looking for the formation of an antigen–antibody precipitate when further antibody or antigen solution respectively is added to a portion of the supernatant. The results of an experiment in which increasing amounts of an antigen are added to a fixed amount of antibody are illustrated in Fig. 4.3. As expected, initially the amount of antigen–antibody precipitate increases with increasing antigen added. However, a sharp plateau indicating that all of the antibody had been precipitated is not obtained and furthermore the precipitate apparently dissolves at higher concentrations of antigen. This is due to the solubility of antigen–antibody complexes containing a single antigen molecule, even if antibody molecules are bound to every antigenic determinant. The resulting curve of Fig. 4.3 may be divided into three zones:

(i) *zone of antibody excess* where the addition of further antigen leads to a substantial increase in the amount of precipitate;
(ii) *zone of equivalence* when the maximum antigen-antibody precipitate is formed;
(iii) *zone of antigen excess* when the precipitate dissolves.

Thus valid data can only be obtained for the precipitin reaction in the zone of equivalence.

4.3.2 Qualitative analysis of antigen

It is possible to detect the presence of an antigen by adding a range of concentrations of the suspected antigen to a fixed amount of monospecific

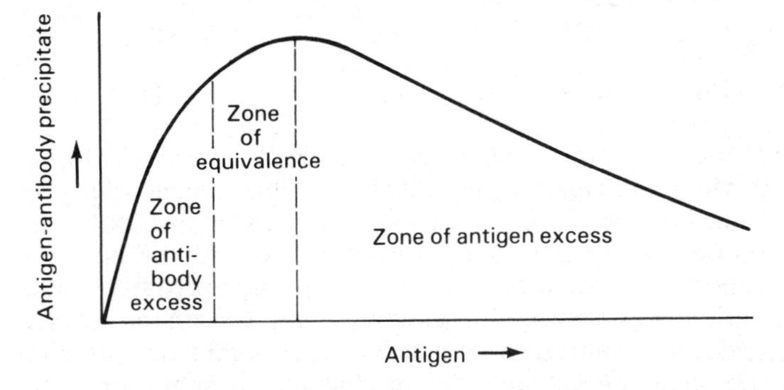

Fig. 4.3 Typical precipitin curve for a protein antigen reacting with rabbit antiserum.

antiserum (containing antibodies to one antigen only) and looking for precipitation. In this way it is possible to ascertain the presence of any one particular antigen (e.g. specific protein or carbohydrate) in the presence of many related compounds, for example during the isolation of any one component of serum or a cell extract. A major disadvantage of this procedure is that false negative data may be obtained if the ratios of antigen to antibody in the experiment do not fall within those pertaining to the zone of equivalence.

4.3.3 Quantitative analysis of antigen

Provided that the zone of equivalence can be determined, the precipitin reaction provides a very accurate method for the quantitative assay of antigens. Usually the total reaction volume is not more than 100 mm^3 and precipitation is allowed to take place overnight at 4 °C. The antigen–antibody complex is washed and its amount determined by any sensitive protein or nitrogen assay (Section 3.2.2) or by liquid scintillation counting if the antibody is labelled with a negatron emitter or by γ-counting if the antibody is labelled with a γ-emitting isotope, for example ^{131}I. It should be noted that all of these methods require the construction of a calibration curve.

The precipitin reaction in free solution is an important historical technique which is generally superceded by more simple methods such as single radial immunodiffusion (Section 4.4.2). It is, however, particularly useful during the quantitative assay of a protein synthesised *in vitro* under the direction of an isolated or chemically synthesised messenger RNA. For example the rate of synthesis of haemoglobin in the presence of mRNA extracted from chick erythrocytes, *E.coli* ribosomes and cofactors, and a radioactive amino acid, may be determined by measuring the rate of radioactivity incorporation into the synthesised protein that is precipitable by an anti-haemoglobin serum. Homologous protein (chick haemoglobin) is mixed into the system as a carrier, followed by the equivalent amount of specific antibody, for example rabbit anti-chick haemoglobin. This co-precipitates virtually all of the *in vitro* synthesised protein which may be assayed by liquid scintillation counting after separation by centrifugation.

4.3.4 Hapten inhibition test

The specificity of the antigen-antibody reaction can be used to elucidate the chemical nature of the antigenic determinants in a macromolecule. For example, an insight into the sequence of monosaccharides in a bacterial heteropolysaccharide may be gained by mixing aliquots of the polysaccharides with monospecific antisera to known small oligosaccharides and from the precipitation of polysaccharide–antibody complex with the specific anti-oligosaccharide serum, noting which antigenic determinants are present. Unfortunately, this requires the laborious production of an antiserum to

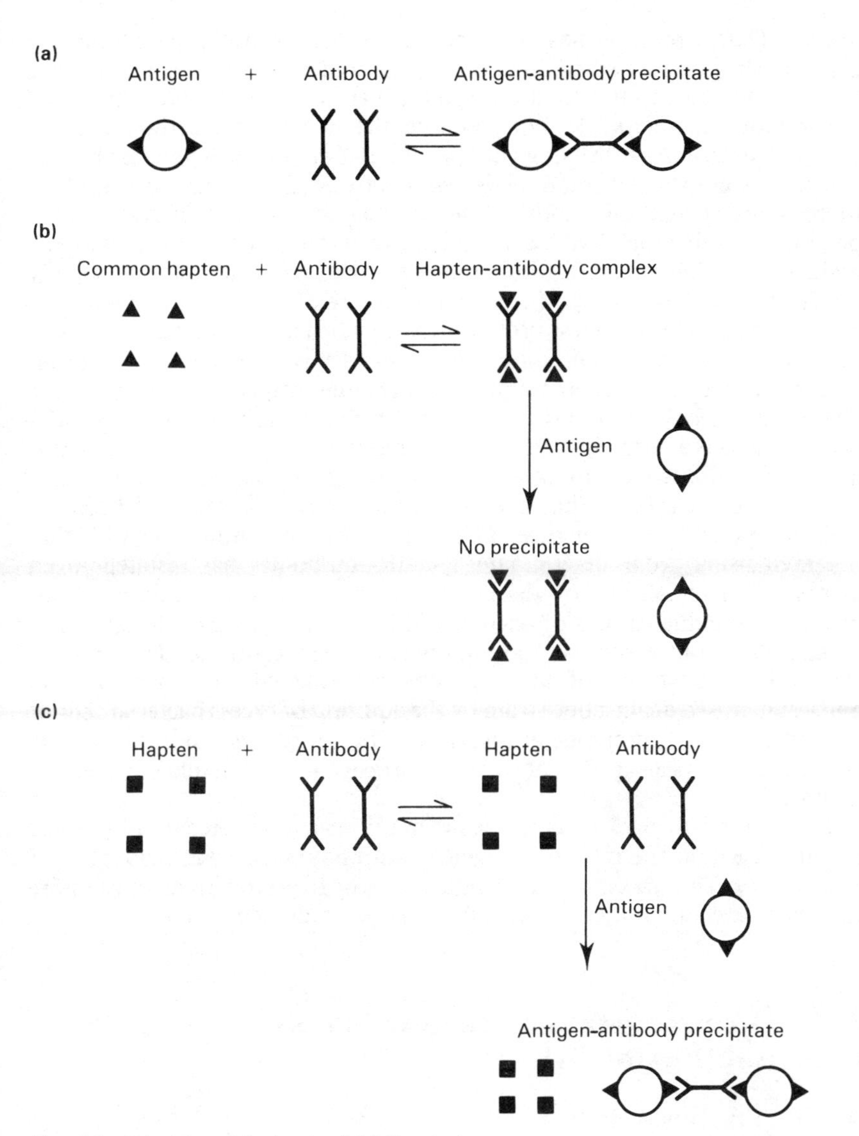

Fig. 4.4 Principles of the hapten inhibition test.

each oligosaccharide which, since they are haptens and hence not antigenic *per se*, also requires their prior coupling to a carrier protein. The *hapten inhibition test* saves most of this effort and is illustrated in Fig. 4.4.

Mixing an antigen which contains repeat determinants with antisera containing antibodies specific to those determinants results in a visible precipitate (Fig. 4.4a). Mixing the antiserum with a hapten which shares a determinant with the antigen also leads to a reaction, but no precipitate results. That a reaction has occurred can be demonstrated by adding the original antigen to the mixture of hapten and antiserum, since no free antibody remains to precipitate the antigen (Fig. 4.4b). Therefore, lack of precipitation at this stage indicates that the hapten and antigen share a common determinant. Mixing the specific anti-serum with another hapten which does not share a common determinant with the antigen does not lead to complex formation. Addition of antigen to this system results in a precipitate, indicating that this hapten does not share a determinant with the antigen (Fig. 4.4c). Thus a whole range of small molecules may be rapidly screened to give evidence of the chemical nature of a complex antigen.

The hapten inhibition test provided early evidence that the maximum size of most polysaccharide and protein antigenic determinants was of the order of 5 to 6 monosaccharides or amino acid residues respectively. The test has been used to show that the major difference between the blood group A and B determinants is a terminal *N*-acetylgalactosamine and galactose respectively, and it is still used to investigate blood and tissue antigens. The major contribution of this test has been to increase our knowledge of bacterial heteropolysaccharide structure. This knowledge, in conjunction with that obtained using mono-determinant specific antibodies has resulted in an extensive understanding of the complexities of cross antigenic reactions between individual strains of bacteria, which allows typing of the strains of clinical importance such as the Salmonellae. Antibodies to determinants shared by two strains of bacteria may be removed from a polyclonal antiserum raised against one strain by the addition of excess bacterial cells of the other strain. A mono-determinant specific antiserum may eventually be obtained by repeating this absorption process with a number of related strains.

The major drawback in using the precipitin reaction in free solution is the requirement that the ratio of antigen to antibody must be in the region of equivalence. This makes such techniques lengthy to perform and may require relatively large quantities of valuable antisera or antigens.

4.4 The precipitation reaction in gels: immunodiffusion (ID)

4.4.1 Principles

When molecules such as soluble antigens diffuse from a homogeneous solution into an agar gel, the concentration falls from a maximum at the solution–gel interface to zero at the leading edge of the region penetrated. Thus the system rapidly adjusts to give a complete antigen concentration gradient. Somewhere along this concentration gradient will be an antigen concentration that will give equivalence with almost any given concentration of antibody. This fact has been used since 1905 and developed into a range of highly sophisticated techniques.

4.4.2 Single (simple) immunodiffusion

This technique usually involves the diffusion of antigen from a solution into a gel containing antibody.

In One Dimension (The Oudin Tube Technique) This is a qualitative technique in which a solution or gel containing antigen(s) is layered onto a layer of agar containing the antiserum in a tube (Fig. 4.5a). As the antigen diffuses into the antibody/agar gel the antigen–antibody complexes initially formed are soluble in the excess antibody. In a simple system containing a single antigen a precipitate is formed where the ratio of antigen and antibody is at equivalence. As more antigen diffuses into the agar layer this precipitate dissolves in excess antigen and the precipitation line appears to move slowly down the tube. In more complicated systems one line is formed for each antigen present, provided that the antiserum contains the corresponding antibodies. Thus the technique may be used to detect the presence of an antigen or antigens in, for example, sera, culture filtrates, tissue extracts, or gradient centrifugation fractions.

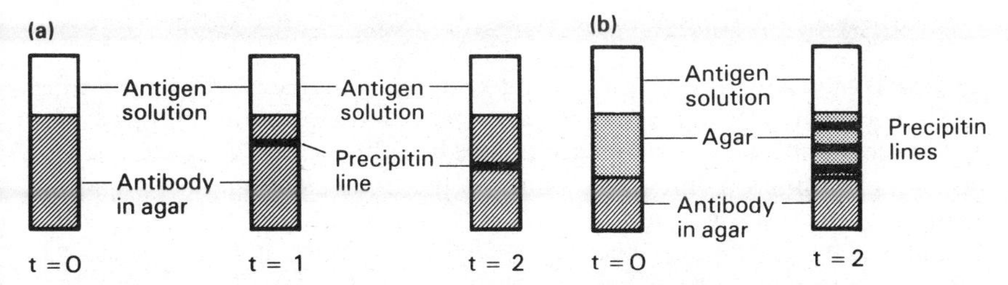

Fig. 4.5 Immunodiffusion in one dimension (t in days). **(a)** Time course of the Oudin tube technique, using a single antigen and its corresponding antibody. **(b)** Double diffusion using a mixture of antigens and the corresponding antiserum.

In Two Dimensions: Single Radial Immunodiffusion (SRID) In this (*Mancini*) technique a range of antigen concentrations is placed in wells cut in agar containing the corresponding antiserum on a microscope slide. As the antigen diffuses out radially from the wells a ring of precipitation forms and appears to move outwards, eventually becoming stationary at equivalence (Fig. 4.6a). The precipitate ring diameter at equivalence is a function of the antigen concentration. By plotting ring diameter or ring area at equivalence against antigen concentration, a calibration curve may be obtained for determining the concentration of antigen in unknown solutions (Fig. 4.6b).

This technique is commonly used for estimating the concentration of various plasma proteins such as IgG and IgM in patients suspected of suffering from agammaglobulinaemia and multiple myeloma respectively.

4.4.3 Double immunodiffusion

This technique involves the diffusion of both antigen and antibody towards each other.

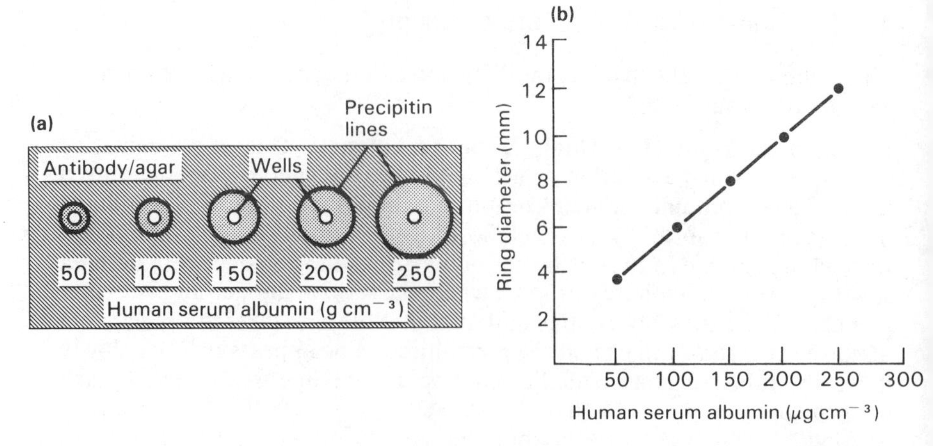

Fig. 4.6 Single radial immunodiffusion (SRID) for the estimation of human serum albumin. **(a)** Microscope slide covered with agar containing rabbit anti-human serum albumin. Wells of 4 mm diameter were filled with the indicated human serum albumin solutions. Precipitin rings formed at 4°C overnight. **(b)** Calibration curve for estimation of human serum albumin.

In One Dimension This rarely used technique is similar to the Oudin tube method except that a layer of agar is placed between the antigens and antibodies, and both reagents are allowed to diffuse towards each other. Again, precipitation lines are found at the equivalence points for each complementary antigen–antibody in the system (Fig. 4.5b). Such tube techniques have generally been superseded by diffusion in two dimensions, which produce more useful information.

In Two Dimensions: Ouchterlony Technique This is probably the most widely used immunochemical technique. It may, for example, be used for detecting which sera or chromatographic or cell fractions, contain a particular antigen and whether or not two antigens are identical, different or share antigenic determinants. Originally 5 to 10 mm diameter wells were cut with a cork borer and removed from 1 to 2 mm thick layers of agar in a petri dish. A typical pattern used for comparing different antigen preparations is shown in Fig. 4.7a. The petri dishes were stored in a humid chamber. The wells were observed and refilled with the appropriate solutions daily. Nowadays patterns are cut in agar layers on microscope slides with stainless steel cutters (Fig. 4.7b). This allows much smaller quantities of antigens and antisera to be used and gives results in hours rather than days. Two-dimensional double diffusion may be carried out through a very thin layer of agar by sticking a small plastic template containing holes cut in a conventional Ouchterlony pattern to a microscope slide with only one drop of agar, which serves as both the adhesive and the diffusion medium. The results are obtained even more rapidly and relatively dilute antigen solutions may be used because the distance to be diffused between wells is small and the ratio of well volume to agar volume large (Fig. 4.7c).

A *reaction of identity* occurs between an antibody and antigens containing

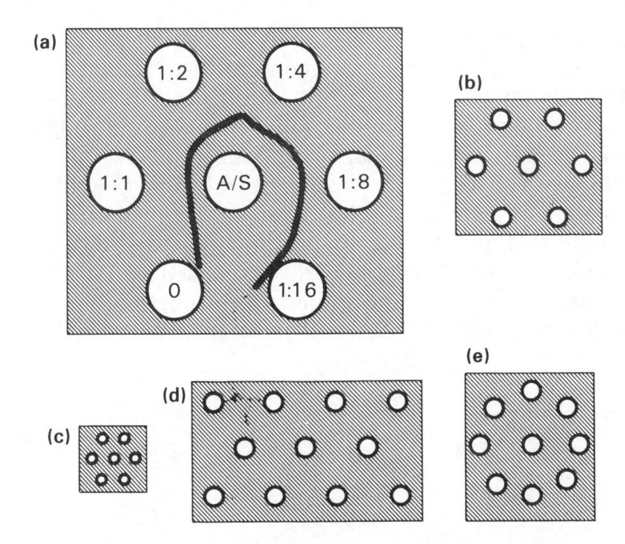

Fig. 4.7 Double immunodiffusion in two dimensions (drawn to same scale). **(a)** Macroscale as used in petri dishes, showing effect of antigen concentration on position of precipitin lines relative to wells; **(b)** commonest pattern used on microscope slide scale; **(c)**commonest pattern using only one drop of agar and a plastic template; **(d)**, **(e)** other common patterns.

identical antigenic determinants, and results in smoothly fused precipitation lines (Fig. 4.8a). The *reaction of non-identity* occurs when the antiserum contains antibodies to both antigens but the two antigens do not share a common determinant. The two lines are formed independently with different antibody molecules and cross without interaction (Fig. 4.8b). A *reaction of partial identity* occurs when two antigens have at least one common determinant, but where the antisera contains antibodies to a determinant in one antigen that is absent from the other (Fig. 4.8c).

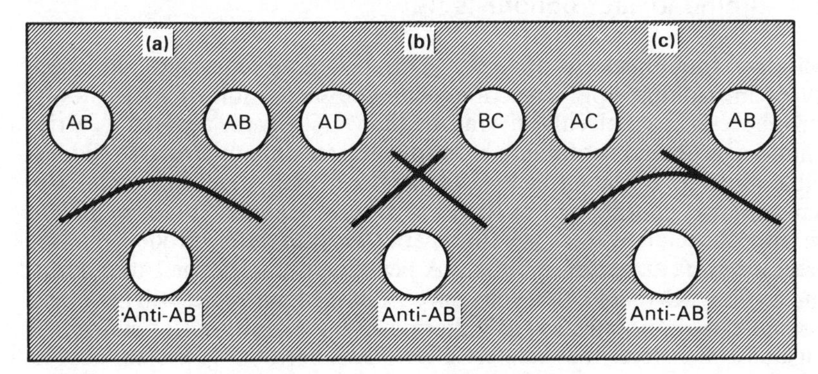

Fig. 4.8 The basic patterns of precipitation formed during double diffusion in two dimensions (Ouchterlony method). **(a)** Reaction of identity lines fuse; **(b)** reaction of non-identity lines cross; **(c)** reaction of partial identity lines spur. A, B, C, D represent antigenic determinants and the antiserum used in each example is anti AB.

The relative position of a precipitation line yields a semi-quantitative estimate of antigen concentration, i.e. the more concentrated the antigen solutions the further from the antigen well the precipitation line is formed, as illustrated in Fig. 4.7a. The shape of a precipitation line can give a rough estimate of the molecular weight of a globular protein antigen. The major antibody is usually IgG, with a molecular weight of 150 000. Globular proteins with molecular weights significantly less than 150 000 will diffuse through agar more rapidly and give an arc of precipitation as shown in Fig. 4.9a. Those with molecular weights significantly greater than 150 000 will diffuse more slowly and give an arc in the opposite direction (Fig. 4.9c). Antigens with similar molecular weights to IgG will give a straight precipitation line (Fig. 4.9b).

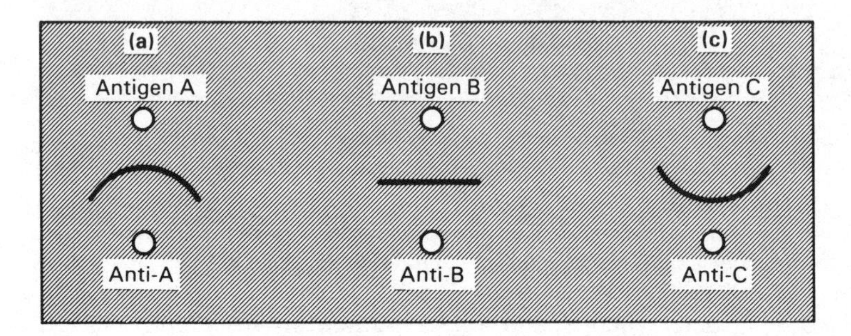

Fig. 4.9 The effect of molecular weight of an antigen on the shape of the precipitin lines formed in the Ouchterlony method. **(a)** Molecular weight antigen A ≪ molecular weight IgG (e.g. human serum albumin, 68 000 daltons); **(b)** molecular weight antigen B = molecular weight IgG (e.g. IgA); **(c)** molecular weight antigen C ≫ molecular weight IgG (e.g. Keyhole limpet haemocyanin, 3×10^6 daltons).

4.4.4 Immunoelectrophoresis (IE)

Qualitative Immunoelectrophoresis This technique combines the specificity of immunoprecipitin reactions with the separation of molecules by electrophoresis in a molecular sieving medium. Usually the analysis is carried out in an agarose gel containing barbitone buffer on a microscope slide. A suitable pattern, as shown in Fig. 4.10a, is cut with a gel punch and 1 to 5 mm^3 solutions containing 1 to 100 μg of antigen are added to the wells. The slides are connected by thick wet filter paper wicks to the electrode wells, and a direct electric current of about 8 mA per slide is passed for 1 to 2 hours, giving a voltage drop of 4 to 8 volts cm^{-1}. Charged molecules will have been separated electrophoretically, as indicated in Fig. 4.10b, but of course will not usually be visible. Immediately after disconnecting the voltage supply, the troughs are filled with appropriate antisera and incubated overnight at room temperature in a humid chamber. The antigens diffuse radially and the antibodies diffuse laterally as shown in Fig. 4.10c, resulting in the antigen–antibody precipitation arcs shown in Fig. 4.10d. Despite using

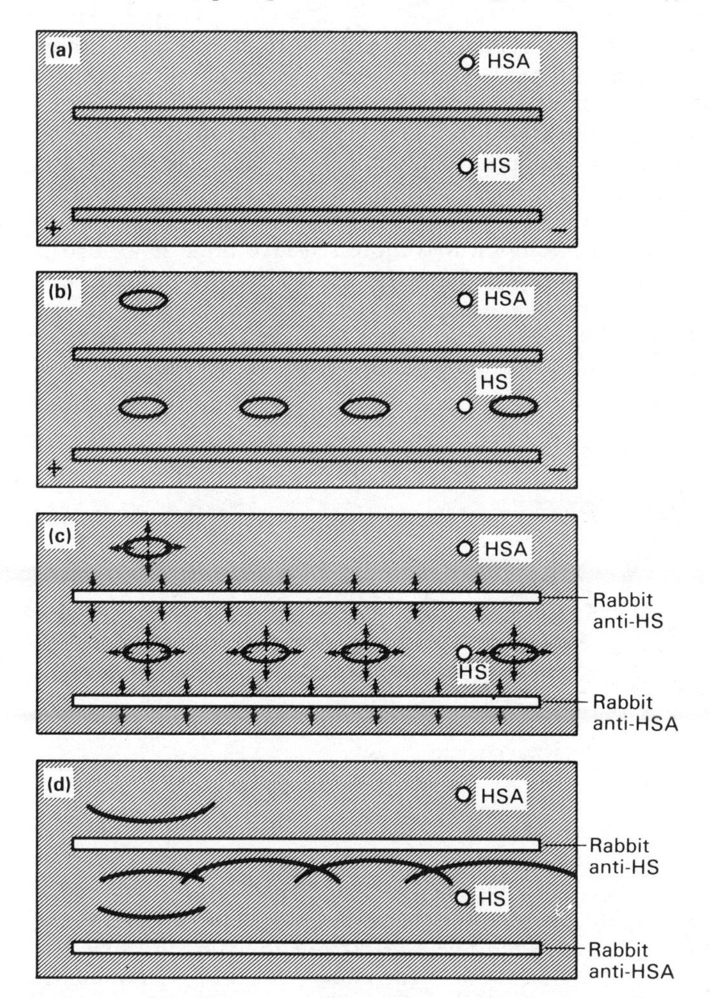

Fig. 4.10 Stages in microimmunoelectrophoresis. **(a)** Pattern cut in agar. Agar removed from wells and wells filled with antigens, human serum albumin (HSA) and human serum (HS) respectively. **(b)** After electrophoresis the proteins will have migrated as indicated but will not be visible. Agar removed from troughs. **(c)** On adding antisera to troughs diffusion of antigens and antibodies occurs as shown. **(d)** The final precipitin lines expected for the systems described.

agarose, which acquires a smaller charge then agar, the electro-osmotic flow of water during electrophoresis (Section 7.2.4) moves all of the antigens towards the cathode. This results in the apparent cathodic migration of γ-globulins including IgG antibodies, although they migrate very little on electrophoresis on uncharged support media. This technique may be used to investigate the purity of, or to detect, particular antigens in sera, culture filtrates, tissue or cell extracts, or fractions from any preparative procedure.

Cross-over Electrophoresis As illustrated in Fig. 4.10, most proteins show anodic migration at pH 8.0 but γ-globulins are exceptional in apparently migrating towards the cathode, due to electro-osmosis (Section 7.2.4). Cross-over electrophoresis, as illustrated in Fig. 4.11, takes advantage of this by moving IgG antibodies (γ-globulins) and antigens towards each other and hence precipitation lines are formed. This technique is therefore more rapid (15 to 20 minutes) than the Ouchterlony method which may take days to produce a clear result. In addition it is more sensitive because all of the molecules migrate towards each other rather than diffusing radially. The technique is particularly useful in forensic science for establishing the species of origin of body fluids such as blood, semen and saliva. Up to 12 samples may be tested on one agar-covered microscope slide, saving time, valuable reagents and samples.

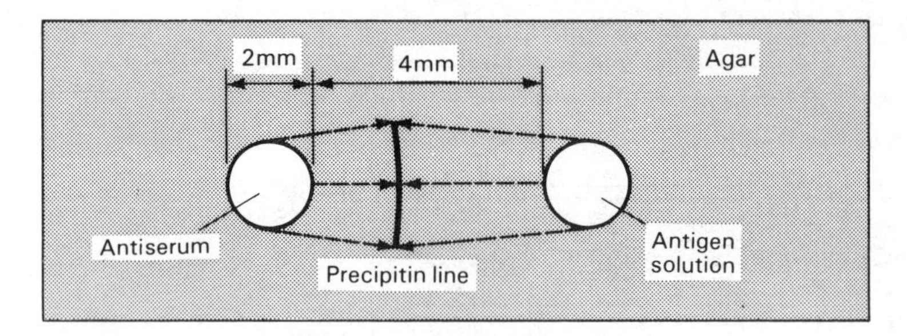

Fig. 4.11 The principles of cross-over electrophoresis.

Quantitative Immunoelectrophoresis *Laurell's 'rocket' electrophoresis* is related to single radial immunodiffusion in the same way that cross-over electrophoresis is related to the Ouchterlony technique. As in single radial immunodiffusion, the antigen sample is placed in wells cut in agar containing specific antiserum to the antigen to be assayed. On applying direct electric current most antigens migrate towards the anode and the IgG antibodies migrate towards the cathode. Initially soluble antigen–antibody complexes are formed in antigen excess. When all of the antigen has migrated into the gel, equivalence is reached and antigen–antibody complexes precipitate as shown in Fig. 4.12. The area under the rocket shape is directly proportional to the antigen concentration. When the precipitation arcs have become stationary (1 to 10 hours) a plot of rocket height against concentration will be linear. Thus, by the use of standards and the preparation of a calibration curve, the concentration of antigen solutions may be determined. Obviously, unlike single radial immunodiffusion, the technique cannot be used to determine the concentration of IgG because of its cathodic migration.

Two Dimensional Immunoelectrophoresis This technique combines electrophoretic separation in a molecular sieving medium with the specificity and speed of rocket electrophoresis. Initially antigens are separated by agar electrophoresis in one dimension (Fig. 4.10b). A suitable slice of the gel is

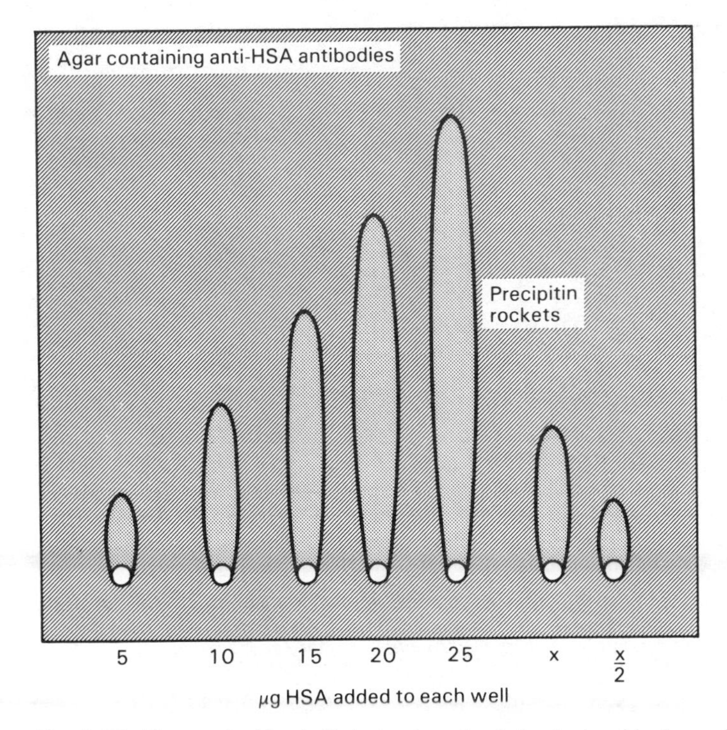

Fig. 4.12 The result of 'rocket' electrophoresis of standard and 'unknown' (*x*) solutions of human serum albumin (HSA).

transferred onto a square glass plate and a layer of agar containing a suitable antiserum is solidified against it over the rest of the plate. After rocket electrophoresis in the second dimension, precipitin arcs are formed as illustrated in Fig. 4.13. By comparison of the area under the arcs with those of standard systems a semi-quantitative estimate of the amount of individual antigens present may be made.

4.4.5 Visualisation and recording of precipitin lines in gels

Precipitin lines in gels should always be developed in a humid atmosphere and wells should not be allowed to dry out. Usually the precipitin lines can be clearly seen with the naked eye (Fig. 4.14), particularly if illuminated from the side and viewed against a dark background. If staining is desired, excess proteins are first removed from the gels by washing for 12 to 24 hours in several changes of phosphate-buffered saline. Precipitin arcs may be stained before or after drying with most protein stains, for example Coomassie Blue. Gels are dried overnight by covering with a good quality filter paper. The filter paper is removed after slight dampening, leaving a permanent record easily stored in a slide box. Alternatively, permanent records may be made

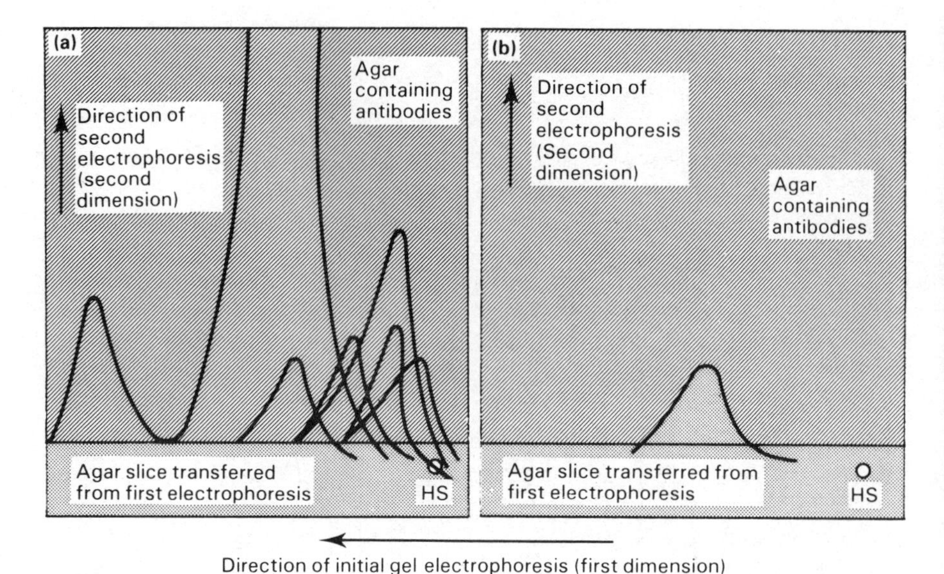

Fig. 4.13 The results of two-dimensional electrophoresis of human serum (HS). **(a)** Using agar containing rabbit anti-human serum for electrophoresis in the second dimension; **(b)** using agar containing rabbit anti-human IgM for electrophoresis in the second dimension.

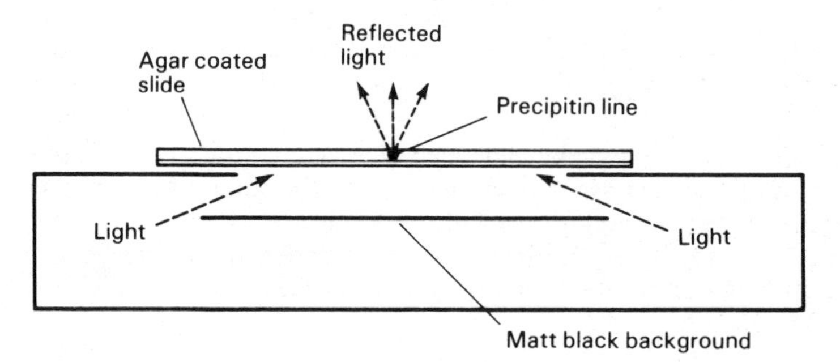

Fig. 4.14 Principle of the dark background viewer for studying precipitin lines.

by drawing or photography without drying. Polaroid cameras enable a series of records to be made and stored as precipitation progresses.

4.5 Radioimmunoassay (RIA)

4.5.1 Principles

Radioimmunoassay is one of the most important techniques in the clinical and biochemical fields for the quantitative analysis of hormones, steroids and drugs. It combines the specificity of the immune reaction with the sensitivity of radioisotope techniques. Alternative names used for RIA include *saturation analysis, displacement analysis* and *competitive radioassay.*

The technique is based on the competition between unlabelled antigen and a finite amount of the corresponding radio-labelled antigen for a limited number of antibody binding sites in a fixed amount of antiserum. At equilibrium in antigen excess there will be both free antigen and antigen bound to the antibody. Under standard conditions the amount of labelled antigen bound to antibody will decrease as the amount of unlabelled antigen in the sample increases, for example:

$$4Ag^* + 4Ab \rightleftharpoons 4Ag^*Ab$$
$$4Ag + 4Ag^* + 4Ab \rightleftharpoons 2Ag^*Ab + 2AgAb + 2Ag^* + 2Ag$$
$$12Ag + 4Ag^* + 4Ab \rightleftharpoons Ag^*Ab + 3AgAb + 3Ag^* + 9Ag$$

where Ab, Ag, Ag* and AgAb represent one equivalent of antibody, unlabelled antigen, labelled antigen and antigen–antibody complex respectively.

By using known amounts of unlabelled antigen and a fixed amount of antibody and labelled antigen, the amount of labelled antigen bound as a function of the total antigen added is measured and a calibration curve constructed (Fig. 4.15). This calibration curve may then be used to determine the amount of antigen in samples treated similarly.

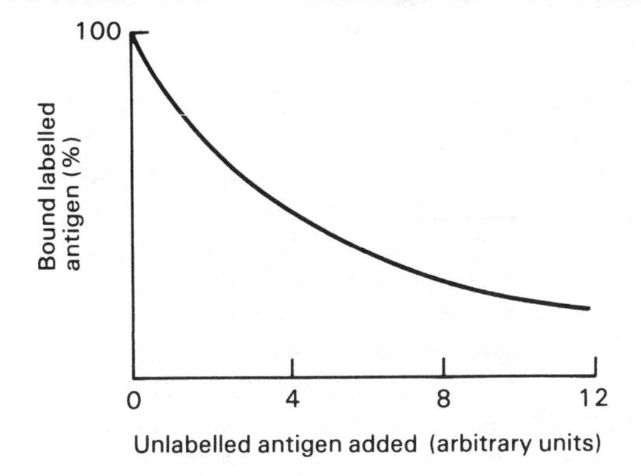

Fig. 4.15 A typical radioimmunoassay calibration curve.

4.5.2 Practical aspects

Pure antigen, labelled antigen and a suitable antiserum are all required for RIA. Pure antigen is required for standard samples, production of labelled antigen and the production of a specific antiserum. Labelled antigen must be produced with minimal alteration of the immunoreactivity of the molecule. It should also have a high *specific activity* (Section 9.1.6) in order to yield an assay with maximum sensitivity. Usually the phenolic group of a tyrosine residue in protein antigens is labelled with ^{125}I by either the chloramine T or the less drastic lactoperoxidase methods as illustrated in Fig. 4.16. The labelled protein is immediately separated from unreacted iodide, and any denatured protein removed by gel filtration. Non-protein haptens are usually labelled with tritium. The availability of a suitable antiserum is probably the most important single factor for satisfactory RIA. Therefore care must be taken to ensure the specificity and avidity of the antiserum produced. It is usual to work with an antiserum dilution that binds 30 to 60% of the labelled antigen.

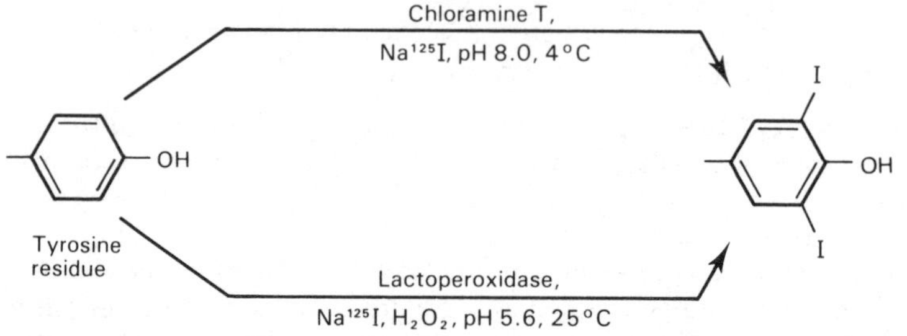

Chloramine T,
$Na^{125}I$, pH 8.0, 4 °C
OH
Tyrosine residue
I
OH
I
Lactoperoxidase,
$Na^{125}I$, H_2O_2, pH 5.6, 25 °C

Fig. 4.16 Methods of labelling the tyrosine residues of protein antigens with ^{125}I.

Antibody-bound and free antigen are separated in order that the radioactivity in one or both fractions may be measured. This enables the proportion of labelled antigen bound to be estimated and hence the amount of unlabelled antigen in the system to be determined from the calibration curve. Separation methods based on the removal of free antigen include ion-exchange chromatography and adsorption onto charcoal or silica. Separation methods based on the removal of antibody-bound antigen include the double antibody technique in which an antiserum capable of reacting specifically with the first antibody to form a separable precipitate is used; and the non-specific precipitation of antibody-bound antigen with salt or organic solvent, for example saturated ammonium sulphate solution, dioxane or ethanol. Alternatively, if the antibodies used in RIA are initially attached to Sephadex beads or test tubes it is possible to centrifuge, decant and wash away the unbound antigen.

Whenever RIA is being carried out it is essential that a number of standards are assayed under identical conditions because environmental conditions are almost never identical. Standards and samples should always be prepared at least in duplicate. Many radioimmunoassays are carried out at room temperature, but, if incubation times exceed 6 hours, 4°C is used in order to prevent proteolysis and microbial growth.

The major advantages of RIA include:

(i) the ability to assay any compound that is immunogenic, available in a pure form, and can be radio-labelled;

(ii) its high sensitivity – some compounds may be detected at a level of pg cm^{-3};
(iii) its high specificity;
(iv) its precision which is comparable to that of other physico-chemical techniques and far better than bioassays;
(v) its ease of automation so that a minimum of manual handling and data processing is necessary. This allows a large number of samples to be processed at minimal cost.

The major disadvantages of RIA include:

(i) the relatively high cost of equipment and reagents. Gamma scintillation counters are expensive to buy and maintain, and radioiodine is not a cheap reagent;
(ii) the short shelf-life of reagents – the half lives of ^{125}I and ^{131}I are 60 days and 8 days respectively, necessitating relatively frequent labelling of antisera;
(iii) the radiological hazards of using radioiodine, particularly during the labelling of antisera, which must be repeated fairly regularly. Staff should have regular thyroid scans and be rested if the level of radioactivity increases significantly;
(iv) assays usually take days rather than hours.

4.5.3 Immunoradiometric assay (IRMA)

This utilises purified radio-labelled antibody and is often more sensitive than RIA, for example thyroid stimulating hormone (TSH) may be assayed by allowing it to bind to an antibody complementary to the α-subunit of TSH attached to Sephadex and then reacting it with radio-labelled antibody against the TSH β-subunit (Fig. 4.17).

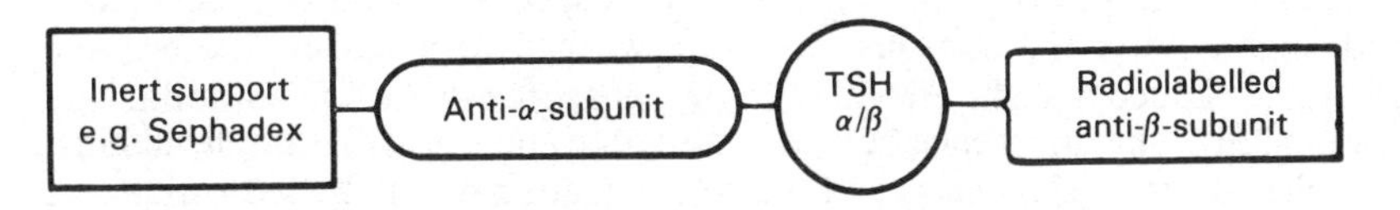

Fig. 4.17 The immunoradiometric assay of thyroid stimulating hormone.

4.6 Enzyme immunoassay (ELISA)

4.6.1 Principles

Enzyme immunoassay, sometimes called the *enzyme-linked immunosorbant assay* or ELISA, combines the specificity of antibodies with the sensitivity of simple spectrophotometric enzyme assays by using antibodies or antigens conjugated (coupled) to an easily assayed enzyme which also possesses a high

Table 4.2 A comparison of enzyme immunoassay (ELISA), radioimmunoassay (RIA), and fluorescent immunoassay (FIA).

	ELISA	RIA	FIA
Specificity	+	+	+
Sensitivity	+	+	+
Reproducibility	+	+	+
Cost	+	−	±
Reagent shelf life	+	−	+
Result reading			
(a) objective	+	+	+
(b) automated	±	+	±
(c) subjective	+	−	−
Safety	+	−	+
Usefulness			
(a) large central laboratories	+	+	+
(b) small diagnostic laboratories	+	−	±

+ favourable; − unfavourable

turnover number. ELISA is replacing RIA, despite the latter already being established, extensively automated and sometimes more sensitive. This is because ELISA is relatively cheap to operate, lacks the radiological hazards of RIA and is suitable for use in small laboratories lacking gamma radioactive counting facilities such as some hospital diagnostic laboratories and the laboratories of the developing nations. A comparison of these two techniques with FIA (Section 4.7) is shown in Table 4.2.

ELISA may be used for assaying antigens by either a competitive method or a double antibody method and for assaying a specific antibody by an indirect method. All these methods require the preparation of a calibration curve during the assay.

The *competitive method* is summarised in Fig. 4.18. A mixture of a known amount of enzyme-labelled antigen (Section 4.6.2) and an unknown amount of unlabelled antigen is allowed to react with a specific antibody attached to a solid phase. After the complex has been washed with buffer, the enzyme substrate is added and the enzyme activity measured in the conventional way (Section 3.4). The difference between this value and that of a sample lacking unlabelled antigen is a measure of the concentration of unlabelled antigen. A major disadvantage of this method is that each antigen may require a different method to couple it to the enzyme; this is not so for the double antibody method.

The *double-antibody method* for ELISA is summarised in Fig. 4.19. The unknown antigen solution is reacted with specific antibody attached to a solid phase, washed and treated with enzyme-labelled antibody. After a further wash the enzyme substrate is added. The amount of enzyme activity measured under standard conditions is directly proportional to the amount of antigen present (compare immunoradiometric assay, Section 4.5.3). An advantage of this method is that only one procedure is required to couple the enzyme to all antibody preparations.

An *indirect* ELISA method may be used to measure antibody levels, as

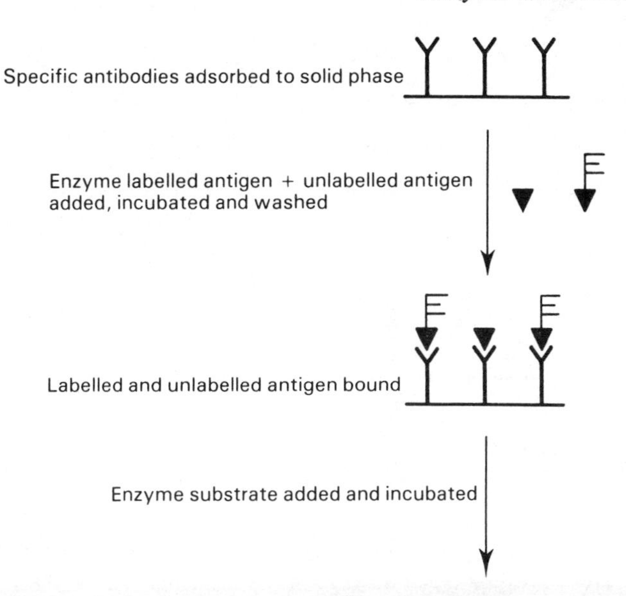

Fig. 4.18 Principles of the competitive method for enzyme immunoassay.

summarised in Fig. 4.20 (compare indirect immunofluorescence, Fig. 4.21b). The putative antiserum is reacted with specific antigen attached to a solid phase. Any specific antibody molecules bind to the antigen and all other material is washed away. Exposure of the complex to enzyme-labelled anti-immunoglobulin antibody results in binding to any specific antibody molecules adsorbed from the original serum. The complex is washed and the substrate for the enzyme added, resulting in activity proportional to the amount of specific antibody in the original serum.

The sensitivity of any type of ELISA may be greatly enhanced by means of *enzyme amplification*. In its simplest form the primary enzyme product is used to trigger a secondary enzyme system that can generate a large quantity of coloured product. Since the product of the first enzyme need not be measurable but only acts catalytically on the second system; enzymes not currently used for ELISA may become important in these systems e.g. aldolase or glucose-6-phosphatase. Enzyme amplification of a double antibody assay in which alkaline phosphatase, the primary enzyme, degrades the substrate $NADP^+$ to NAD^+ can be achieved by using alcohol dehydrogenase as the amplifying enzyme and a tetrazolium dye as a redox acceptor. On triggering the alcohol dehydrogenase the tetrazolium dye is reduced to a coloured formazan product, which can be assayed spectrophotometrically e.g. iodonitrotetrazolium violet may be reduced to a red

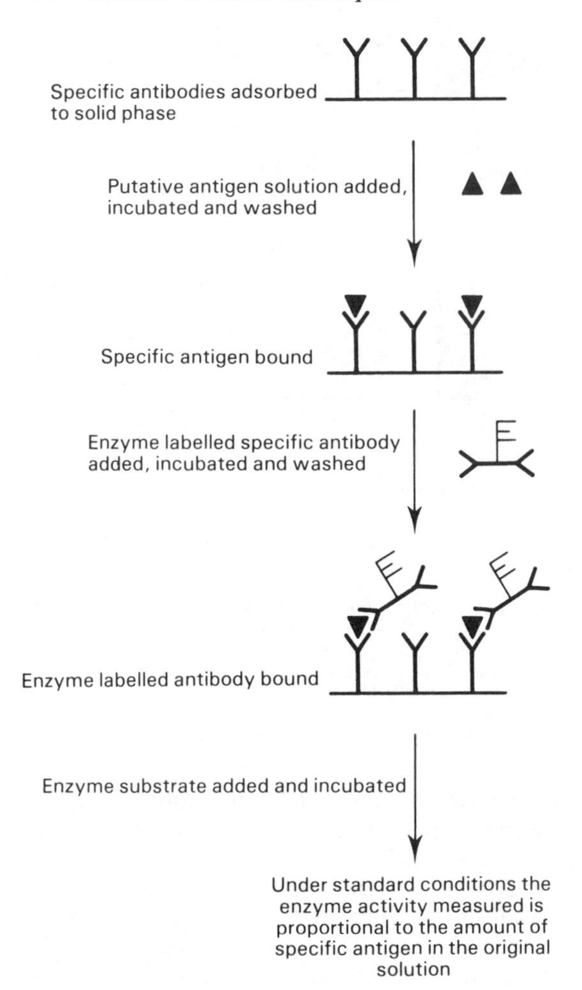

Fig. 4.19 The double antibody method for enzyme immunoassay.

formazan and the yellow thiazolyl blue may be reduced to a blue formazan. Under appropriate conditions 500 molecules of formazan are produced per minute by such a redox amplifier for each molecule of NAD^+ produced by the primary enzyme system. Enzyme amplification can be carried out as either a one-step amplification, in which both enzymes and substrates are reacting at the same time or as a two-step amplification in which the first enzyme is inhibited before or during the addition of the second enzyme and substrate.

Enzyme amplification immunoassays have already been developed for hormones, pathogenic viruses and bacteria, and tumour markers. A possible significant advantage of enzyme amplification immunoassays over the

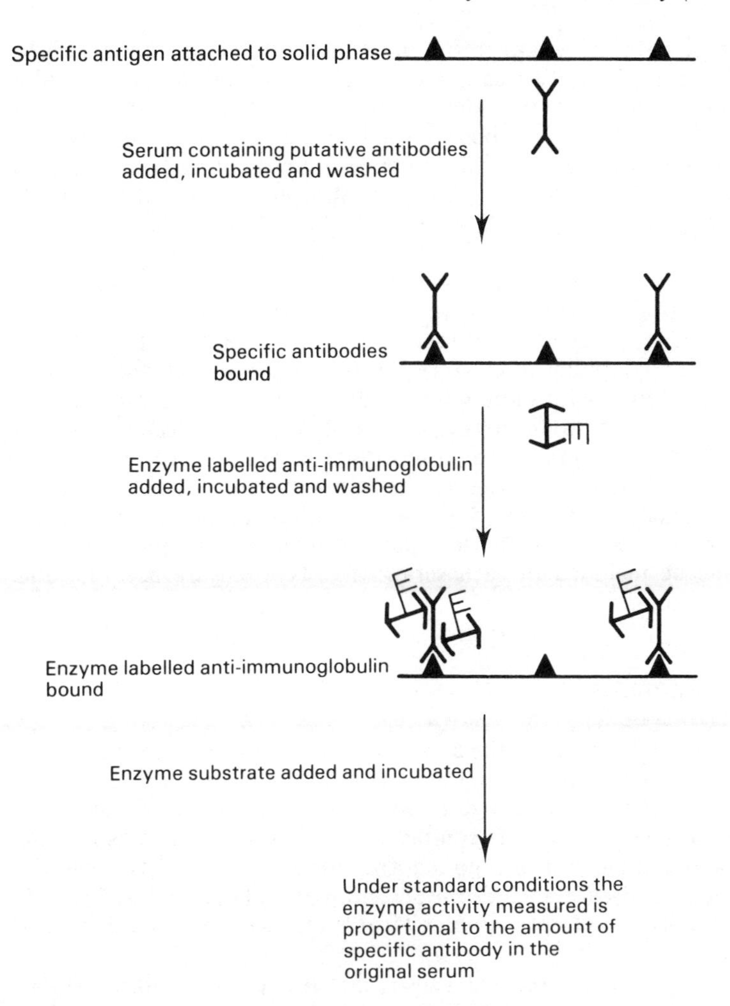

Fig. 4.20 The indirect method of enzyme immunoassay.

technological demands of RIA (Section 4.5.2) and FIA (Section 4.7) is that they may be assayed with a simple spectrophotometer or even by visual inspection.

4.6.2 Practical aspects

Solid phases used in ELISA include cross-linked dextran or polyacrylamide beads, filter paper (cellulose) discs or polypropylene tubes, but disposable polystyrene microtitration plates are particularly convenient for large numbers of samples. The appropriate antigen or antibody may be attached to the solid phase by passive adsorption or covalent coupling with cyanogen bromide (Section 6.7.2). Test samples of antigen or antibody are usually diluted with the same buffer containing a wetting agent such as Triton X-100 used for washing the antigen-bound solid phase. The antibody–enzyme conjugate used must contain a highly reactive antibody coupled to an enzyme with a high turnover number. Alkaline phosphatase and horseradish peroxidase are commonly used enzymes. Glutaraldehyde is a popular coupling agent for conjugating the enzyme to the antibody but requires reactive amino groups to be part of the enzyme as well as the antibody. Periodate oxidation of the extensive carbohydrate group of horseradish peroxidase produces aldehyde groups which will combine with the amino groups of immunoglobulins to produce Schiff base conjugates which can be stabilised by reduction with sodium borohydride. Enzyme substrates should ideally be stable, safe and inexpensive. Colourless substrates that are converted to a coloured product by the enzyme are popular; for example *p*-nitrophenylphosphate is converted to the yellow *p*-nitrophenol by alkaline phosphatase. Diaminobenzidine, 5-aminosalicylate and *o*-phenylenediamine have been used as substrates with peroxidase. Reference positive and negative samples must be included in each series of tests to ensure accurate and reproducible results.

4.6.3 Applications

Whilst ELISA can be used for the assay of virtually any antigen, hapten or antibody, it is predominantly used in clinical biochemistry laboratories to measure, for example, immunoglobulins G and E, oncofoetal proteins, haematological factors, immune complexes and hormones such as insulin, oestrogens and human chorionic gonadotrophin. Examples of its use in the study of infectious diseases include the detection of bacterial toxins, *Candida albicans*, Rotaviruses, Herpes simplex viruses and the Hepatitis B surface antigen.

ELISA has also been used extensively for the assay of antibodies in infectious diseases including: antiviral antibodies, e.g. Epstein Barr Virus and Rubella Virus; antibacterial antibodies, e.g. *Brucella, Rickettsia* and *Salmonella* spp; antifungal antibodies, e.g. *Aspergillus* and *Candida* spp; antiparasite antibodies, e.g. *Plasmodium, Schistosoma* and *Trypanosoma* spp; and autoantibodies e.g. anti-DNA and anti-thyroglobulin.

4.7 Fluorescence immunoassays (FIA)

4.7.1 Immunofluorescence (IF)

Immunofluorescence provides sensitive assays for the detection of antigens in frozen or fixed tissue sections or viable cells. By staining with specific antibody conjugated to a fluorescent chromophore and illuminating with ultraviolet light the location of specific antigens in a tissue or cell preparation may be studied. Fluoroscein emits green light and rhodamine orange light, whereas the weak natural fluorescence of some biological materials occurs in the blue region of the spectrum. Immunofluorescence is used in a number of modified forms, illustrated in Fig. 4.21. The more complicated methods are more sensitive as they amplify the amount of fluorescent antibody bound. Samples must be very carefully washed between the incubation with each reagent, and both positive and negative controls should be included. Immunofluorescence is particularly useful for detecting autoantibodies to thyroglobulin, DNA, mitochondria, nuclei and nucleoli, immune complexes and cell receptors such as the immunoglobulin receptors on the surface of lymphocytes. Monoclonal antibodies have vastly increased the range and effectiveness of immunofluorescent measurements e.g. identification of helper and suppressor subpopulations of T lymphocytes.

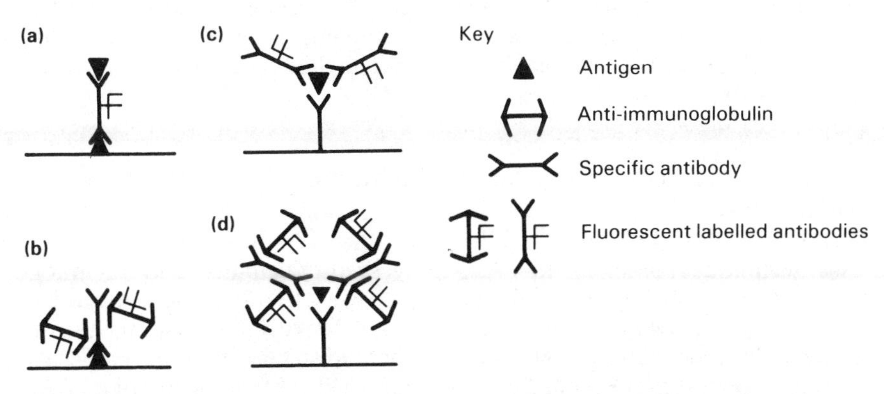

Fig. 4.21 The principles of various immunofluorescence assays. **(a)** Direct immunofluorescence test of single layer technique; **(b)** indirect immunofluorescence test or double layer technique; **(c)** simple sandwich technique for the detection of specific antibodies; **(d)** multiple sandwich technique.

4.7.2 The homogeneous substrate-labelled fluorescent immunoassay (SLFIA)

SLFIA uses the principles of competive binding reactions to measure the concentration of low molecular mass compounds, such as drugs, in body fluids. The compound is covalently linked to a fluorogenic enzyme substrate (e.g. galactosylumbelliferone) to produce a non-fluorescent reagent which can be hydrolysed by an appropriate enzyme (e.g. β-galactosidase) to produce a fluorescent product. It is essential that the binding of antibody to the non-fluorescent reagent prevents the enzymic hydrolysis of the reagent. Thus the enzyme can produce fluorescence from unbound reagent but not from antibody-bound reagent. In the presence of the original compound and of limited amounts of reagent and specific antibody the enzyme will produce fluorescence as a function of the concentration of the compound (compare radioimmunoassay, Section 4.5). Unlike radioimmunoassay it is not necessary to separate the free and antibody-bound forms of the reagent because only the free reagent can be enzymically hydrolysed to produce fluorescence. A calibration curve of percent fluorescence against concentration may be constructed by incubating known amounts of compound with appropriate fixed amounts of reagent, adding enzyme and measuring the fluorescence produced. The antiserum and the enzyme can be mixed and added together because the affinity of the antibody for the reagent is usually very much greater than the affinity of the enzyme for the reagent.

As an example, under appropriate conditions SLFIA produces a linear relationship between the antibiotic gentamicin and fluorescence over the range 0 to 24 ng gentamicin with a mid-range coefficient of variation of less than 3%. This means that human sera being monitored for drug level can be diluted 10^3 times and hence it is unnecessary to carry out controls for any native fluorescence of the sera. Other compounds routinely assayed by SLFIA include tobramycin, kanamycin, theophylline, phenobarbitone, IgG and IgM.

4.7.3 Delayed enhanced lanthanide fluorescence immunoassay (DELFIA)

This technique combines the sensitivity, reproducibility and accuracy normally associated with radioimmunoassay with the speed and versatility usually associated with enzyme immunoassays. Usually it involves the use of antibody labelled with a chelate of the lanthanide metal europium. The weak fluorescence of the europium chelate is intensified 10^6 fold on addition of an *enhancing solution* which releases the europium and re-chelates it into micelles which protect it from the quenching effects of water molecules. The lanthanide chelates have long fluorescent decay times (10 to 1000 μs) after excitation. This allows *time resolved fluorescent immunoassays* (Fig. 4.22a) to measure the lanthanide fluorescence after interference from scattered light and the short-lived (1–20 ns) natural fluorescence of biological materials has disappeared. Usually the fluorescence emitted by the lanthanide chelate is measured 400 to 500 μs after it has been irradiated for 1 μs at 340 nm with a xenon lamp. Other advantages of using lanthanide chelates include: a very large Stokes' shift (Section 8.5.1) e.g. from 340 nm to 613 nm, as illustrated in Fig. 4.22b and a very narrow emission peak with a half intensity bandwidth of about 10 nm (Fig. 4.22b) which together allow great spectral selectivity in the assay; a broad excitation region with a half-intensity bandwidth of about 50 nm (Fig. 4.22b), resulting in great sensitivity of the assay and great stability and biological inertness, unlike radioisotopes and enzymes respectively.

DELFIA is usually carried out by means of a solid phase two site direct sandwich technique analagous to the double antibody ELISA illustrated in Fig. 4.19 but using excess europium-labelled second antibody in place of the

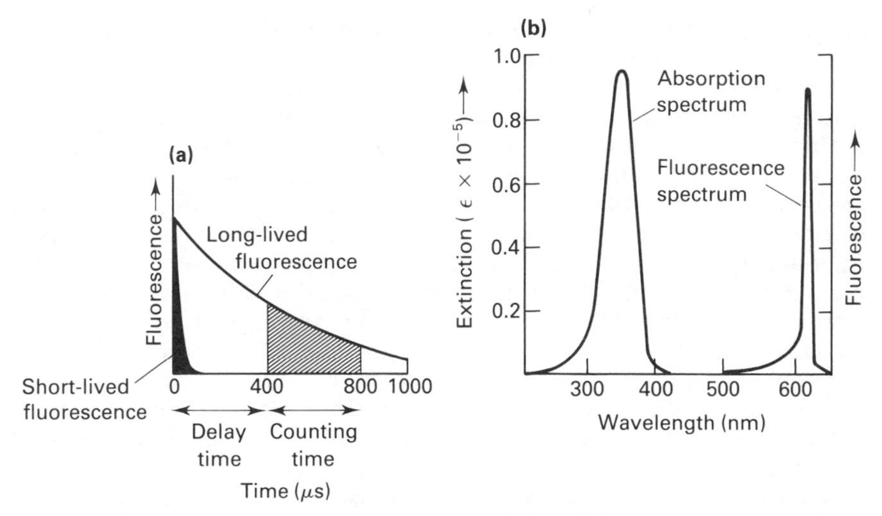

Fig. 4.22 Delayed enhanced lanthanide fluorescence. **(a)** Time course of fluorescence at 613 nm after excitation by a 1 μs flash at 340 nm; **(b)** absorption and fluorescence spectra of a europeum chelate.

enzyme labelled antibody and *enhancer solution* in place of the enzyme substrate. Under standard conditions the fluorescence produced is directly proportional to the amount of specific antigen in the original solution. As little as 10^{-16} moles europium per sample can be measured and the precision of DELFIA is better than 5% up to 10^{-12} moles europium. Commercial kits utilising microtitre plates are already available for DELFIA of Hepatitis B antigen, digoxin, thyroxine, testosterone, progesterone, oestradiol, cortisol, human follicle stimulating hormone and human luteinising hormone. Results are usually available within 1 to 5 hours.

4.7.4 Flow cytofluorimetry and fluorescence activated cell sorting (FACS)

The cytofluorograph or flow cytofluorimeter measures the fluorescence emitted by individual cells in a mixed cell population (Section 1.6.4). Monoclonal antibodies have vastly increased the effectiveness of flow cytofluorimetry and FACS. In particular the availability of a vast array of monoclonal antibodies to leucocyte surface antigens is greatly facilitating studies on the different lineages of leucocytes and their numerous roles in the immune response.

4.8 Particle counting immunoassays (PACIA)

PACIA relies on the principles that the number of free particles coated with antigen will decrease during agglutination and that the angles at which a beam of light is scattered is a function of particle size.

Cell agglutination has been used to detect cell surface antigens for decades. The relative concentrations of antibodies to cell surface antigens can be determined by using doubling dilutions of the antiserum, carefully standardised conditions and noting the end point of cell agglutination. The technique has been extended to other antibodies by binding the corresponding antigen to erythrocytes. The relative concentration of an antigen solution can be determined by agglutination inhibition assays, using doubling dilutions of the antigen. However, because the erythrocytes used are labile they cannot be stored indefinitely, and these techniques are laborious and difficult to automate. Polystyrene beads (usually 0.8 μm in diameter and called latex) are stable and can be coated with antibodies and used to assay the corresponding antigen by agglutination (*latex fixation test*). Antibodies are either physically adsorbed to plain latex or covalently coupled with carbodiimide to carboxylated latex. Similar techniques can be used to assay antibodies and immune complexes. An optical size counter is set to count only nonagglutinated particles and ignore those with diameters smaller than 0.6 μm and greater than 1.2 μm. Alternatively an electrical resistance particle counting system (Section 1.6.5) can be used to count nonagglutinated particles. Both of these techniques of counting unagglutinated particles give a sensitive and precise evaluation of the extent of the agglutination reaction by simply mixing the antigen and antibody coated latex and diluting to give a count of 3000 to 4000 particles s^{-1}.

The advantages of PACIA include the ease of attaching proteins to the latex; the relative stability of the latex suspension; the sensitivity of the assay (0.1 to 1 ng antigen cm^{-3}); and the ready automation of the technique. The limitations to PACIA are its susceptibility to non-specific agglutination and agglutination inhibition, but these phenomena can be minimised by the use of Fab_2 fragments of antibody in place of whole Ig molecules, high ionic strength media and control serum. Antibodies and haptens can be assayed by PACIA using agglutination inhibition.

4.9 Complement fixation

This sensitive semiquantitative technique for detecting very small amounts of antigen (less than 1 μg), although still popular in microbiology, is being superseded by RIA and ELISA, so it will be described only briefly here.

4.9.1 Principles

Most antibodies fix (bind) complement on reacting with the corresponding antigen. In the presence of complement, rabbit antibodies raised against sheep erythrocytes will lyse such cells, a reaction which is easily detected by means of the haemoglobin liberated. Thus the presence of a complement system may be detected by mixing it with sheep erythrocytes sensitised with

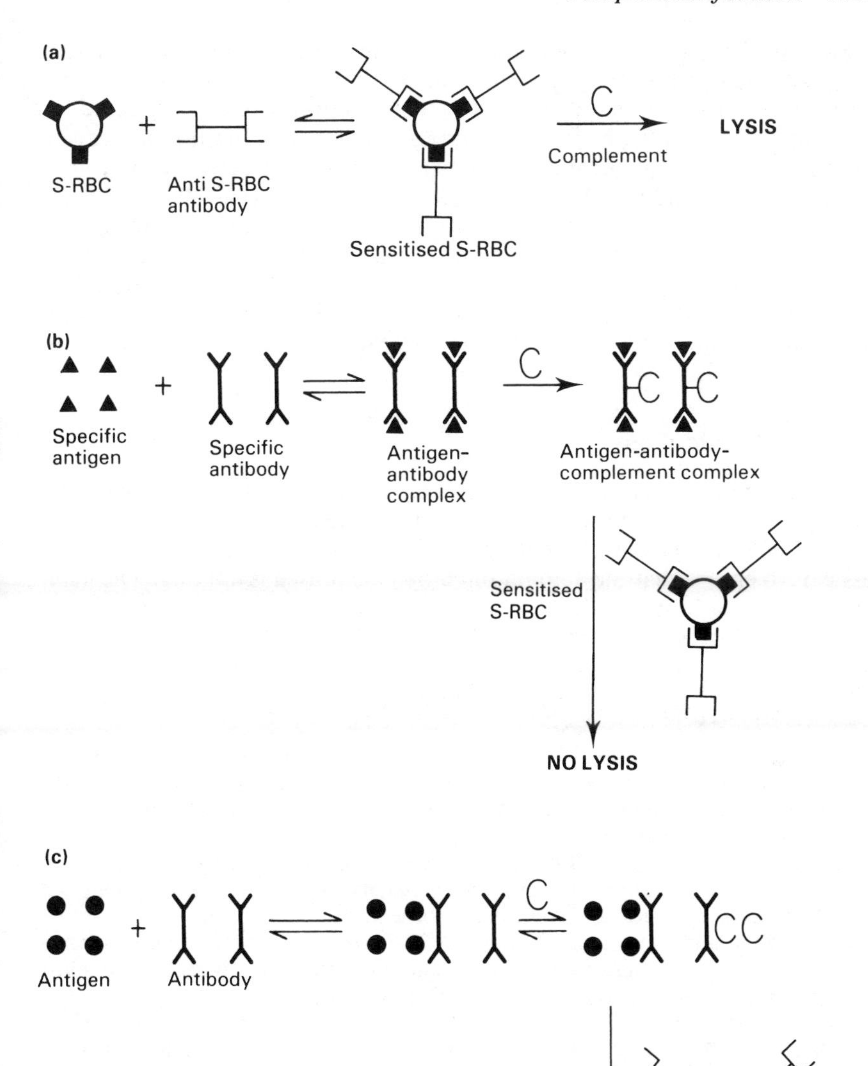

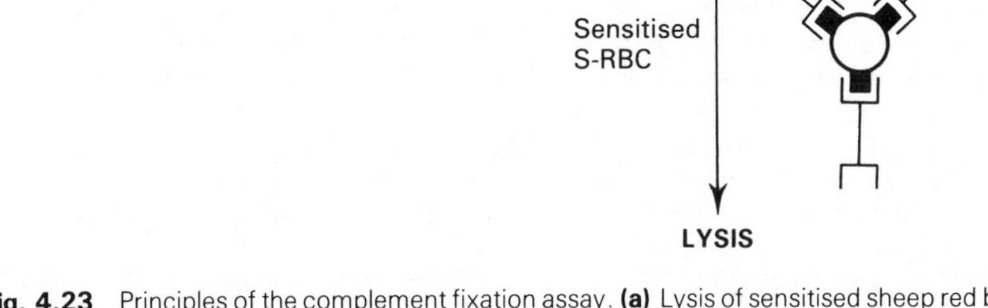

Fig. 4.23 Principles of the complement fixation assay. **(a)** Lysis of sensitised sheep red blood cells (S-RBC) by complement; **(b)** the fixing of complement by antigen/antibody complexes and its detection by the lack of lysis of the sensitised sheep red blood cells; **(c)** failure to fix complement by antibody in the absence of its specific antigen and the detection of this failure by the lysis of sensitised red blood cells.

rabbit antibodies (i.e. exposed to rabbit antibody, which binds to the erythrocytes but does not lyse them in the absence of complement). An antigen that reacts with a complement-fixing antibody may be detected, as illustrated in Fig. 4.23, by first incubating them together in the presence of a limiting amount of guinea-pig complement. If they react, complement will be fixed and on the addition of the sensitised sheep erythrocytes cell lysis will not occur, indicating that the original solution contained the putative antigen. If specific antigen was absent, complement will not have been fixed and the sensitised erythrocytes will be lysed. By performing the assay on doubling dilutions of standard and unknown antigen the concentration of the latter may be ascertained. Obviously the technique may also be used to detect and measure complement-fixing antibodies.

4.9.2 Practical aspects

All sera used as a source of antibodies in complement-fixation assays must have been heated at 56°C for 30 minutes to destroy essential but labile components of the complement system. The most laborious aspect of the assay is to ensure that the amount of complement used is limiting. The main clinical applications include the Wasserman test for syphilis and the detection of autoantibodies such as those to thyroid antigens.

4.10 Suggestions for further reading

Colowick, S.P. and Kaplan, N.O. (Eds)(1980–84). *Immunochemical techniques* in *Methods in Enzymology*, vols 70, 73, 74, 84, 92, 93, 108, Academic Press, London. (Comprehensive and exhaustive accounts of all the techniques discussed in this chapter).

Hudson, L. and Hay, F.C. (1981). *Practical Immunology*. 2nd edition. Blackwell Scientific Publications, Oxford. (An excellent introduction to the practice of many immunological and some immunochemical techniques.)

Hunter, W.M. and Corrie, J.E.T. (Eds)(1983). *Immunoassays for Clinical Chemistry*. Churchill Livingstone, London. (Contains a comprehensive account of most immunochemical assays used in clinical chemistry.)

Lefkovits, I. (Ed.)(1979). *Immunological Methods*. Academic Press, London. (Contains useful detailed, individual articles on the classical immunochemical techniques.)

Marchalonis, J.J. and Warr, G.W. (Eds)(1982). *Antibody as a Tool – The Applications of Immunochemistry*. J. Wiley and Sons, New York. (A readable comprehensive account of most techniques discussed in this Chapter, including detailed practical applications.)

Roit, S.M. (1985). *Essential Immunology*, 5th edition. Blackwell Scientific Publications, Oxford. (An excellent textbook on immunology.)

Steward, M.W. (1974). *Immunochemistry*. Chapman and Hall, London. (An excellent introduction to immunochemistry.)

Weir, D.M. (Ed.)(1986). *Handbook of Experimental Immunology*, 4th edition. Blackwell Scientific Publications, Oxford. (The standard reference work on all immunological and immunochemical techniques, containing comprehensive details on each.)

5
Molecular biology techniques

5.1 Introduction

The shapes and functions of cells are determined by their proteins. These molecules catalyse the reactions which synthesise membranes, cell walls and pigments, and they are needed to extract energy from substrates. Proteins in membranes are responsible for the transport of molecules from one cellular compartment to another, and between the inside and outside of the cell. Synthesis of proteins is itself catalysed by proteins, but this process is directed by DNA, which carries all the information needed to specify the structure of every protein the cell can make. The realisation that DNA lies behind all the cell's activities led to the development of molecular biology, which aims to explain biological processes in terms of the structures and interactions between nucleic acids and proteins. Although it is a relatively young discipline, it has already transformed our understanding of the way in which cells store and express their genetic information, and has had an enormous impact on many fields of study, notably immunology and medicine.

Molecular biology has also led directly to the immensely powerful, and potentially profitable techniques of genetic engineering. A great deal of effort (and money) is being directed into the *genetic manipulation,* or *genetic engineering*, of microorganisms, to make them produce a range of valuable polypeptides such as insulin, blood clotting factor VIII, growth hormone, or interferons, which they would not normally produce, and which are expensive to prepare by conventional biochemical means. Since it is easy and cheap to grow microorganisms on a large scale, they are very attractive as potential sources of these polypeptides. Genetic manipulation of plants and livestock should also permit the introduction of beneficial characteristics, such as resistance to herbicides or diseases, which could not readily be obtained by conventional breeding.

5.2 Structure of nucleic acids

5.2.1 Components and primary structure of nucleic acids

In order to appreciate the ways in which nucleic acids can be analysed and

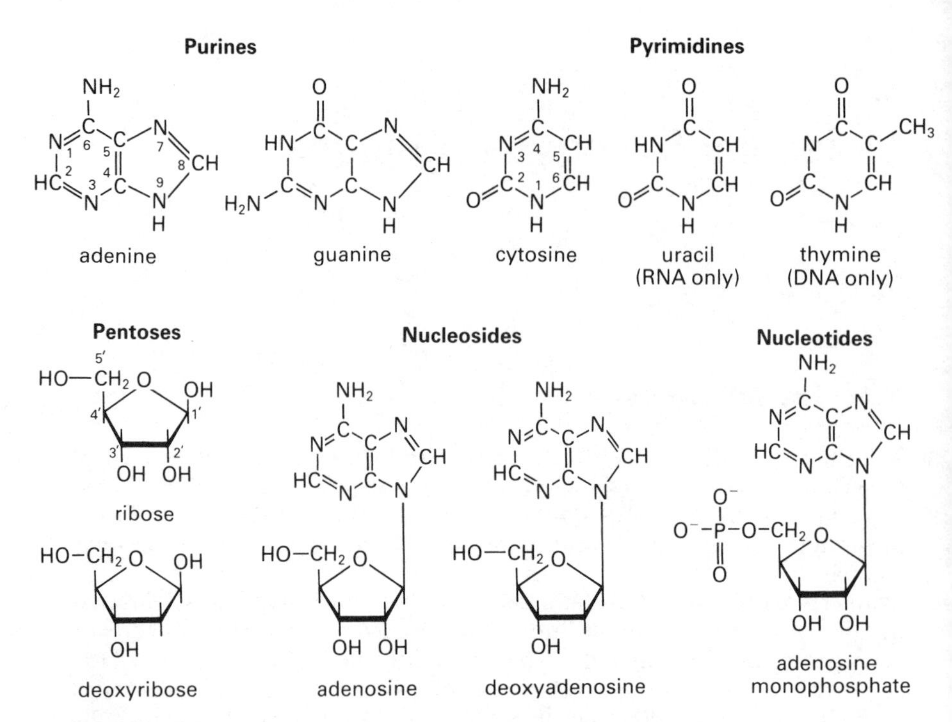

Fig. 5.1 Structures of bases, nucleosides and nucleotides.

manipulated, it is essential to have a basic understanding of their structures and functions. Such information can be found in any general biochemistry textbook, but the most important features will be summarised here.

In spite of their ultimate complexity, ribonucleic acid (RNA) and deoxyribonucleic acid (DNA) are made up of relatively few components. Both contain a pentose sugar (ribose in RNA, 2′-deoxyribose in DNA) to which is attached a purine or pyrimidine base, forming a *nucleoside*. The pentose sugar carbon atoms are numbered as shown in Fig. 5.1, using a prime (′) to indicate that the carbon is part of the sugar rather than of the purine or pyrimidine base, and it can be seen that the base is attached to the 1′ position of the pentose. A *nucleotide*, or *nucleoside phosphate*, is formed by the attachment of a phosphate to the 5′ position of a nucleoside by an ester linkage. Such nucleotides can be joined together by the formation of a second ester bond by reaction between the phosphate of one nucleotide and the 3′ hydroxyl of another, thus generating a *5′ to 3′ phosphodiester bond* between adjacent sugars; this process can be repeated indefinitely to give long polynucleotide molecules. Each polynucleotide will have a free phosphate at one of its ends, and a free 3′ hydroxyl at its other end; thus the molecule has *polarity*, and we can refer to its *3′ and 5′ ends* (Fig. 5.2).

The purine bases adenine and guanine are found in both RNA and DNA, as is the pyrimidine cytosine. The other pyrimidines are each restricted to one

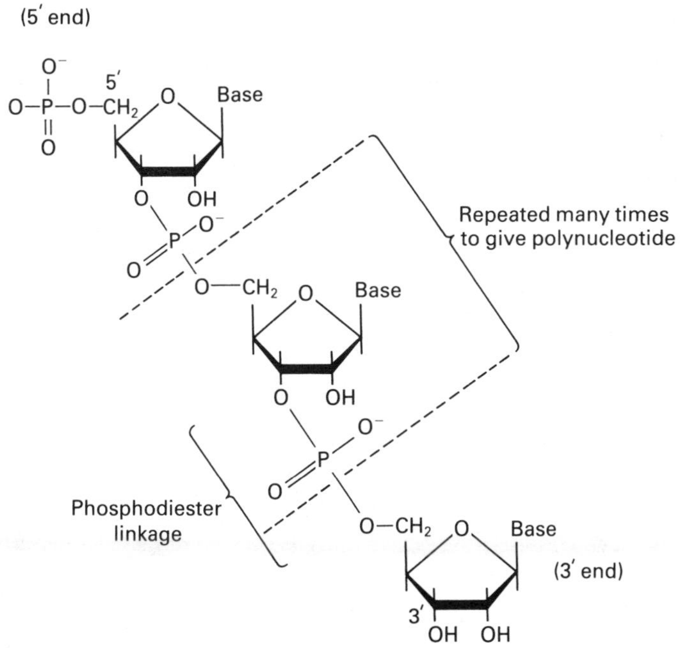

Fig. 5.2 Polynucleotide structure.

type of nucleic acid: uracil occurs exclusively in RNA, whilst thymine is limited to DNA. Thus we can distinguish between RNA and DNA on the basis of the presence of ribose and uracil in RNA, and deoxyribose and thymine in DNA. However, it is the *sequence* of bases along a molecule which distinguishes one RNA (or DNA) from another. It is conventional to write a nucleic acid sequence starting at the 5′ end of the molecule, using single capital letters to represent each of the bases, e.g. CGGATCT. Note that there is usually no point in including the sugar or phosphate groups, since these are identical throughout the length of the molecule. Terminal phosphate groups can, when necessary, be indicated by use of a letter p; thus 5′ pCGGATCT 3′ indicates the presence of a phosphate on the 5′ end of the molecule.

5.2.2 Secondary structure of nucleic acids

DNA is not usually found in a single-stranded form, but as a double-stranded molecule in the shape of a *double helix*, in which the bases of the two strands lie in the centre of the molecule, with the sugar-phosphate backbones on the outside (Fig. 5.3). A crucial feature of this structure is that it depends on the sequence of bases in one strand being *complementary* to that in the other. As shown in Fig. 5.4, thymine can form hydrogen bonds with adenine, and cytosine with guanine, in such a way that the distance between the 1′ carbons

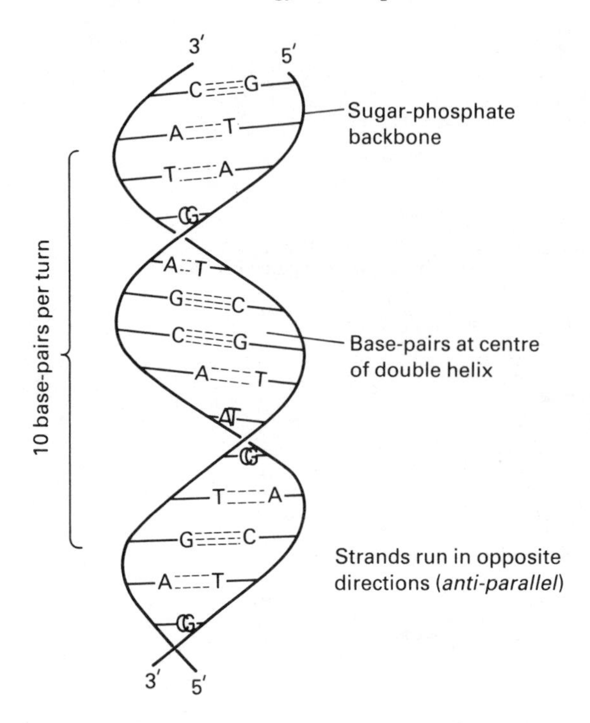

Fig. 5.3 DNA double helix.

of their respective deoxyriboses is the same. Thus, only if adenine is always *base-paired* with thymine, and cytosine with guanine, will a stable double helix result, in which the backbones of the two strands are a constant distance apart. Consequently, if the sequence of one strand is known, that of the other strand can be predicted. The strands are designated as *plus* (+) and *minus* (−), depending on which is copied during transcription (Section 5.3.3). Another important feature of DNA is that the two strands are *anti-parallel*, i.e. they run in opposite directions. For example:

5′ C G G T A A C T 3′
3′ G C C A T T G A 5′

It should be noted that the two strands of DNA are held together only by the weak forces of hydrogen bonding between complementary bases, and by hydrophobic interactions between adjacent, stacked base pairs. Little energy is needed to separate a few base pairs, and so, at any instant, as a result of the kinetic energy of the molecules, a few short stretches of DNA will be opened up to the single-stranded conformation. However, such stretches immediately pair up again at room temperature, so the molecule as a whole remains predominantly double-stranded. If, however, a DNA solution is heated to about 90°C there will be enough kinetic energy to *denature* the

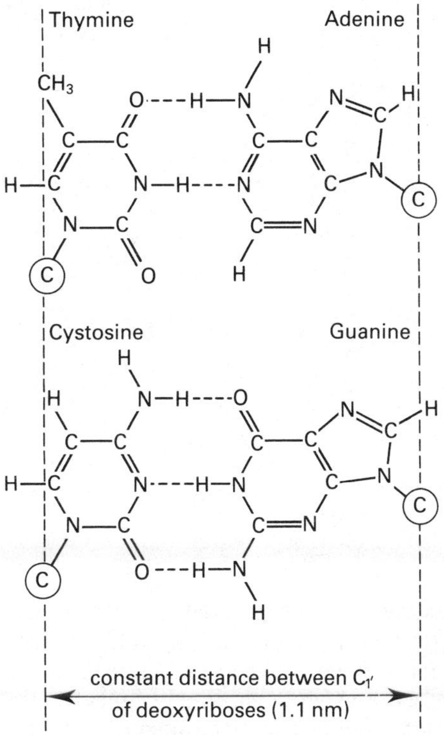

Fig. 5.4 Base pairing in DNA. © represents carbon at 1′ position of deoxyribose.

DNA completely, causing it to separate into single strands. This denaturation can be followed spectrophotometrically by monitoring the absorbance of light at 260 nm. The stacked bases of double-stranded DNA are less able to absorb light than the less constrained bases of single-stranded molecules, and so the absorbance of DNA at 260 nm increases as the DNA becomes denatured, a phenomenon known as the *hyperchromic effect*.

If absorbance at 260 nm is plotted against temperature of a DNA solution, to give a *melting curve* (Fig. 5.5), it is seen that little denaturation occurs below about 70°C, but further increases in temperature result in a marked increase in the extent of denaturation. Eventually a temperature is reached at which the sample is totally denatured, or *melted*. The temperature at which the DNA is 50% melted is called the *melting temperature* or T_m, and will depend on the nature of the DNA. If several different samples of DNA are melted, it is found that T_m is highest for those DNAs which contain the highest proportion of cytosine and guanine, and T_m can actually be used to estimate the percentage (C + G) in a DNA sample. This relationship between T_m and (C + G) content arises because cytosine and guanine form three hydrogen bonds when base-pairing, whereas thymine and adenine form only two (Fig. 5.4); consequently, more energy is needed to separate C – G pairs.

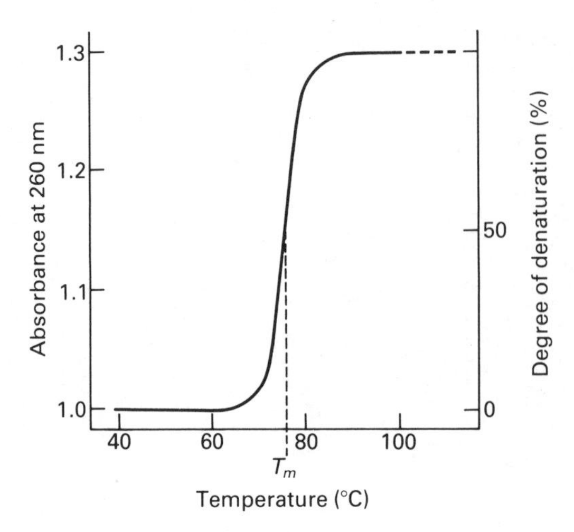

Fig. 5.5 Melting curve of DNA.

If melted DNA is cooled it is possible for the separated strands to reassociate, a process known as *renaturation*. However, a stable double-stranded molecule will only be formed if the complementary strands collide in such a way that their bases are paired precisely, which is an unlikely event if the DNA is very long and *complex* (i.e. if it contains a large number of different genes). Measurements of the rate of renaturation can give information about the complexity of a DNA preparation (Section 5.5.4).

Although RNA almost always exists as a single strand, it often contains sequences within the same strand which are complementary to each other, and which can therefore base-pair if brought together by suitable folding of the molecule. This is most obvious in the case of *transfer RNAs* (tRNA), which have four pairs of complementary sequences within their 70 to 80 nucleotide lengths (Fig. 5.6); consequently the single strand folds up to give a clover leaf secondary structure. As with DNA, cytosine pairs with guanine, but in RNA adenine pairs with uracil.

Strands of RNA and DNA will associate with each other, if their sequences are complementary, to give double-stranded, *hybrid* molecules (uracil of RNA base-pairs with adenine of DNA). Similarly, strands of radioactively labelled RNA or DNA, when added to a denatured DNA preparation, will act as *probes* for DNA molecules to which they are complementary. This *hybridisation* of complementary strands of nucleic acids is very useful for pulling a specific piece of DNA out of a complex mixture (Section 5.9).

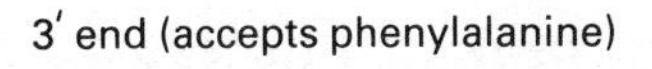

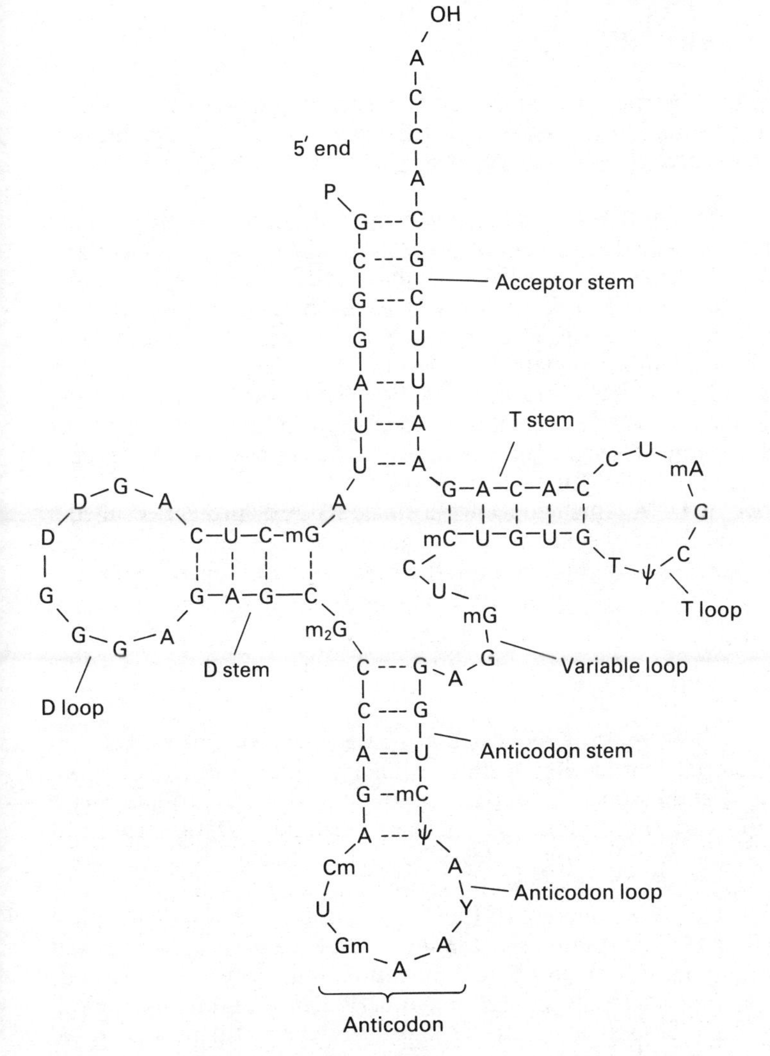

Fig. 5.6 Secondary structure of yeast $tRNA^{Phe}$. A single strand of 76 ribonucleotides forms four double stranded 'stem' regions by base-pairing between complementary sequences. The anticodon will base-pair with UUU or UUC (both are codons for phenylalanine), and phenylalanine is attached to the 3′ end by a specific amino-acyl tRNA synthetase. Several 'unusual' bases are present: D, dihydrouridine; T, ribothymidine; ψ, pseudouridine; Y, very highly modified, unlike any 'normal' base. 'mX' indicates methylation of base X (m_2X shows dimethyation); 'Xm' indicates methylation of ribose, on the 2′ position.

5.3 Functions of nucleic acids

5.3.1 Classes of RNA

The genetic information of cells and most viruses is stored in the form of DNA. This information is used to direct the synthesis of RNA molecules, which fall into three classes, messenger RNA, ribosomal RNA and transfer RNA.

Messenger RNA (mRNA) contains sequences of ribonucleotides which code for the amino acid sequences of proteins. A single mRNA codes for a single polypeptide chain in eukaryotes, but may code for several polypeptides in prokaryotes. *Ribosomal RNA* (rRNA) forms part of the structure of *ribosomes*, which are the sites of protein synthesis. Each ribosome contains only three or four different rRNA molecules, complexed with a total of between 55 and 75 proteins. *Transfer RNA* (tRNA) molecules carry amino acids to the ribosomes, and interact with the mRNA in such a way that their amino acids are joined together in the order specified by the mRNA. There is at least one type of tRNA for each amino acid.

Each block of DNA which codes for a single RNA or protein is called a *gene*, and the entire set of genes in a cell, organelle or virus forms its *genome*. Cells and organelles may contain more than one copy of their genome.

5.3.2 DNA replication

Chromosomal DNA must be replicated at a rate which will at least keep up with the rate of cell division. Replication begins at a sequence called the *origin of replication*, and involves the separation of the two DNA strands over a short length, and the binding of enzymes, including *DNA and RNA polymerases* (Fig. 5.7). In prokaryotes, RNA polymerase synthesises a short, complementary RNA chain on each exposed strand, using the DNA as a template. Then DNA polymerase III (polIII) also uses the DNA as a template for synthesis of a DNA strand, using the short RNA as a *primer*. Synthesis of the DNA strand occurs only in a 5′ to 3′ direction, but, since the two strands of DNA are antiparallel, only one can be synthesised in a continuous fashion. The other is synthesised in relatively short stretches, still in a 5′ to 3′ direction, using an RNA primer for each stretch. These RNA primers are then removed by DNA polI, acting as a 5′ to 3′ exonuclease, the gaps are filled by the same enzyme acting as polymerase, and the separate fragments are joined together by DNA ligase to give a continuous strand of DNA. The replication of eukaryotic DNA is less well characterised, and is certainly more complex than that of prokaryotes; however, both processes involve 5′ to 3′ synthesis of new DNA strands.

The net result of the replication is that the original DNA is replaced by two molecules, each containing one old and one new strand; the process is therefore known as *semi-conservative replication*.

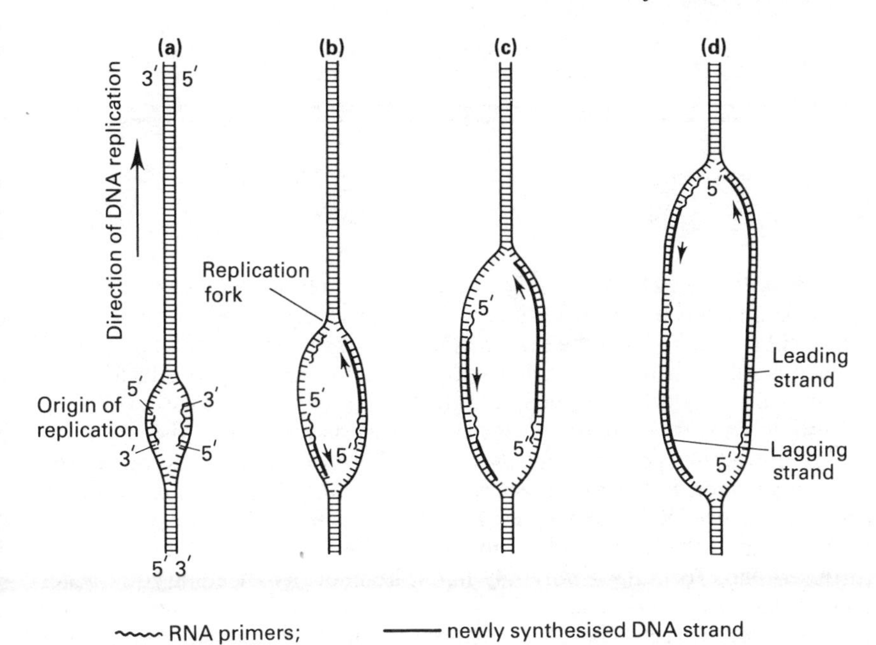

Fig. 5.7 DNA replication. **(a)** Double stranded DNA separates at *origin of replication*. RNA polymerase synthesises short RNA primer strands complementary to both DNA strands. **(b)** DNA polymerase III (pol III) synthesises new DNA strands in a 5′ to 3′ direction, complementary to the exposed, old DNA strands, and continuing from the 3′ end of each RNA primer. Consequently DNA synthesis is in the same direction as DNA replication for one strand (the *leading strand*), and in the opposite direction for the other (the *lagging strand*). RNA primer synthesis occurs repeatedly to allow the synthesis of fragments of lagging strand. **(c)** As the *replication fork* moves away from the origin of replication, pol III continues the synthesis of the leading strand, and synthesises DNA between RNA primers of the lagging strand. **(d)** DNA polymerase I (pol I) removes RNA primers from the lagging strand, and fills the resulting gaps with DNA. DNA ligase then joins the resulting fragments, producing a continuous DNA strand.

5.3.3 Transcription

At any instant only a fraction of all the genes in a genome are active. These genes, which are being *expressed*, undergo the process of *transcription*, in which an RNA molecule complementary to one of the gene's DNA strands is synthesised.

Most prokaryotic genes are made up of three regions (Fig. 5.8). At the centre is the sequence which will be copied in the form of RNA, called the *transcription unit*. To the 5′ side (*upstream*) of the strand which will be copied (the + *strand*) lies a region called the *promoter*, and *downstream* of the transcription unit is the *terminator* region. Transcription begins when DNA-dependent RNA polymerase binds to the promoter region and moves along the DNA to the transcription unit. At the start of the transcription unit the polymerase begins to synthesise an RNA molecule complementary to

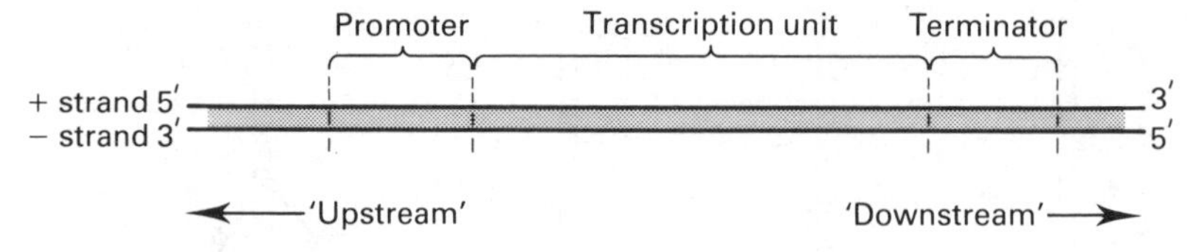

Fig. 5.8 Gene structure.

the − *strand* of the DNA, moving along this strand in a 3′ to 5′ direction, and synthesising RNA in a 5′ to 3′ direction, using nucleoside triphosphates. The RNA will therefore have the same sequence as the + strand of DNA, apart from the substitution of uracil for thymine. On reaching the terminator region, transcription is stopped, and the RNA molecule is released.

Both tRNA and rRNA are produced as *precursors*, which are then trimmed to their correct sizes by *ribonucleases*. The mRNA of prokaryotes can be used without any modification to direct protein synthesis, but *post-transcriptional processing* is needed in eukaryotes. In this processing a *cap* sequence is added to the 5′ end of the RNA, and about 150 to 200 adenosine residues are added to the 3′ end, forming a *poly (A) tail*. The majority of eukaryotic genes contain lengths of non-coding DNA, called *introns*, which interrupt their coding regions (*exons*) (Fig. 5.9). Transcription of these genes results in the production of *heterogeneous nuclear RNA*, or hnRNA, which must have its introns excised, leaving the exons spliced together to form mRNA.

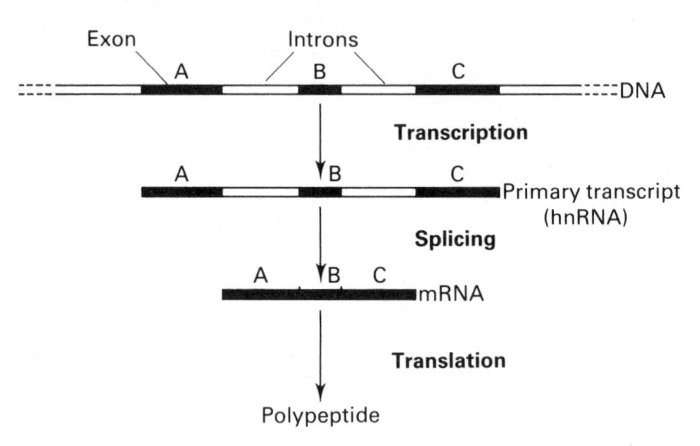

Fig. 5.9 Splicing of RNA to remove introns. Regions A, B and C, the exons, are sequences which, when joined together, code for the polypeptide product. The primary transcript still contains introns, and is called *heterogeneous nuclear RNA* (hnRNA).

5.3.4 Translation

Each mRNA codes for the primary amino-acid sequence of a protein, using a *triplet* of nucleotides to represent each of the amino acids. Each triplet is known as a *codon*, and since there are 64 possible triplet codons but only 20 different amino acids (plus start and stop codons), most amino acids are coded for by more than one codon. The *genetic code* has been worked out, and is found to be universal for all chromosomal and chloroplast DNAs or RNAs so far examined; but a few differences have been found in the codes used by mitochondria.

The mRNAs are read, and proteins assembled, on the *ribosomes*, which are structures formed of a complex of rRNAs and proteins. Each ribosome consists of a large and a small subunit, which associate during the process of protein synthesis, or *translation*. The ribosomes of prokaryotes and of organelles have a *sedimentation coefficient* (Section 2.2) of 70S, whilst those of the eukaryotic cytoplasm are 80S. Transfer RNA molecules (Fig. 5.6) are also needed for translation. Each of these can be covalently linked to a specific amino acid, forming an *amino-acyl tRNA*, and each has a triplet of bases exposed which is complementary to the codon for that amino acid. This exposed triplet is known as the *anti-codon*, and allows the tRNA to act as an adapter molecule, bringing together a codon and its corresponding amino acid.

After binding to a specific sequence at the 5′ end of the mRNA, known as a *Shine-Dalgarno sequence* in prokaryotes, the ribosome moves towards the 3′ end, allowing an amino-acyl tRNA molecule to base-pair with each successive codon, thereby carrying in amino acids in the correct order for protein synthesis. The ribosome forms peptide bonds between these amino acids as it moves along the mRNA, and releases a completed polypeptide chain when it reaches a *termination codon* (UAA, UGA or UAG) (Fig. 5.10). The first codon after the ribosome binding site, the *initiation codon*, is always AUG, which acts as a start signal.

Since the mRNA is read in triplets, an error of one or two nucleotides in positioning of the ribosome will result in the synthesis of an incorrect polypeptide. Thus it is essential for the correct *reading frame* to be used during translation. This is ensured in prokaryotes by base-pairing between the Shine-Dalgarno sequence and a complementary sequence of one of the ribosome's rRNAs, thus establishing the correct starting point for movement of the ribosome along the mRNA.

5.4 Isolation of nucleic acids

5.4.1 DNA

Before nucleic acids can be cut or otherwise manipulated, they must be isolated and purified to some extent, the degree of purity depending on the intended use of the preparation.

DNA is very easily damaged by *shear forces*; even rapid stirring of a solution can break high molecular weight DNA into much shorter fragments. The other main threat comes from digestion by *deoxyribonucleases* (*DNases*), which are found in most cells, and may also be present in dust

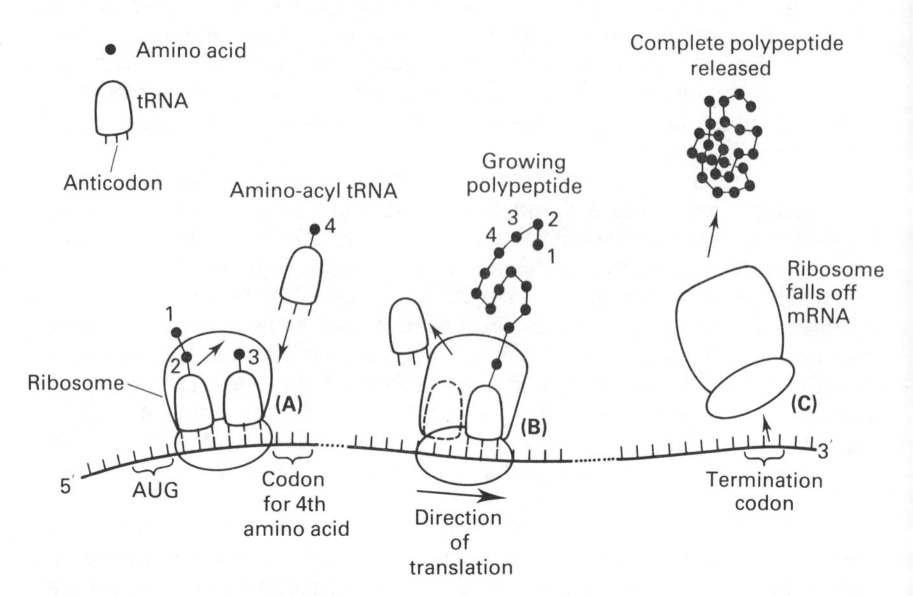

Fig. 5.10 Translation. Ribosome (A) has moved only a short way from the 5′ end of the mRNA, and has built up a dipeptide (on one tRNA), which is about to be transferred onto the third amino acid (still attached to tRNA). Ribosome (B) has moved much further along the mRNA, and has built up an oligopeptide which has just been transferred onto the most recent amino-acyl tRNA. The resulting free tRNA leaves the ribosome, and will receive another amino acid. The ribosome moves towards the 3′ end of the mRNA by a distance of three nucleotides, so that the next codon can be aligned with its corresponding amino-acyl tRNA on the ribosome. Ribosome (C) has reached a termination codon, so has released the completed polypeptide, and has fallen off the mRNA.

which could contaminate laboratory glassware. Consequently, DNA is recovered from cells by the gentlest possible method of cell rupture, in the presence of EDTA to chelate the Mg^{2+} ions needed for DNase activity. Ideally, cell walls, if present, should be digested enzymatically (e.g. lysozyme treatment of bacteria), and the cell membrane should be solubilised using detergent. If physical disruption is necessary, it should be kept to a minimum, and should involve cutting or squashing of cells, rather than the use of shear forces. Cell disruption (and most subsequent steps) should be performed at 4°C, using glassware and solutions which have been autoclaved to destroy DNase activity.

After release of nucleic acids from the cells, RNA can be removed by treatment with ribonuclease (RNase) which has been heat treated to inactivate any DNase contaminants; RNase is relatively stable to heat as a result of its disulphide bonds, which ensure rapid renaturation of the molecule on cooling. The other major contaminant, protein, is removed by shaking the solution gently with water-saturated phenol, or with a phenol/chloroform mixture, either of which will *denature* proteins (Section 3.1) but not nucleic acids. Centrifugation of the emulsion formed by this mixing produces a lower, organic phase, separated from the upper, aqueous phase by an interface of denatured protein. The aqueous solution is recovered and deproteinised repeatedly, until no more material is seen at the interface. Finally, the deproteinised DNA preparation is mixed with two volumes of absolute ethanol, and allowed to precipitate out of solution in a freezer. After centrifugation, the DNA pellet is redissolved in a buffer containing EDTA for protection against DNases, and this solution can be stored at 4°C for at least a month. DNA solutions can be stored frozen, but repeated freezing and thawing tends to damage long molecules by shearing; so preparations in frequent use are normally stored at 4°C.

The procedure described above is suitable for total cellular DNA. If the DNA from a specific organelle or viral particle is needed, it is best to isolate the organelle or virus before trying to extract its DNA, since the recovery of a particular type of DNA from a mixture is usually rather difficult. The isolation of *plasmids* (Section 5.8.1) will be described later.

5.4.2 RNA

The methods used for RNA isolation are very similar to those described above for DNA; however, RNA molecules are relatively short, and therefore less vulnerable to damage by shearing, so cell disruption can be rather more violent. RNA is, however, very vulnerable to digestion by RNases which are present on fingers, so gloves should be worn, and a strong detergent should be included in the isolation medium to denature any RNases immediately. Subsequent deproteinisation should be particularly rigorous, since RNA is often tightly associated with proteins. DNase treatment can be used to remove DNA, and RNA can be precipitated by ethanol.

Frequently one wishes to isolate mRNA, for translation *in vitro*, or for the synthesis of a *cDNA probe* (Section 5.9.1). Since almost all mRNA molecules encoded by chromosomal DNA have lengths of adenosine units at their 3′-ends, forming poly(A) tails, they can be separated from a mixture of RNA molecules by affinity chromatography on oligo(dT)-cellulose columns (Section 6.7). At high salt concentrations (approx. 0.5M NaCl) the poly (A) tails will bind to the complementary oligo(dT) units of the affinity column, and so mRNA will be retained; all other RNA molecules can be washed through the column by further high salt solution. Finally, the bound mRNA can be eluted using a low concentration of salt (less than approx. 50mM NaCl).

5.5 Analysis of DNA

5.5.1 Electrophoresis

Electrophoresis in agarose or polyacrylamide gels is the most usual way to separate DNA molecules according to size. The technique can be used analytically or preparatively, and can be qualitative or quantitative. Electrophoresis is discussed in Chapter 7, so no practical details will be given here. The easiest and most widely applicable method is electrophoresis in horizontal agarose gels, followed by staining with ethidium bromide. This dye binds to DNA by insertion between stacked base pairs (*intercalation*), and it exhibits a strong orange/red fluorescence when illuminated with ultraviolet light (Section 8.5). Very often electrophoresis is used to check that restriction (Section 5.7.1) or ligation (Section 5.7.2) reactions have gone to completion, or to assess the purity and intactness of a DNA preparation. For such checks *mini-gels* are particularly convenient, since they need little sample and give results quickly. Agarose gels (Section 7.5) can be used to separate molecules larger than about 200 base-pairs (bp), but polyacrylamide gels (Section 7.6) must be used for shorter molecules.

By calibration against DNA molecules of known sizes, such as *restriction fragments* (Section 5.7.1), the lengths of molecules in a sample can be measured. This can be used for such purposes as determining the sizes of inserts in *cloning vectors* (Section 5.8), or to map the positions of *restriction sites* on a length of DNA. *Restriction mapping* involves the size analysis of restriction fragments produced by several *restriction enzymes* (Section 5.7.1) individually and in combination. The principle of this mapping is illustrated in Fig. 5.11, in which the restriction sites of two enzymes, A and B, are being mapped. Cleavage with A gives fragments 2 and 7 kilobases (kb) from a 9 kb

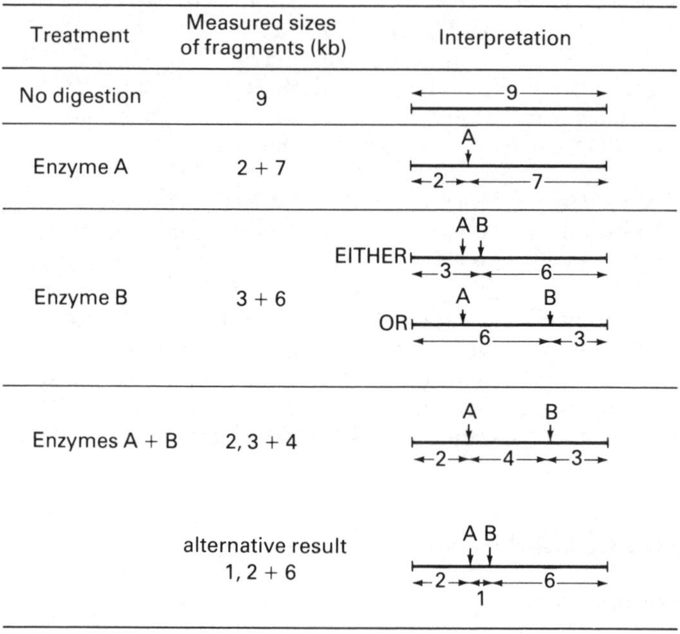

Treatment	Measured sizes of fragments (kb)	Interpretation
No digestion	9	←9→
Enzyme A	2 + 7	A ←2→←7→
Enzyme B	3 + 6	EITHER: A B ←3→←6→ OR: A B ←6→←3→
Enzymes A + B	2, 3 + 4	A B ←2→←4→←3→
	alternative result 1, 2 + 6	A B ←2→←1→←6→

Fig. 5.11 Restriction mapping of DNA. Note that each experimental result and its interpretation should be considered in sequence, thus building up an increasingly unambiguous map.

molecule, hence we can position the single A site 2 kb from one end. Similarly, B gives fragments 3 and 6 kb, so it has a single site 3 kb from one end; but it is not possible at this stage to say if it is near to the site of A, or at the opposite end of the DNA. This can be resolved by a double digestion. If the resultant fragments are 2, 3, and 4 kb, then A and B cut at opposite ends of the molecule; if they are 1, 2 and 6 kb, the sites are near each other. Not surprisingly, the mapping of real molecules is rarely as simple as this, and computer analysis of the restriction fragment lengths is usually needed to construct a map.

When electrophoresis is used *preparatively*, the piece of gel containing the desired DNA fragment is cut out, and the DNA is recovered from it in various ways, including *crushing* with a glass rod in a small volume of buffer, or by *electroelution*. In this method the piece of gel is sealed in a length of dialysis tubing containing buffer, and is then placed between two electrodes in a tank containing more buffer. Passage of an electrical current between the electrodes causes DNA to migrate out of the gel piece, but it remains trapped within the dialysis tubing, and can therefore be recovered easily.

It is, of course, necessary to discover which band on a gel contains the particular piece of DNA which is needed. This can be achieved by transferring the DNA from the intact gel onto a piece of nitrocellulose paper placed in contact with it, using denaturing conditions, so that the DNA becomes bound to the paper in exactly the same pattern as that originally on the gel. This transfer, named a *Southern blot* after its inventor, can be performed electrophoretically, or by drawing large volumes of buffer through both gel and paper, thus transferring DNA from one to the other (Fig. 5.12). The point of this operation is that the paper can now be treated with a radioactive DNA molecule, acting as a *probe*, for example a cDNA (Section 5.9.1), in the same way as in colony hybridisation (Section 5.9.3), to discover which bands of DNA contain sequences complementary to the probe. The same process can be used to transfer RNA from gels onto nitrocellulose paper, for identification of specific sequences by hybridisation, and it is then known as *Northern blotting*. When the procedure is used with proteins separated on polyacrylamide gels it is termed *Western blotting*.

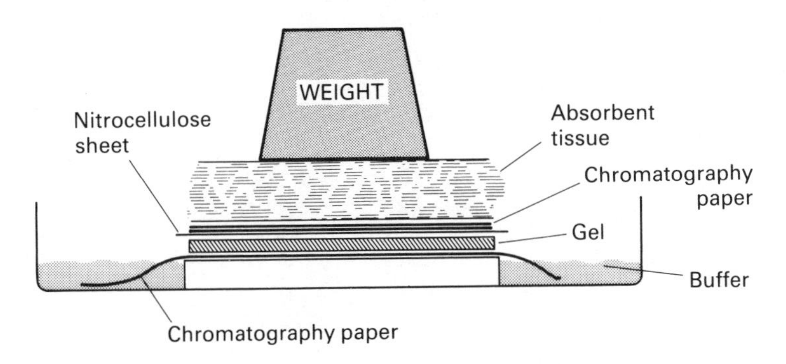

Fig. 5.12 Southern blot apparatus.

5.5.2 Sequencing of DNA

The advent of methods for the *sequencing of DNA* has revolutionised our understanding of gene structure, and it is now routine to sequence any newly isolated DNA fragment of interest. There are two methods in use, the *dideoxy, or chain termination method of Sanger*, and the *chemical cleavage method of Maxam and Gilbert*. Both methods are based on the high resolution electrophoresis of four sets of radioactive oligonucleotides produced from the DNA to be sequenced, but they differ in the procedures used to generate the oligonucleotides.

For the Sanger method, single stranded DNA is required. This is readily prepared by cloning (Section 5.6) the DNA in M13 bacteriophage, which

Fragment to be sequenced, cloned in M13 phage

3′ – – – AG – – – CT**GCTCGCAT** – – – 5′

TC – – – GA

Primer

DNA polymerase
4 dNTPs (radioactive)
ddGTP

Synthesis of complementary second strands:

5′ TC – – – GA**C**dd**G** 3′

5′ TC – – – GA**CGA**dd**G** 3′

5′ TC – – – GA**CGAGC**dd**G** 3′

Denature to give single strands.

Run on sequencing gel alongside products of ddCTP, ddATP and ddTTP reactions.

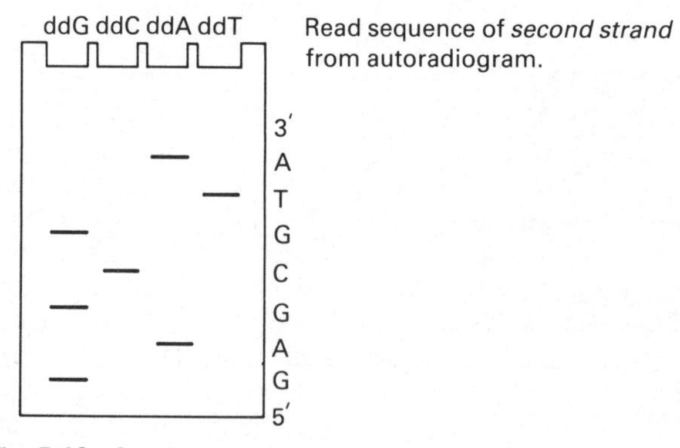

Fig. 5.13 Sanger sequencing of DNA.

packages its DNA in a single stranded form. The DNA is isolated and allowed to hybridise with a short oligonucleotide which is complementary to a sequence in the M13 DNA just to the 3′ side of its inserted DNA (Fig. 5.13). The oligonucleotide will then act as a primer for synthesis of a second strand of DNA, catalysed by DNA polymerase. Since the new strand is synthesised from its 5′ end, the first DNA to be made will be complementary to the inserted DNA, which we wish to sequence. One (or all) of the deoxyribonucleoside triphosphates (dNTPs) which must be provided for DNA synthesis is radioactively labelled with ^{32}P or ^{35}S, and so the newly synthesised strand will be labelled. If a 2′, 3′-dideoxynucleoside triphosphate (ddNTP) is also present in the reaction mixture, there is a chance that it will be added to the growing DNA chain in place of the normal dNTP, since it is identical to its corresponding dNTP, apart from the absence of a 3′-hydroxyl group. Once this has happened, the chain growth terminates, since a 5′ to 3′ phosphodiester linkage cannot be formed without a 3′-hydroxyl. Since the incorporation of ddNTP rather than dNTP is a random event, the reaction will produce new molecules varying widely in length, but all terminating at the same type of base. By splitting the cloned DNA into four aliquots and using a different ddNTP for each, four sets of molecules are generated, each terminating at a different type of base, but all having a common 5′ end (the primer).

The four samples so produced are then denatured and loaded next to each other on a polyacrylamide gel for electrophoresis. Electrophoresis is performed at about 70°C in the presence of urea, to prevent renaturation of the DNA, since even partial renaturation would seriously alter the rates of migration of DNA fragments. Very thin, long gels are used for maximum resolution over a wide range of fragment lengths. After electrophoresis, the positions of radioactive DNA bands on the gel are detemined by *autoradiography* (Section 9.2.4). Bearing in mind that every band in the track from the dideoxyadenosine triphosphate sample must contain molecules which terminate at adenine, and that those in the ddGTP terminate at guanine, etc., it is possible to read the sequence of the newly synthesised strand from the autoradiogram, provided that the gel can resolve differences in length equal to a single nucleotide (Fig. 5.13). Under ideal conditions, sequences up to about 300 bases in length can be read from one gel.

The chemical cleavage method of Maxam and Gilbert usually starts with the enzymic addition of a radioactive label to either the 3′ or the 5′ ends of a double-stranded DNA preparation (Fig. 5.14). The strands are then separated by electrophoresis under denaturing conditions, and analysed separately. DNA labelled at one end is split into four aliquots and each is treated with chemicals which will act on a specific base (or, in some cases, either of two bases) by methylation or removal of the base. Conditions are chosen so that, on average, each molecule is modified at only one position along its length; every base in the DNA strand has an equal chance of being modified. Following the *modification reactions*, the separate samples are cleaved by piperidine, which breaks phosphodiester bonds exclusively on both sides of nucleotides whose base has been modified. The result is similar

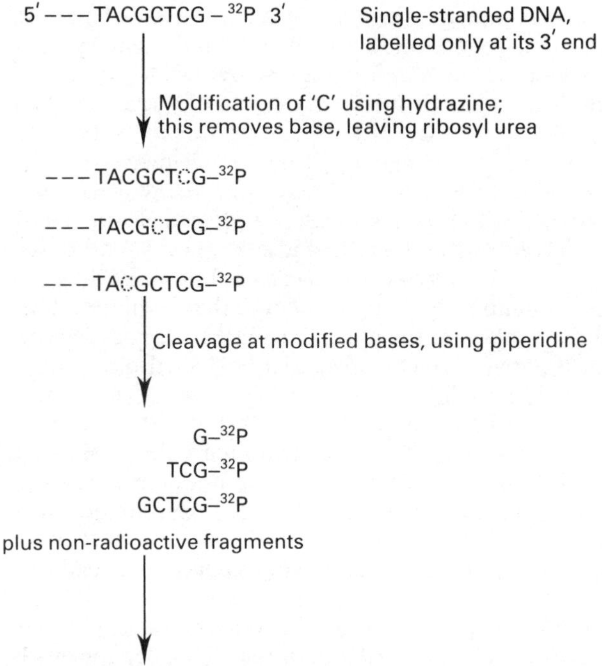

Fig. 5.14 Maxam and Gilbert sequencing of DNA. Only modification and cleavage of deoxycytidine is shown, but three more aliquots of the end-labelled DNA would be modified and cleaved at G, G + A, and T + C, and the products would be separated on the sequencing gel alongside those from the 'C' reactions.

to that produced by the Sanger method, since each sample now contains radioactive molecules of various lengths, all with one end in common (the labelled end), and with the other end cut at the same type of base. Analysis of the reaction products by electrophoresis is as described for the Sanger method.

Because the Sanger method produces oligonucleotides which are radioactively labelled throughout their lengths, rather than only at one end, the molecules can be made a lot more radioactive, and therefore easier to detect; so less DNA is needed for sequencing. Once M13 cloning has been set up in a laboratory, it provides a very convenient and rapid way to obtain single-stranded DNA. For these reasons, dideoxy sequencing of M13-cloned DNA is probably the most commonly used sequencing method, though the chemical procedure is still used by many laboratories.

5.5.3 Protein sequencing

Although protein sequencing may seem out of place in a section dealing with the analysis of DNA, the molecular biologist can often make use of a knowledge of protein sequences when manipulating DNA. If the sequence of a protein is known, a gene coding for it can be synthesised chemically (though this is usually only worth doing for small polypeptides), or an *oligonucleotide probe* can be synthesised for use in recovering the gene for that protein from a *gene library* (Section 5.9.5).

Since it is currently impossible to sequence a polypeptide longer than about 100 amino acids, pure proteins must be fragmented to give polypeptides of a length which can be sequenced, and these polypeptides must be separated from each other prior to sequencing. Fairly specific and limited cleavage can be obtained by chemical means. For example, cyanogen bromide cleaves only at (rare) methionine residues, BNPS-skatole cleaves at tryptophan, and hydroxylamine breaks the linkage between asparagine and glycine. Similarly, several proteolytic enzymes, such as trypsin and V8-Protease, have a fairly specific site of action, and will therefore generate relatively few cleavage products.

The polypeptides so produced are separated from each other prior to sequencing, using such techniques as exclusion chromatography (Section 6.6) or HPLC (Section 6.8). Relative positions of the polypeptides within a protein can be found by looking for overlaps in the sequences of polypeptides generated by different means, and in this way the entire protein sequence may be deduced.

All protein sequencing methods are based on the *Edman degradation* of polypeptides, in which the *N*-terminal amino acid is specifically removed, leaving a polypeptide one amino acid residue shorter. Variations arise in the method of identifying the removed amino acid or the newly exposed *N*-terminal amino acid. By repeated cycles of Edman degradation and identification of product, the polypeptide can be sequenced.

In the *Edman reaction* (Fig. 5.15) the polypeptide is treated with phenylisothiocyanate (PITC), which reacts with the *N*-terminal amino acid to form a phenylthiocarbamyl (PTC) derivative of the polypeptide. Anhydrous trifluoroacetic acid is then used to cleave the molecule, giving the 2-anilino-5-thiazolinone derivative of the *N*-terminal amino acid and also the polypeptide shortened by one residue. The thiazolinone derivative is separated from the polypeptide and converted into the more stable 3-phenyl-2-thiohydantoin (PTH) derivative, which is then identified by HPLC or TLC. By repeating this cycle the polypeptide can be sequenced from its *N*-terminal end. The process has been automated, either by immobilising the protein on an inert, solid support (*solid-phase sequencers*), or by keeping the protein spread out in a thin film for maximum exposure to reagents (*spinning cup sequencers*). Such instruments can, under ideal conditions, sequence up to 100 residues of a protein.

The alternative *Dansyl-Edman* procedure (Fig. 5.16) is highly sensitive, allowing as little as 1 nmole of polypeptide to be sequenced, and it is therefore well suited to manual determination of sequences. It uses cycles of the Edman reaction to remove *N*-terminal amino acids sequentially, but, instead of identifying the released PTH derivatives, it identifies the newly exposed *N*-terminal amino acids. This is achieved by adding a dansyl group to the *N*-terminal of a very small sample of the polypeptide after each cycle of the

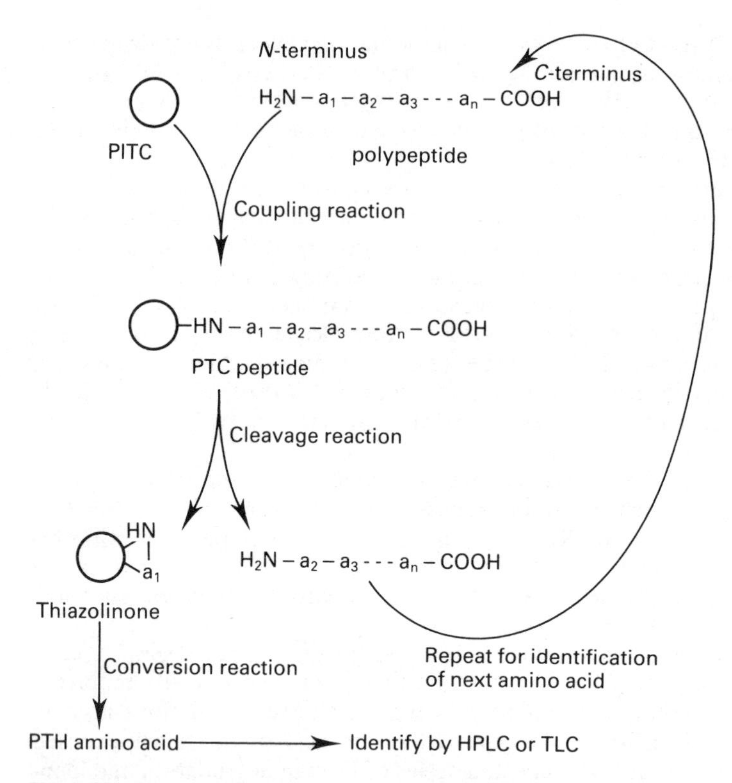

Fig. 5.15 Edman reactions. PITC phenylisothiocyanate; PTC, phenylthiocarbamyl; PTH 3-phenyl-2-thiohydantoin. Note that each cycle of reactions removes one amino acid from the *N*-terminus of the polypeptide.

Edman reaction, followed by cleavage with hydrochloric acid to release a dansyl amino acid plus free amino acids. The dansyl derivative can be identified by two-dimensional TLC on polyamide plates (Section 6.1.3). Up to about 15 amino acids can be sequenced before the cumulative effects of incomplete reactions and side reactions make impossible the unambiguous identification of the dansyl amino acid.

Given the nucleotide sequence of a gene, and our knowledge of the genetic code, it is easy to read off the amino acid sequence for which the gene codes, provided the correct reading frame is used, and the sequence is not interrupted by introns. Ironically, DNA sequencing, rather than protein sequencing, has sometimes been used to obtain amino acid sequences of proteins, especially when the pure protein has not been obtainable in sufficient quantities for direct sequencing. However, it should be remembered that a lot of effort is involved in the isolation of a specific gene, and this may more than offset the rapidity of DNA sequencing. The pace of sequencing is such that computers are now used by some laboratories for the analysis of sequencing gels, and sequence data banks have been set up to cope with the massive flow of information. In spite of this, it will be some time before the human genome

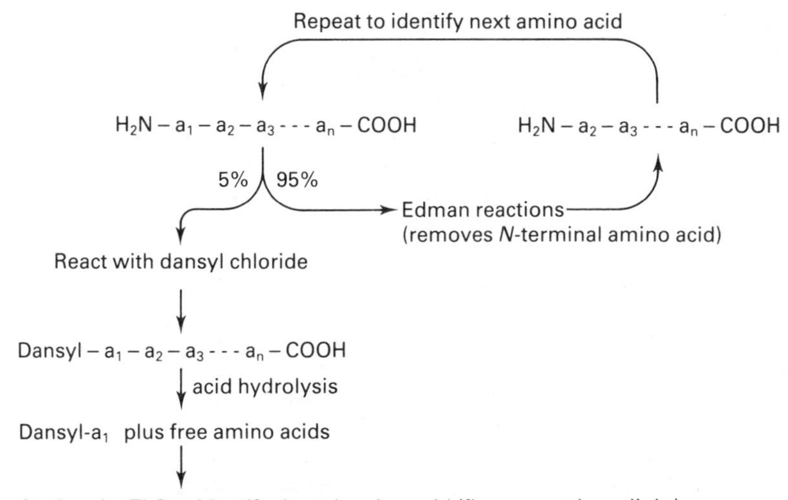

Fig. 5.16 Dansyl-Edman procedure. Only the *N*-terminal amino acid becomes dansylated, and can therefore be identified by TLC. The Edman degradation is used to remove *N*-terminal amino acids one-by-one from the polypeptide, and dansylation allows the identification of each newly exposed *N*-terminal residue.

is completely sequenced. Even at a rate of one base per second, the 3×10^6 kb of the haploid genome would take more than 100 years to be sequenced.

5.5.4 Renaturation kinetics

When preparations of double-stranded DNA are denatured by heat or alkali, and then allowed to renature, measurement of the *rate of renaturation* can give valuable information about the *complexity* of the DNA, i.e. how much information it contains (measured in base-pairs). The complexity of a molecule may be much less than its total length if some sequences are *repetitive*, but complexity will equal total length if all sequences are *unique*, appearing only once in the genome. In practice, the DNA is first cut randomly into fragments about 1 kb in length (Section 5.9.2), and is then completely denatured by heating above its T_m. Renaturation at a temperature about 10°C below the T_m is monitored either by decrease in absorbance at 260 nm (the *hypochromic effect*), or by passing samples at intervals through a column of hydroxyapatite, which will adsorb only double-stranded DNA, and measuring how much of the sample is bound. The degree of renaturation after a given time will depend on C_0, the concentration (in nucleotides per unit volume) of double-stranded DNA prior to denaturation, and t, the duration of the renaturation.

For a given C_0, it should be evident that a preparation of λ DNA (genome size 49 kb) will contain many more copies of the same sequence per unit volume than a preparation of human DNA (haploid genome size 3×10^6 kb), and will therefore renature far more rapidly, since there will be more molecules complementary to each other per unit volume in the case of λ DNA, and therefore more chance of two complementary strands colliding with each other. In order to compare the rates of renaturation of different DNA samples it is usual to measure C_0 and the time taken for renaturation to proceed half way to completion, $t_{1/2}$, and to multiply these values together to give a $C_0t_{1/2}$ value. The larger $C_0t_{1/2}$, the greater the complexity of the DNA; hence λ DNA has a far lower $C_0t_{1/2}$ than does human DNA.

In fact, the human genome does not renature in a uniform fashion. If the extent of renaturation is plotted against log C_0t (this is known as a *Cot curve*), it is seen that part of the DNA renatures quite rapidly, whilst the remainder is very slow to renature (Fig. 5.17). This must mean that some sequences have a higher concentration than others; in other words, part of the genome consists of *repetitive sequences*. These repetitive sequences can be separated from the single-copy DNA by passing the renaturing sample through a hydroxyapatite column early in the renaturation process, at a time which gives a low value of C_0t. At this stage only the rapidly renaturing sequences will be double-stranded, and they will therefore be the only ones able to bind to the column.

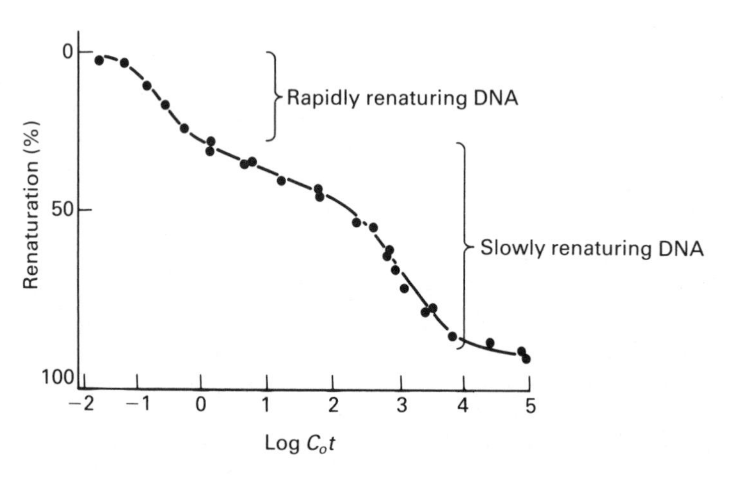

Fig. 5.17 *Cot* curve of human DNA. DNA was allowed to renature at 60°C after being completely dissociated by heat. Samples were taken at intervals and passed through a hydroxyapatite column to determine the percentage of double-stranded DNA present. This percentage was plotted against log *Cot* (original concentration of DNA × time of sampling).

5.6 Outline of genetic manipulation

The molecular biologist has a wide range of techniques available for the investigation and alteration of gene structure and function, but the key steps are those of cutting and joining lengths of DNA in a precise way, using restriction endonucleases and ligases.

By such cutting and joining it is possible to divide a complex genome into a large number of small fragments, each about the size of a single gene, and to insert each of those fragments into a carrier (or *vector*) DNA molecule, which can then be replicated indefinitely within bacterial cells. In this way genes can be *cloned* to provide sufficient material either for detailed analysis, or for insertion into the genome of a cell which is to be genetically engineered. The principle steps involved in gene cloning are illustrated in Fig. 5.18. Each step will be discussed in more detail below.

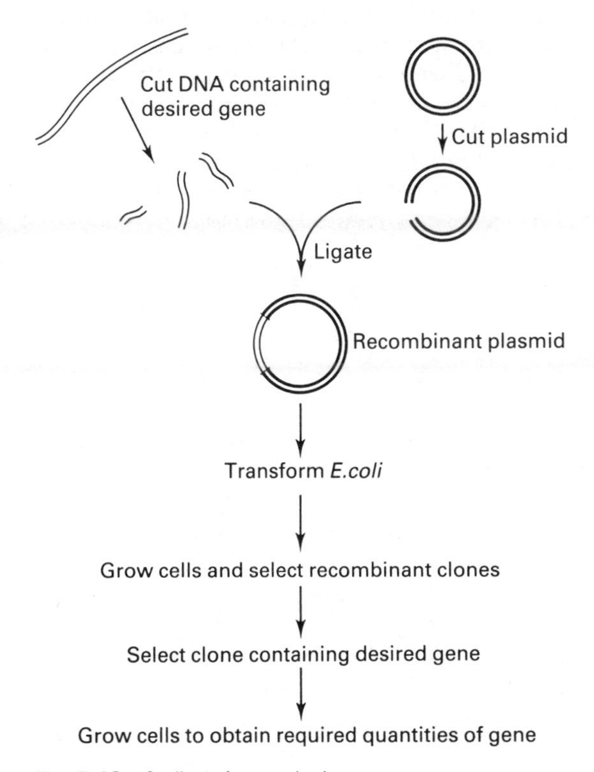

Fig. 5.18 Outline of gene cloning.

5.7 Enzymes used in genetic manipulation

5.7.1 Restriction endonucleases

It was as recently as 1970 that the first enzyme was isolated which would recognise a specific sequence of DNA, and cut the molecule within that sequence. Such enzymes are known as *restriction endonucleases*, and they are used by bacteria as a defence mechanism against *foreign DNA*, e.g. viral DNA, since they recognise and digest such invading molecules. Bacterial DNA is protected from digestion by the cell's own enzymes as a result of methylation of bases within vulnerable sequences. A large number of restriction enzymes have been isolated, and the class which is most useful to the molecular biologist is known as Type II. These enzymes recognise a specific sequence of four or six nucleotides, and cleave the DNA within this *restriction site*. Clearly a tetranucleotide sequence will occur more frequently in a given molecule than a hexanucleotide, so more fragments will be generated by an enzyme which recognises a tetranucleotide sequence.

Some enzymes cut straight across the DNA to give *blunt ends*, whilst others make staggered *single-strand cuts*, producing short single-stranded projections at each end of the cleaved DNA (Fig. 5.19). Since the restriction sites are symmetrical, so that both strands have the same sequence when read in the 5′ to 3′ direction, such staggered cuts will generate identical single-stranded projections on either side of the cut. These ends are not only identical, but complementary, and will base-pair with each other; they are

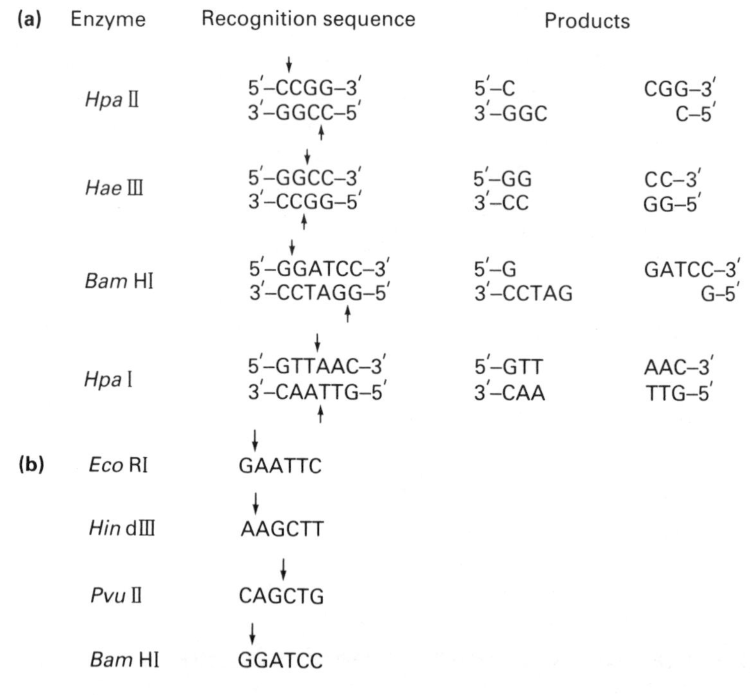

Fig. 5.19 Recognition sequences of some restriction enzymes showing **(a)** full descriptions and **(b)** conventional representations. Arrows indicate positions of cleavage. Note that all the information in **(a)** can be derived from knowledge of a single strand of the DNA whilst in **(b)** only one strand is shown, drawn from 5′ to 3′; this is the conventional way of representing restriction sites.

therefore known as *cohesive* or *sticky ends*. It is most important to realise that, because of the specificity of restriction enzymes, every copy of a given DNA molecule will give the same set of fragments when cleaved with a particular enzyme, and different DNA molecules will, in general, give different sets of fragments when treated with the same enzyme. Well over 400 Type II enzymes, recognising nearly 100 different *restriction sites*, have been characterised, and the list is growing steadily.

5.7.2 Ligases

Although cutting DNA precisely is very useful for DNA analysis, its full potential is only revealed when the fragments produced are joined together to give a new structure, known as *recombinant DNA*. This joining, or *ligation* is achieved by the use of a DNA ligase enzyme, the most common being that isolated from the bacterial virus known as T4 phage.

If two different DNA preparations are treated with the same restriction enzyme to give fragments with sticky ends, these ends will be identical in both preparations. Thus, when the two sets of fragments are mixed, base pairing between sticky ends will result in the coming together of fragments which were derived from different molecules. There will, of course, also be pairing of fragments derived from the same molecule. Such pairings are transient, owing to the weakness of hydrogen bonding between the few bases in the sticky ends, but they can be stabilised by use of DNA ligase, which forms a covalent bond between the 5′-phosphoryl at the end of one strand and the 3′-hydroxyl of the adjacent strand (Fig. 5.20). This reaction, which is driven by ATP, is often carried out at 4°C to lower the kinetic energy of molecules, and so reduce the chances of base-paired sticky ends parting before they have

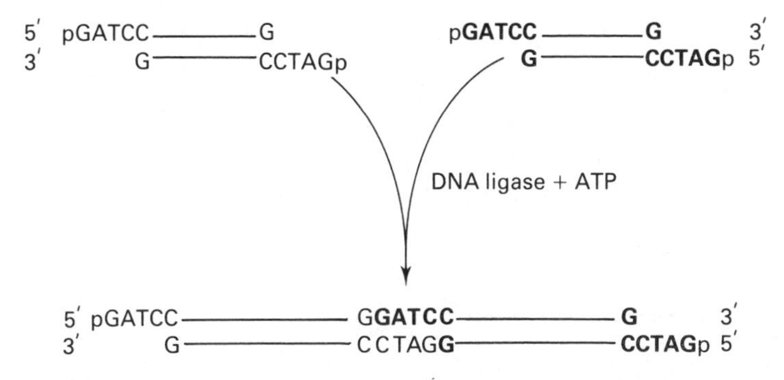

Fig. 5.20 Ligation of molecules with cohesive ends. Complementary cohesive ends base-pair, forming a temporary link between two DNA fragments. This association of fragments is stabilised by the formation of 3′ to 5′ phosphodiester linkages between cohesive ends, a reaction catalysed by DNA ligase.

been stabilised by ligation. However, long reaction times are needed to compensate for the low activity of DNA ligase in the cold.

Since ligation reconstructs the site of cleavage, recombinant molecules produced by ligation of sticky ends can be cleaved again at the joins, using the same restriction enzyme that was used to generate the fragments initially. Consequently a fragment can be inserted into a vector DNA, and recovered again after cloning of the recombinant molecule.

Lengths of blunt-ended DNA can be ligated, but, since there is no base-pairing to hold fragments together temporarily, concentrations of DNA and ligase must be high. However, *blunt-ended ligation* is a useful way of joining together DNA fragments which have not been produced by the same restriction enzyme, and which therefore probably have incompatible sticky ends. These ends are removed prior to ligation, using the enzyme S1 nuclease, which digests single-stranded DNA (Fig. 5.21). In such cases a restriction site will not be regenerated, and this may prevent recovery of a fragment after cloning. For this reason molecules called *linkers* are frequently used for joining DNA. Linkers are short, double-stranded oligonucleotides, with blunt ends, containing at least one restriction site within their sequence (Fig. 5.22). These linkers can be joined to one preparation of DNA by blunt-ended ligation, and then sticky ends can be created by cleavage of the linkers with a suitable restriction enzyme. The linker is chosen so that the sticky end it produces is identical to that on the other DNA preparation, consequently, the two can then be joined by ligation of their sticky ends. Some very versatile

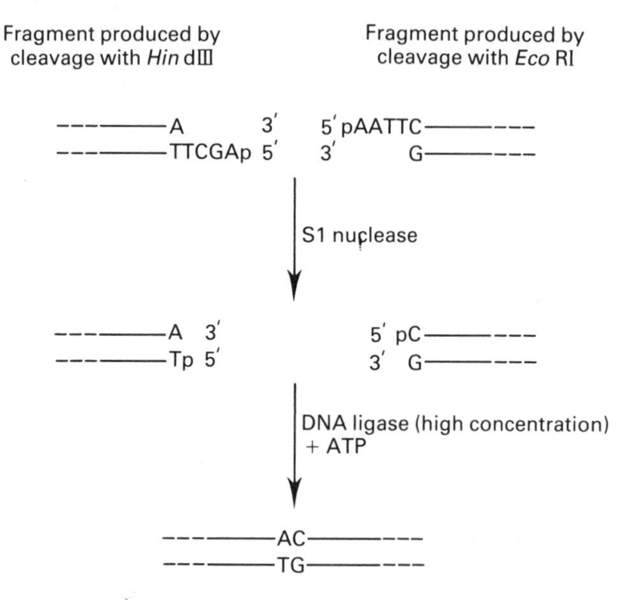

Fig. 5.21 Production and ligation of blunt ends. Different blunt-ended fragments cannot be held together temporarily by base-pairing between cohesive ends, and so both DNA and ligase must be used at high concentrations to increase the chances of two DNA fragments occupying the ligase active site simultaneously.

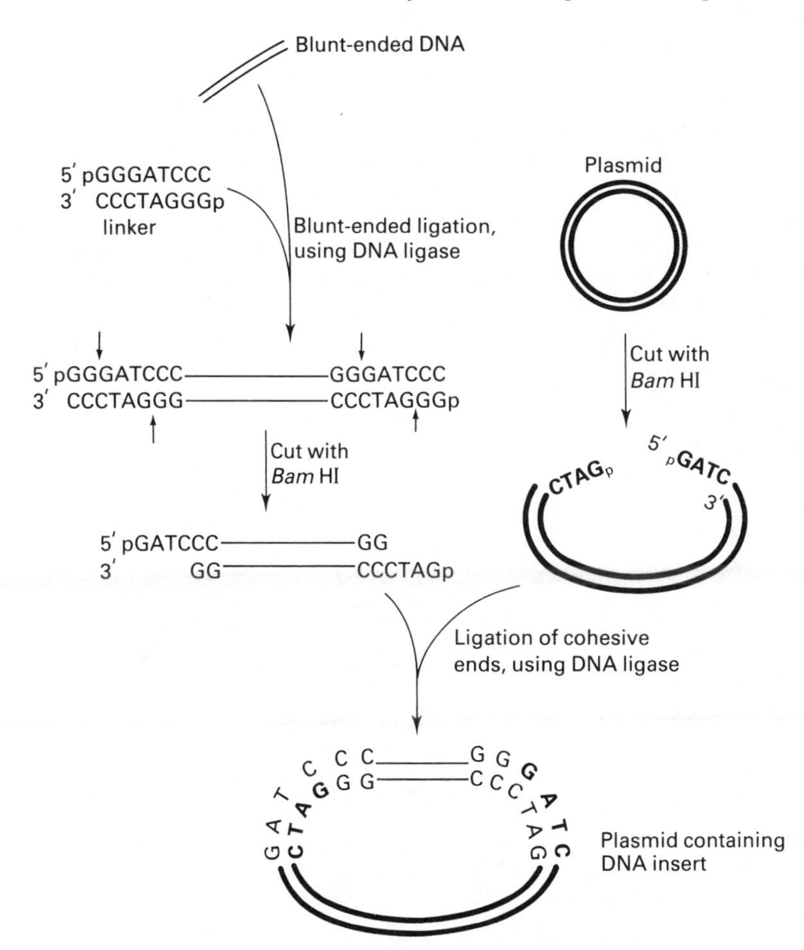

Fig. 5.22 Use of linkers. In this example blunt-ended DNA is inserted into a specific restriction site on a plasmid, after ligation to a linker containing the same restriction site.

linkers are available which contain restriction sites for several different enzymes within a sequence of only eight to ten nucleotides (Fig. 5.23).

Using a technique called *homopolymer tailing*, sticky ends can be built up on blunt-ended molecules (Fig. 5.24). For example, one preparation of DNA could be treated with the enzyme terminal transferase in the presence of dATP, resulting in the addition of a poly (dA) chain to the 3′ end of each strand. The other preparation would then be given 3′ tails of poly (T), using the same enzyme with TTP. On mixing, there would be base pairing between complementary sticky ends, which could then be ligated. An attractive feature of this method is that ligation will not occur between fragments from the same preparation.

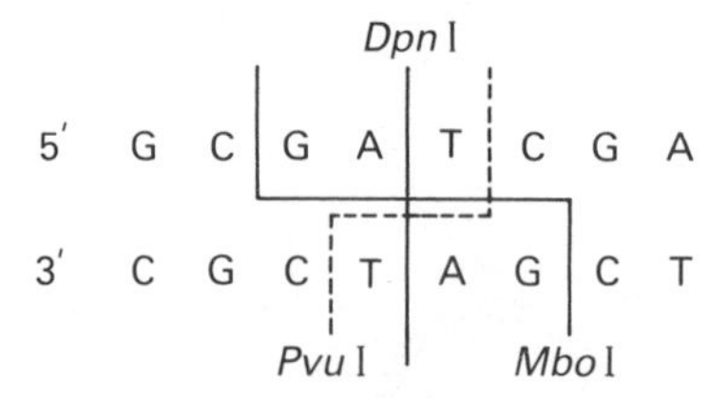

Fig. 5.23 A versatile linker. Cleavage sites of three different restriction enzymes are marked. Recognition sequences are: *Dpn* I GA↓TC; *Mbo* I ↓GATC; *Pvu* I CGAT↓CG. Thus *Dpn* I generates blunt-ended fragments, whilst *Mbo* I and *Pvu* I produce fragments with cohesive ends.

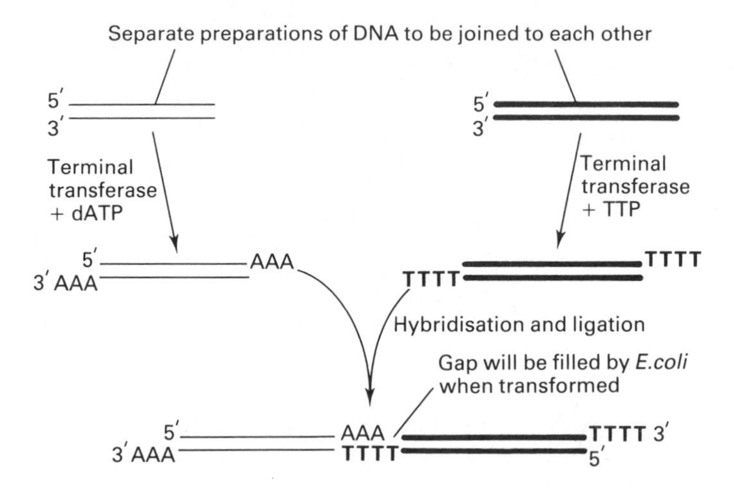

Fig. 5.24 Homopolymer tailing. One DNA preparation is given poly(dA) tails at its 3′ ends, the other receives poly(T) tails. These tails will hybridise with each other when the two preparations are mixed, and ligation can be used to stabilise this association. If the tails are long enough, hybridisation will be so stable that ligation is not needed.

5.8 Cloning vectors

5.8.1 Plasmids

By cloning, we can produce unlimited amounts of any particular fragment of DNA. In principle, the DNA is introduced into a suitable *host cell*, most usually a bacterium such as *Escherichia coli*, where it is replicated as the cell grows and divides. However, replication will only occur if the DNA contains a sequence which is recognised by the cell as an *origin of replication*. Since such sequences are infrequent, this will rarely be so, and therefore the DNA to be cloned has to be attached to a *carrier*, or *vector* DNA which does contain an origin of replication. Many bacteria contain such a piece of DNA, called a *plasmid*, which is a relatively small, circular, extrachromosomal molecule, carrying genes for such properties as antibiotic resistance, conjugation and the metabolism of unusual substrates. Some of these plasmids are replicated at a high rate by bacteria, and so they are excellent potential vectors. Starting from a selection of natural plasmids, *artificial plasmids* have been constructed as vectors, by a complex series of cutting and joining reactions.

One of the most widely used plasmids, called pBR322, illustrates the desirable features that have been incorporated into these vectors (Fig. 5.25).

(i) The plasmid is much smaller than a natural plasmid, since this makes it more resistant to damage by shearing, and increases the efficiency of uptake by bacteria during *transformation* (see below).

(ii) A bacterial origin of DNA replication ensures that the plasmid will be replicated by the host cell. Some replication origins display *stringent* regulation of replication, in which rounds of replication are initiated at the same frequency as cell division. Most plasmids, including pBR322, have a *relaxed* origin of replication, whose activity is not tightly linked to cell division, and so plasmid replication can be initiated far more frequently than chromosomal replication. Under suitable conditions a large number of plasmid molecules will be produced per cell.

(iii) Two genes coding for resistance to antibiotics have been included. One of these allows the selection of cells which contain plasmid: if cells are plated on medium containing an appropriate antibiotic, only those which contain plasmid will grow to form colonies. The other resistance gene can be used, as described below, for detection of those plasmids which contain inserted DNA.

(iv) There are single restriction sites for a number of enzymes, scattered

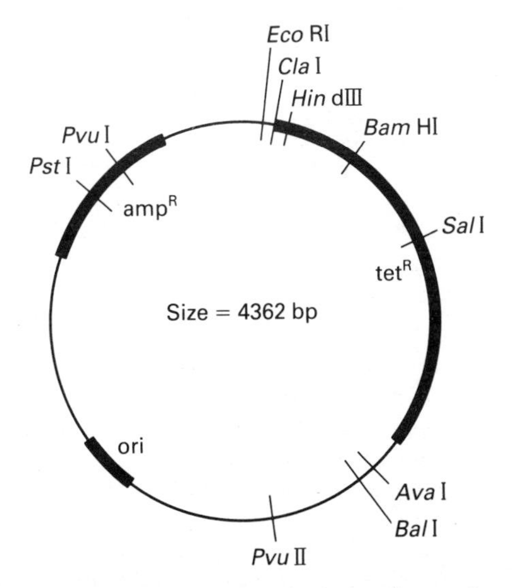

Fig. 5.25 The plasmid pBR322: amp^R and tet^R are genes for resistance to ampicillin and tetracycline, respectively; ori is the origin of DNA replication. Some unique sites of cleavage by a selection of restriction endonucleases are indicated (e.g. *Eco*R I, *Cla* I).

around the plasmid, which can be used to open the circle at a specific point prior to insertion of a piece of DNA to be cloned. The variety of sites not only makes it easier to find a restriction enzyme which is suitable for both vector and inserted DNA, but, since some of the sites are placed within an antibiotic resistance gene, the presence of an insert can be detected by loss of resistance to that antibiotic.

The value of pBR322 can be illustrated by considering its use for cloning a fragment of DNA. It is assumed that this fragment has been produced by cleavage of a larger molecule with the restriction enzyme *Bam* HI, followed by separation of the different fragments using gel electrophoresis, and recovery of the desired fragment from the gel by crushing or electroelution (Section 5.5.1). The plasmid would also be cut at a single site, using *Bam* HI, and both samples would then be deproteinised to inactivate the restriction enzyme. *Bam* HI cleaves to give sticky ends, and so it is easy to obtain ligation between plasmid and fragment, using T4 DNA ligase. The products of this ligation will include plasmid containing a single fragment of the DNA as an insert, but there will also be unwanted products, such as plasmid which has recircularised without an insert, dimers of plasmid, fragments joined to each

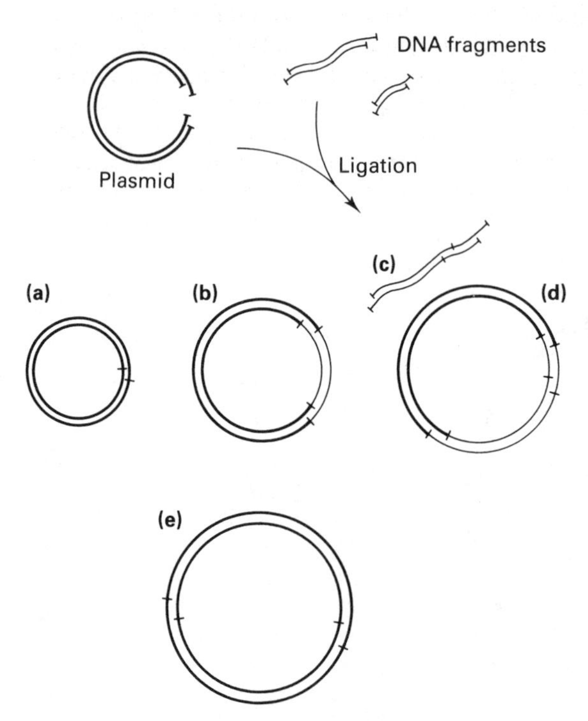

Fig. 5.26 Products of ligation. When linearised plasmid is mixed with DNA fragments in the presence of DNA ligase, there will be a mixture of products consisting of **(a)** recircularised plasmid; **(b)** recircularised plasmid containing a single DNA insert; **(c)** fragments of DNA joined together; **(d)** recircularised plasmid containing an insert made up of more than one fragment; **(e)** two plasmids joined together and recircularised to give a dimer.

other, and plasmid with an insert composed of more than one fragment (Fig. 5.26). Most of these unwanted molecules can be eliminated during subsequent steps.

The ligated DNA must now be used to *transform E. coli*. Bacteria do not normally take up DNA from their surroundings, but can be induced to do so by prior treatment with Ca^{2+} in the cold; they are then said to be *competent*, since DNA added to the suspension of competent cells will be taken up during a mild heat shock. Small, circular molecules are taken up most efficiently, whereas long, linear molecules will not enter the bacteria.

After a brief incubation, the cells are plated onto medium containing the antibiotic ampicillin. Colonies which grow on these plates must be derived from cells which contain plasmid, since this carries the gene for resistance to ampicillin; but it is not possible to distinguish between those containing inserts and those which are simply recircularised. To do this, the colonies are *replica plated*, using a sterile velvet pad, onto plates containing tetracycline in their medium (Fig. 5.27). Since the *Bam* HI site lies within the tetracycline resistance gene, this gene will be inactivated by the presence of insert, but will be intact in those plasmids which have merely recircularised. Thus colonies which grow on ampicillin but not on tetracycline must contain plasmids with inserts. Because replica plating gives an identical pattern of colonies on both sets of plates, it is easy to recognise the colonies with inserts, and to recover them from the ampicillin plate for further growth. This illustrates the importance of a second gene for antibiotic resistance in a vector.

Although recircularised plasmid can be selected against, its presence decreases the yield of recombinant plasmid containing inserts. If the cut plasmid is treated with alkaline phosphatase prior to ligation, recircularisation

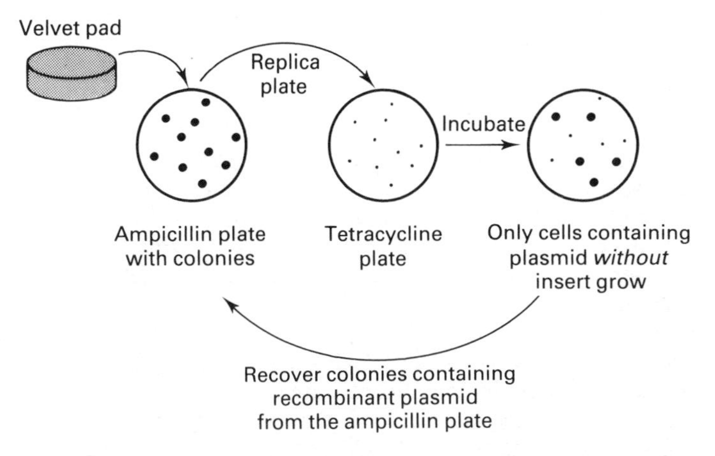

Fig. 5.27 Replica plating to detect recombinant plasmids. A sterile velvet pad is pressed onto the surface of an agar plate, picking up some cells from each colony growing on that plate. The pad is then pressed onto a fresh agar plate, thus inoculating it with cells in an identical pattern to that of the original colonies. Clones of cells which fail to grow on the second plate (e.g. owing to the loss of antibiotic resistance) can be recovered from their corresponding colonies on the first plate.

will be prevented, since the enzyme removes the 5′-phosphoryl groups which are essential for ligation. Links can still be made between the 5′-phosphoryl of insert and the 3′-hydroxyl of plasmid, so only recombinant plasmids and chains of linked fragments will be formed. It does not matter that only one strand is ligated, since the nick will be repaired by bacteria transformed with these molecules.

Both prior to, and after cloning, plasmid must be isolated from the bacteria. It is possible to obtain a DNA preparation highly enriched for plasmid by very gentle cell lysis, using lysozyme and then detergent, followed by *clearing of the lysate* by centrifugation. Centrifugation sediments the high molecular weight DNA (predominantly chromosomal) and cell debris, leaving the small plasmid molecules and RNA in the supernatant. Undamaged plasmid is particularly compact, since it is *supercoiled* as a consequence of having slightly too few turns of the double helix per unit length. Such supercoiling can easily be demonstrated by attempting to unwind a piece of string which is clamped at one end. Other methods rely on the preferential denaturation of linear DNA by heat or alkali, using centrifugation to remove it from the circular plasmid.

Further purification of the plasmid can be achieved by caesium chloride density gradient ultracentrifugation of the nucleic acid preparation in the presence of ethidium bromide (Section 2.7.2). Ethidium bromide causes unwinding of DNA as it binds to it, simultaneously producing a decrease in its buoyant density. Since the supercoiled plasmid DNA can unwind to only a very limited extent, it will not bind as much dye as will the linear and open circle forms of DNA; hence supercoiled plasmid will have a higher density than the other types of DNA in the presence of saturating levels of ethidium bromide. Because of this density difference, plasmid DNA can be separated from other DNA by isopycnic ultracentrifugation (Section 2.4.6).

5.8.2 Viral DNA

The cloning of single genes is usually best carried out using plasmids, since the insert will rarely be larger than about 2 kb. As will be shown later, there are several reasons for wanting to clone much larger pieces of DNA, particularly when constructing *gene libraries* (Section 5.9.2) of higher eukaryote genomes. Large inserts increase plasmid size to the point at which efficient transformation cannot occur, and so another way must be found to get recombinant DNA into the bacterial cells.

One way of doing this is by using the DNA of *bacteriophages* (bacterial viruses) as a vector, since these large molecules are injected into bacteria by the viral particles. A commonly used vector is that of the phage λ, which is 49 kb in length. For the cloning of long DNA fragments, up to about 20 kb much of the inessential λDNA is removed and replaced by the insert. The recombinant DNA is then packaged within viral particles *in vitro*, and these are allowed to infect bacterial cells which have been plated out on agar (Fig. 5.28). Since the DNA is injected into the cells, a very high efficiency of

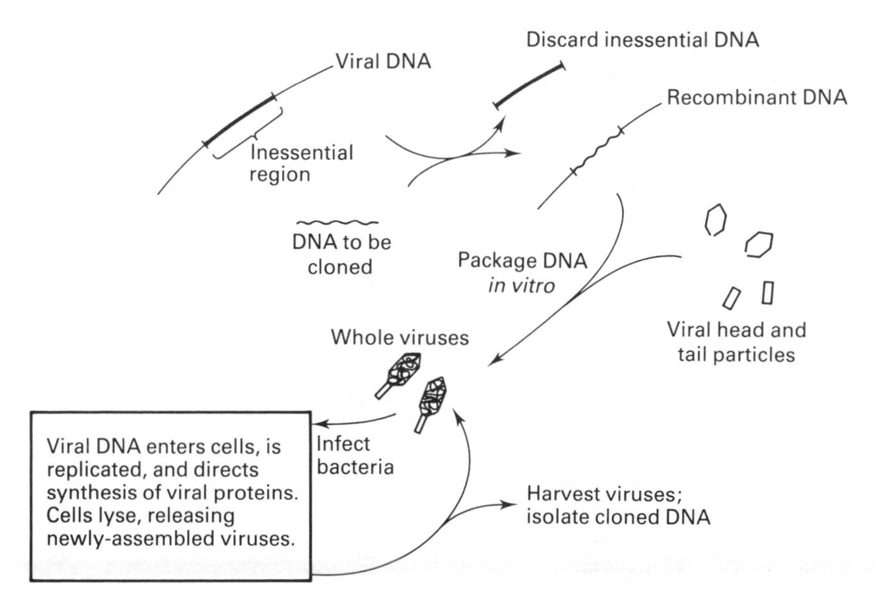

Fig. 5.28 Cloning DNA in the bacteriophage λ. Inessential DNA is cut out of the λ DNA using restriction enzymes, and is separated from essential DNA fragments by electrophoresis. Essential DNA and DNA to be cloned are ligated and packaged in empty head particles, and tails are added to give infective viruses. See Fig. 5.29 for details of packaging.

transformation can be obtained. Once inside the cells, the recombinant viral DNA is replicated. All the genes needed for normal lytic growth are still present in the DNA, and so multiplication of the virus takes place by cycles of cell lysis and infection of surrounding cells, giving rise to *plaques* of lysed cells on a background, or *lawn*, of bacterial cells. Cloned DNA can be recovered from the viruses in these plaques.

5.8.3 Cosmids

Even longer fragments of DNA must be cloned for the analysis of highly complex genomes, and in order to understand how this can be achieved, it is necessary to know, in outline, how viral DNA is packaged. Viral DNA is injected into the cell as a linear molecule, at each end of which are cohesive ends, complementary to each other, 12 bases in length. Once inside the cell, these ends base pair and become permanently joined by ligation to form a region known as the *cos* site, giving a circular DNA molecule (Fig. 5.29). Replication of the DNA by a rolling circle mechanism gives rise to a *concatamer*, which is a long molecule made up of many copies of the viral DNA, linked end to end through the *cos* sites. DNA is packaged by looping regions between *cos* sites into the precursor of the viral head; when a head is full, the *cos* sites should be at the mouth of the head, where they will be

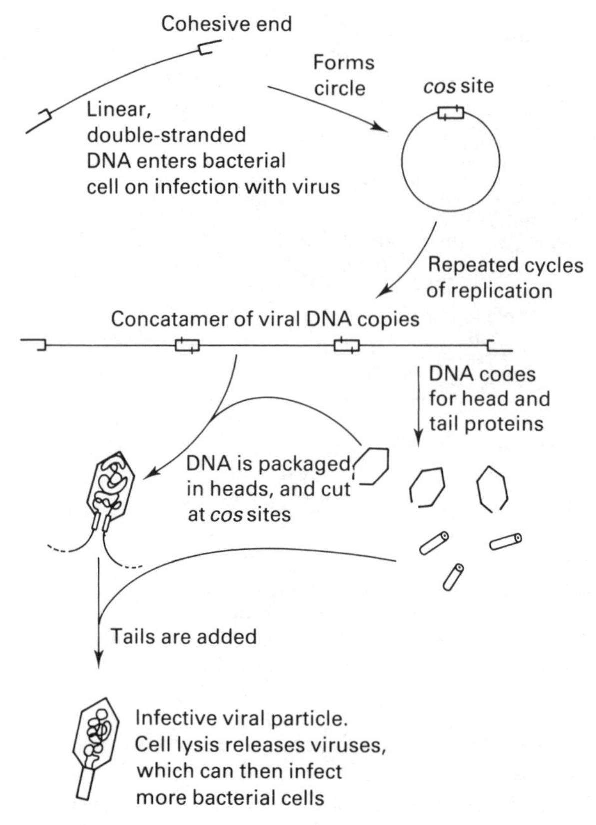

Fig. 5.29 Packaging of λDNA.

cleaved to generate a linear molecule with cohesive ends. Subsequently, tail proteins are added to give infective particles.

Thus the only requirement for a length of DNA to be packaged into viral heads is that it should contain *cos* sites spaced the correct distance apart; in practice this spacing can range between 37 and 52 kb. Consequently vectors called *cosmids* have been constructed, which contain a *cos* site plus essential features of a plasmid, such as the plasmid origin of replication, a gene for drug resistance, and several unique restriction sites for insertion of DNA to be cloned (Fig. 5.30). When a cosmid preparation is linearised by restriction, and ligated to DNA for cloning, the products will include concatamers of alternating cosmid and insert. Such DNA can be packaged *in vitro* if bacteriophage head precursors, tails and packaging proteins are provided, so long as the *cos* sites are spaced between 37 and 52 kb apart. Since the cosmid is very small, inserts about 40 kb in length will be most readily packaged. Once inside the cell, the DNA recircularises through its *cos* site, and from then onwards behaves exactly like a plasmid.

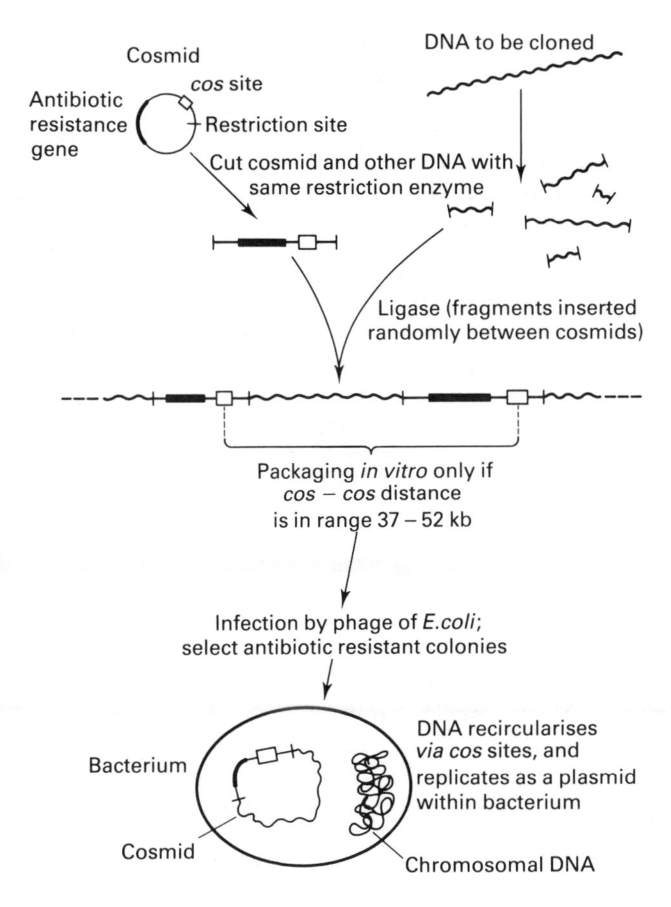

Fig. 5.30 Cloning in a cosmid. Note that in eukaryotic systems upper case letters are used to denote sites (i.e. *COS*).

5.8.4 Vectors used in eukaryotes

Plasmids are also used for cloning DNA in eukaryotic cells, but they need a eukaryotic origin of replication and marker genes which will be expressed by eukaryotic cells. At present the two most important applications of plasmids to eukaryotic cells are for cloning in yeast and in plants.

Although yeast has a natural plasmid, called the *2μ circle*, this is too large for use in cloning. Plasmids have been created by genetic manipulation, sometimes using replication origins from the 2μ circle, and usually incorporating a gene which will complement a defective gene in the host yeast cell. If, for example, a strain of yeast is used which has a defective gene for the biosynthesis of an amino acid, an active copy of that gene on a yeast plasmid can be used as a selectable marker for the presence of that plasmid. Yeast, like bacteria, can be grown rapidly, and it is therefore well suited for use in cloning. Moreover, it is not pathogenic, and can carry out such *post-translational modifications* of the polypeptide as glycosylation and limited proteolysis. Such modifications are sometimes required for the activation or export from the cell of a polypeptide, and so yeast is particularly attractive for the expression of cloned genes on an industrial scale.

The bacterium *Agrobacterium tumefaciens* infects plants which have been damaged near soil level, and this infection is often followed by the formation of plant tumours in the vicinity of the infected region. It is now known that *A. tumefaciens* contains a plasmid called the *Ti plasmid*, part of which is transferred into the nuclei of plant cells which are infected by the bacterium. Once in the nucleus, this DNA is maintained by integrating with the chromosomal DNA. The integrated DNA carries genes for the synthesis of opines (which are metabolised by the bacteria, but not by the plants) and for *tumour induction* (hence Ti). DNA inserted into the correct region of the Ti plasmid will be transferred to infected plant cells, and in this way it has been possible to clone and express foreign genes in plants. This is an essential prerequisite for the genetic engineering of crops.

5.9 Isolation of specific nucleic acid sequences

5.9.1 Complementary DNA

The most difficult part of genetic manipulation is not the cloning of DNA, but isolation of the particular piece of DNA to be cloned. If the aim is to clone a gene, then it is enormously helpful to have as much information as possible about the gene product. Usually this product is a protein, and ideally antibodies to the protein should be available for detection or precipitation of the protein or its precursors (see below); but even knowledge of its molecular weight can be of use.

Frequently an attempt is made to isolate the mRNA transcribed from a desired gene. If this codes for a major protein of the cell, it should form a major fraction of the total mRNA, as in the case of B cells of the pancreas, which contain high levels of proinsulin mRNA. It is sometimes possible to precipitate polysomes which are translating the mRNA, by using antibodies to its protein; mRNA can then be dissociated from the precipitated ribosomes. More usually, the mRNA is only a minor component of the total cellular mRNA. In such cases, total mRNA is fractionated by size, using sucrose density gradient centrifugation (Section 2.6). Then each fraction is used to direct the synthesis of proteins, using an *in vitro translation system* derived from lysates of rabbit reticulocytes or from wheat germ extracts. Immunoprecipitation (Section 4.3) or polyacrylamide gel electrophoresis (Section 7.6) can then be used to detect the target protein amongst the many other products.

When a fraction containing the desired mRNA has been identified, it is used to direct the synthesis of DNA molecules complementary to all of the mRNAs in that fraction. This cDNA (*complementary DNA*) is made using

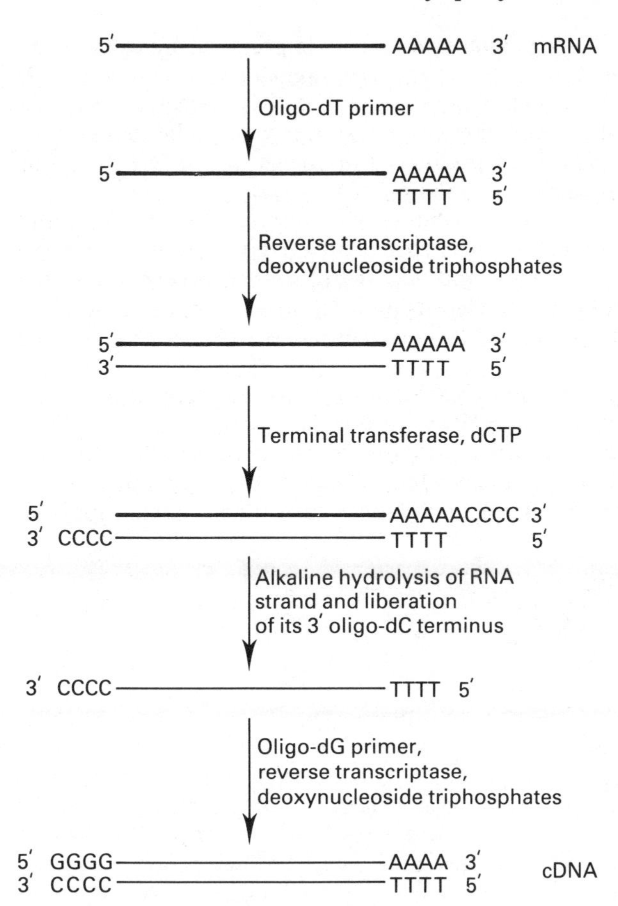

Fig. 5.31 Synthesis of cDNA. Although each mRNA molecule results in the formation of only one cDNA molecule, the cDNA can be cloned, using a plasmid vector, to give an unlimited number of copies.

the enzyme reverse transcriptase, as shown in Fig. 5.31. Reverse transcriptase will synthesise a DNA strand complementary to an mRNA template, using a mixture of the four deoxyribonucleoside triphosphates, provided that a short length of 'primer' is base-paired with the 3′ end of the RNA. Since mRNA has a poly(A) tail at its 3′ end, a short oligo-dT molecule will act as the primer for reverse transcriptase. After synthesis of the first DNA strand, a poly(dC) tail is added to its 3′ end, using terminal transferase and dCTP. This will also, incidentally, put a poly(dC) tail on the poly(A) of mRNA. Alkaline hydrolysis is then used to remove the RNA strand, leaving single-stranded DNA which can be used, like the mRNA, to direct the synthesis of a complementary DNA strand. The second-strand synthesis requires an oligo-dG primer, base-paired with the poly(dC) tail, and it can be catalysed by the

Klenow fragment of DNA polymerase I. This is prepared by cleavage of DNA polymerase with subtilisin, giving a large fragment which has no 5′ to 3′ exonuclease activity, but which still acts as a 5′ to 3′ polymerase. Surprisingly, since the template is now DNA, the reaction can also be catalysed by reverse transcriptase. The final product is double-stranded DNA, one of whose strands is complementary to the mRNA.

The mixture of cDNA molecules can now be inserted into plasmids and used to transform bacteria, which are then grown to give colonies. Since the cloned cDNAs lack promoters, and will therefore not be expressed, the wanted sequence cannot be detected using antibodies to its corresponding protein. Consequently, a rather devious method is used to pick out the sequence.

Plasmid is extracted from part of each colony, and each preparation is then denatured and immobilised on a nitrocellulose filter (Fig. 5.32). The filters are soaked in total cellular mRNA, under *stringent conditions* (temperature only a few degrees below T_m) in which hybridisation will occur only between complementary strands of nucleic acid. Hence each filter will bind just one

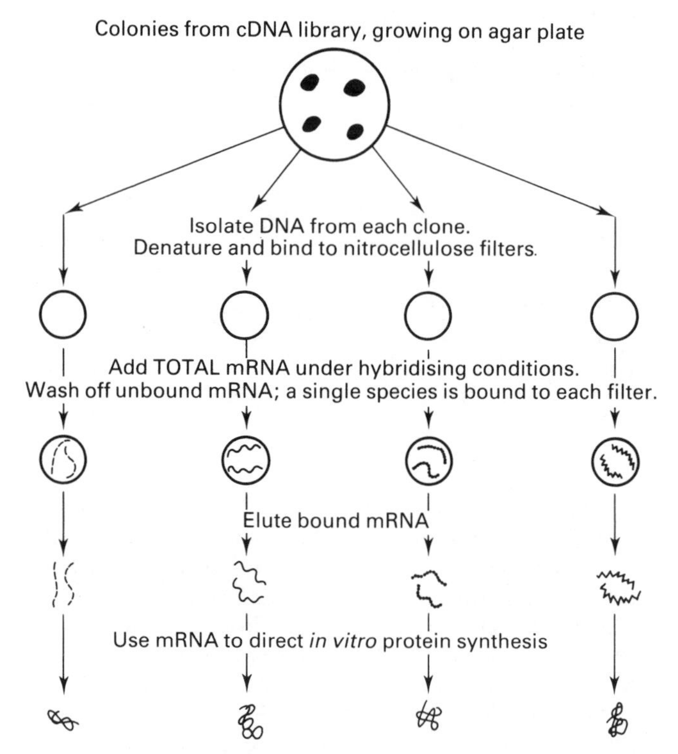

Fig. 5.32 Hybrid release translation.

species of mRNA, since it has only one type of cDNA immobilised on it. Unbound mRNA is washed off the filters, and then the bound mRNA is eluted and used to direct translation *in vitro*. By immunoprecipitation or electrophoresis of the translation products, the mRNA coding for a particular protein can be detected, and the clone containing its corresponding cDNA isolated. This technique is known as *hybrid release translation*. In a related method called *hybrid arrested translation* a positive result is indicated by the *absence* of a particular translation product when total mRNA is hybridised with cDNA. This is a consequence of the fact that mRNA cannot be translated when it is hybridised to another molecule.

In some cases the cDNA is all that needs to be cloned; as, for example, when the aim is to persuade bacteria to synthesise a foreign protein. The cDNA sequence can be inserted into an *expression vector* (Section 5.10) and should then result in production of the desired protein. It should be noted that cDNA will often not be identical to the original gene, if the gene is eukaryotic, since the majority of such genes contain introns (Section 5.3.3) which interrupt their coding regions (exons). During maturation of mRNA the introns are excised from the molecule, leaving only the exons spliced together; it is this spliced molecule which is used to make cDNA. Genes also have regions flanking them which are of importance in the regulation of their expression, and which are not transcribed as part of the mRNA. When a complete gene must be isolated, cDNA can be used as a *probe* to search through a *gene library* for the desired gene.

5.9.2 Gene libraries

Gene libraries are constructed by isolating the complete genomic DNA from a cell, and cutting it almost randomly into fragments of the desired average length. This can be achieved by partial restriction with an enzyme which recognises tetranucleotide sequences. Complete restriction with such an enzyme would produce a large number of very short fragments (Section 5.7.1), but, if the enzyme is allowed to cleave only a few of its potential restriction sites before the reaction is stopped, each DNA molecule will be cut almost randomly into relatively large fragments. Average fragment size will depend on the concentrations of DNA and restriction enzyme, and on the conditions and length of incubation.

The mixture of fragments is ligated with a vector, and cloned. If enough clones are produced there will be a very high chance that any particular gene will be present in at least one of the clones. Such a collection of clones is known as a gene library. To keep the number of clones to a manageable size, fragments about 10 kb in length are needed for prokaryotic libraries, but the length must be increased to about 40 kb for mammalian libraries.

5.9.3 Colony hybridisation

The technique of *colony hybridisation* is used to pull a particular gene out of a gene library (Fig. 5.33). A large number of clones is grown up to form

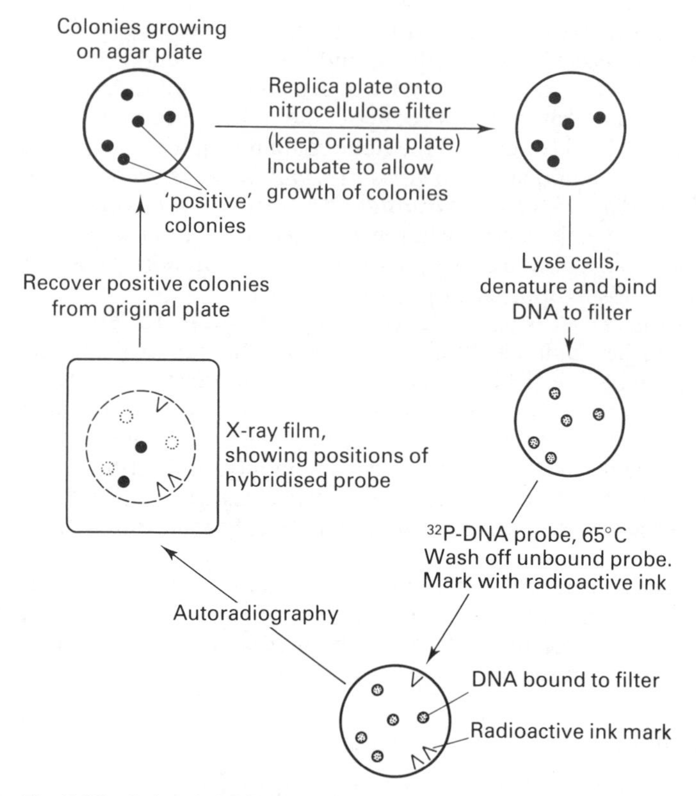

Fig. 5.33 Colony hybridisation.

colonies on one or more plates, and these are then replica plated onto nitrocellulose filters placed on solid agar medium. Nutrients diffuse through the filters and allow colonies to grow on them. The colonies are then lysed, and liberated DNA is denatured and bound to the filters, so that the pattern of colonies is replaced by an identical pattern of bound DNA. If the filters are incubated with denatured, radioactive cDNA under hybridising conditions, the cDNA will bind only to cloned fragments containing at least part of its corresponding gene. Such binding can be detected by autoradiography of the washed filters (Section 9.2.4). It is worth noting that the introns of a gene do not interfere with its hybridisation to cDNA, since there is a strong interaction between the exons and the cDNA. By comparison of the autoradiograms with the original plates of colonies, those which contain the desired gene (or part of it) can be identified and used for gene isolation.

5.9.4 Nick translation

Radioactive labelling of cDNA is most easily carried out by *nick translation*, in which DNA polymerase I is used to make single strand nicks in the DNA,

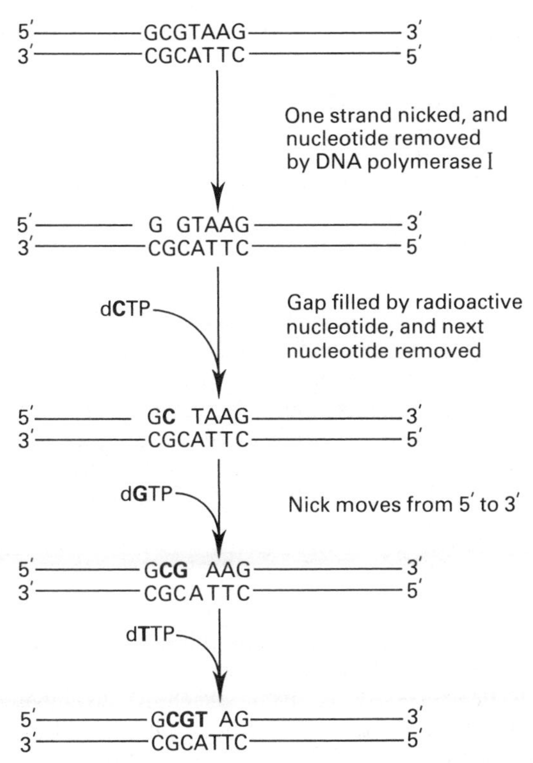

Fig. 5.34 Nick translation. If one (or more) of the deoxynucleoside triphosphates supplied is radioactive, the DNA will become progressively more highly radioactive as nick translation proceeds.

which it then fills in, using an appropriate deoxynucleoside triphosphate (dNTP), at the same time making a new nick to the 3′ side of the previous one (Fig. 5.34). In this way the nick is *translated* along the DNA. If radioactive dNTPs are added to the reaction mixture, they will be used to fill the nicks, and so the DNA can be labelled to a very high specific radioactivity.

5.9.5 Oligonucleotide probes

If the amino acid sequence of a protein is known, there is no need to prepare a cDNA probe for its gene. From our knowledge of the genetic code, we can predict the DNA sequences that would code for the protein, and can then synthesise appropriate oligonucleotide sequences chemically. Since most amino acids are coded for by more than one codon, there will be more than one possible nucleotide sequence which could code for a given polypeptide (Fig. 5.35). The longer the polypeptide, the greater the number of possible oligonucleotides which must be synthesised. Fortunately, there is no need to

Polypeptide:		Phe	Met	Pro	Trp	His	
Corresponding		T		T		T	
nucleotide	5′	TTC	ATC	CCC	TGG	CAC	3′
sequences:				A			
				G			

Fig. 5.35 Oligonucleotide probes. Note that only methionine and tryptophan have unique codons. It is impossible to predict which of the indicated codons for phenylalanine, proline and histidine will actually be present in the gene to be probed, and so all possible combinations must be synthesised (16 in the example above).

synthesise a sequence longer than about 20 bases, since this should hybridise efficiently with any complementary sequences, and should be specific for one gene. Ideally, a section of the protein should be chosen which contains as many tryptophan and methionine residues as possible, since these have unique codons, and there will therefore be fewer possible base sequences which could code for that part of the protein. The synthetic oligonucleotides can then be used as probes in a colony hybridisation, as described for cDNA.

5.10 Expression of genes

One of the main purposes of genetic engineering is to obtain the expression within bacterial cells of a *foreign gene* which codes for some valuable polypeptide. However, for a gene to be expressed in a bacterial cell, it must have particular sequences of bases forming a promoter upstream of the coding region, to which the RNA polymerase will bind prior to transcription of the gene (Section 5.3.3). It must also contain a Shine-Dalgarno sequence, placed just before the coding region, which is transcribed and then acts as a ribosome binding site at the start of translation (Section 5.3.4). Unless a cloned gene contains both these sequences, it will not be expressed in its bacterial host cell. If the gene has been produced via cDNA from a eukaryotic cell, then it will certainly not have such sequences.

Consequently, *expression vectors* have been developed which contain promoter and Shine-Dalgarno sequences sited just before one or more restriction sites for the insertion of foreign DNA. These regulatory sequences, such as that from the *lac* operon of *E. coli*, are usually derived from genes which, when induced, are strongly expressed in bacteria. Since the mRNA produced from the gene is read as triplet codons, the inserted sequence must be placed so that its reading frame is in phase with the regulatory sequence. This can be ensured by the use of three vectors which differ only in the number of bases between promoter and insertion site, the second and third vectors being respectively one and two bases longer than the first. If an insert is cloned in all three vectors it is bound to be in the correct reading frame in one of them. The resulting clones can be screened, using antibodies, enzyme assays, etc., for the production of a functional foreign protein; approximately a third of the colonies should be positive.

It is not only possible, but usually essential to use cDNA instead of a eukaryotic gene to direct the production of a functional protein by bacteria. This is because bacteria are not capable of *processing* RNA to remove introns, and so any foreign genes must be *pre-processed* as cDNA if they contain introns. A further problem arises if the protein must be *glycosylated*, by the addition of oligosaccharides at specific sites, in order to become functional. Again, this is something that bacteria are not equipped to carry out. Yeast cells can perform such *post-translational modifications*, producing a glycosylation pattern that is usually adequate, if not indentical to that which would be produced in animal cells. Although yeast genes often contain introns, which are spliced out at the RNA stage, yeast is not able to process the introns of other eukaryotes, and so it is still necessary to use cDNA when expression of foreign genes in yeast is needed.

5.11 Safety of cloning procedures

As soon as DNA cloning became possible, fears were expressed about the dangers of the process. An obvious concern was that the gene for some toxic polypeptide could be cloned in a bacterium, which could then escape from the laboratory and either infect humans itself, or transfer the cloned DNA to other organisms which were pathogenic, and which might express the cloned gene in humans. Even more worrying was the possibility that a piece of DNA which was in some way able to cause cancer might be similarly transferred to humans. Similar fears concerned the inadvertent transfer of harmful genes into animals and crops. Consequently, very strict guidelines were drawn up, which imposed a ban on the cloning of viral or tumour DNA, and which demanded high levels of physical *containment* in most other cases.

Fortunately, it was not long before *safe host cells and plasmids* were developed, with features which made it highly unlikely that the cells could survive outside the laboratory, and which prevented any transfer of DNA into other organisms. Experience has also shown how difficult it is to obtain expression of cloned DNA, and so the chances of expression of a rogue gene occurring spontaneously are very low. Thus it is now accepted that, provided containment of microorganisms is at a level in keeping with their pathogenicity and that of any genes being deliberately cloned, genetic manipulation should not present any new hazard. There is still debate about the release of genetically engineered organisms into the environment, for such purposes as the microbial degradation of pollutants, but it is likely that this will be allowed on a limited scale in the near future.

5.12 Applications of molecular biology

In terms of pure research, all the techniques of molecular biology, from cloning to sequencing, have revolutionised our understanding of gene structure and function. Promoter regions, control sites, ribosome binding sites, introns and other protein binding sites have been sequenced, and secondary structures have been postulated which could help to explain how these regions act. *Oncogenes* have been discovered through the application of molecular biology, and this has changed the pattern of research into the mechanisms and prevention of cancer. Since the physical characteristics of an individual depend largely on their DNA, and only identical twins will have identical genomes, analysis of DNA could be the ultimate method of fingerprinting an individual. Forensic scientists are developing methods, based on restriction analysis and hybridisation of DNA, for matching tissue samples unambiguously. This type of approach is also being used by taxonomists to help them work out the evolutionary relationships between organisms.

Over 500 inherited diseases, such as sickle-cell anaemia and β-thalassemias, are known to be caused by a mutation in a single gene. More than forty of these diseases can be detected early in the development of a foetus by enzyme assays carried out on cells sampled by amniocentesis. This process involves the insertion of a needle through the abdominal wall into the amniotic cavity, and withdrawal of some of the amniotic fluid. Suspended in the fluid are foetal cells, which can be isolated and cultured in the laboratory to provide sufficient material for diagnostic tests. Techniques such as *chorionic sampling* are being developed for even earlier sampling of foetal cells. A positive result allows an informed decision to be taken over abortion of the foetus.

Unfortunately such tests are applicable only to those cases where the gene is normally expressed in the sampled cells. Ideally, one would like to be able to screen the DNA itself for the mutations; but this would involve detecting a single base change in a genome of 6×10^9 bases. Surprisingly, this can be done in some cases. It seems that some diseases are almost always caused by a mutation at the same site in the gene, and this mutation may destroy a restriction site. In such cases, the DNA is restricted, run on a gel, Southern blotted, and the blot probed with radiolabelled, cloned gene or cDNA. This will reveal the positions of restriction fragments containing the gene, and the positions of such fragments will be altered by mutations which destroy a restriction site in the gene. Alternatively, a Southern blot can be hybridised with a chemically synthesised oligonucleotide, complementary to the region of the gene which is susceptible to mutation. Under stringent conditions (Section 5.9.1), strong hybridisation will occur only if the gene is not mutated, and so this provides a way of detecting mutations which do not destroy a restriction site.

Molecular biology has probably received most publicity in connection with genetic engineering. Already it has been possible to clone the genes for such polypeptides as human insulin, growth hormone, interferons, tumour necrosis factor, blood clotting factor VIII, and viral coat proteins (for vaccines) in bacteria in such a way that they are expressed, and the polypeptides can be recovered from the cell cultures. Genes have been altered *in vitro* to produce slightly altered enzymes with increased stability or different reaction kinetics; this is likely to be of great importance for the production of enzymes for use on an industrial scale.

In the long term it is anticipated that genetic engineering will be used by plant and animal breeders to introduce genes for such characteristics as disease resistance and improved yield, into crops and livestock. A major problem here is that we can rarely attribute such characteristics to a single gene, and, until our understanding of the biochemistry and physiology of plants and animals is more profound, we will not really know which genes to 'engineer'.

5.13 Suggestions for further reading

Mainwaring, W.I.P., Parish, J.H., Pickering, J.D. and Mann, N.H. (1982). *Nucleic Acid Biochemistry and Molecular Biology.* Blackwell Scientific Publications, Oxford. (A detailed and comprehensive textbook on molecular biology.)

Maniatis, T., Fritsch, E.F. and Sambrook, J. (1982). *Molecular Cloning.* Cold Spring Harbor, New York. (An invaluable source of practical details and recipes; found on the shelves of most molecular biology laboratories.)

Old, R.W. and Primrose, S.B. (1985). *Principles of Gene Manipulation,* 3rd edition. Blackwell Scientific Publications, Oxford. (An advanced textbook, including explanations of the latest techniques.)

Stryer, L. (1981) *Biochemistry,* 2nd edition. Freeman, San Francisco. (An excellent general biochemistry textbook, superbly illustrated.)

Watson, J.D., Tooze, J. and Kurtz, D.T. (1983). *Recombinant DNA: A Short Course.* Scientific American Books, Freeman, New York. (An exceptionally clear introduction to the subject.)

Walker, J.M. (Ed.) (1984). *Methods in Molecular Biology,* Vols 1 and 2. Humana, Clifton. (An extensive collection of procedures for use in protein and nucleic acid biochemistry.)

Walker, J.M. and Gaastra, W. (Eds) (1983). *Techniques in Molecular Biology.* Croom Helm, London. (Explains the principles behind key techniques in molecular biology.)

6

Chromatographic techniques

6.1 General principles and techniques

6.1.1 General principles

One of the problems continually facing biochemists is the separation, purification and identification of one or more biological compounds from a mixture of such compounds. One of the most convenient methods for achieving such separations is the use of *chromatographic techniques*. Such techniques can be used for the separation of large amounts (several grammes) or small amounts (picogramme quantities) of materials. The selection of a particular form of chromatography to achieve a separation is dependent on the material to be isolated, and often several chromatographic methods may be used sequentially to achieve the complete purification of a compound.

The basis of all forms of chromatography is the *partition* or *distribution coefficient* (K_d) which describes the way in which a compound distributes itself between two immiscible phases. For a compound distributing itself between equal volumes of two immiscible solvents A and B, the value for this coefficient is a constant at a given temperature and is given by the expression:

$$\frac{\text{concentration in solvent A}}{\text{concentration in solvent B}} = K_d$$

The distribution of a compound can, however, be described not only in terms of its distribution between two solvents, but also by its distribution between any two phases, such as solid/liquid or gas/liquid phases. Thus a distribution coefficient of a substance between silicic acid and benzene might be 0.5, which means that the concentration of the substance in benzene is twice that in the silicic acid.

The term *effective distribution coefficient* is defined as the total amount, as distinct from the concentration, of substance present in one phase divided by the total amount present in the other phase. It is in fact the distribution coefficient multiplied by the ratio of the volumes of the two phases present. If the distribution coefficient of a compound between two solvents A and B is 1, then, if this compound is equilibrated between 10 cm^3 of A and 1 cm^3 of B, the concentration in the two phases will be the same, but the total amount of the compound in solvent A will be 10 times the amount in solvent B.

Basically, all chromatographic systems consist of two phases. One is the *stationary phase* which may be solid, gel, liquid or a solid/liquid mixture which is immobilised. The second *mobile phase* may be liquid or gaseous and flows over or through the stationary phase. The choice of stationary or mobile phases is made so that the compounds to be separated have different distribution coefficients. This may be achieved by setting up:

(i) an adsorption equilibrium between a stationary solid and a mobile liquid phase (*adsorption chromatography*);
(ii) a partition equilibrium between a stationary liquid (or semi-liquid) and a mobile liquid phase (*countercurrent chromatography* and *partition chromatography*);
(iii) a partition equilibrium between a stationary liquid and a mobile gaseous phase (*gas-liquid chromatography*);
(iv) an ion-exchange equilibrium between an ion-exchange resin stationary phase and a mobile electrolyte phase (*ion-exchange chromatography*);
(v) an equilibrium between a liquid phase inside and outside a porous structure or molecular sieve (*exclusion chromatography*);
(vi) an equilibrium between a macromolecule and a small molecule for which it has a high biological specificity and hence affinity (*affinity chromatography*).

The principle of separation may be depicted by considering a column packed with a solid granular stationary phase to a height of 5 cm surrounded by the mobile liquid phase of which there is 1 cm^3 per cm of column (Fig. 6.1). If 32 μg of a compound is added to the column in 1 cm^3 of solvent then as this 1 cm^3 moves onto the column to occupy position A, 1 cm^3 of solvent will leave the base of the column. If the compound added has an effective distribution coefficient of 1, it will distribute itself equally between the solid and liquid phases. If a further 1 cm^3 of solvent is introduced onto the column, the solvent in section A will move down to B taking 16 μg of the compound with it, leaving 16 μg at A. At both A and B a redistribution of the compound will occur so that there is 8 μg in the solvent and 8 μg in the solid phase. The addition of a further 1 cm^3 of solvent to the column displaces the solvent in A to B and that in B to C giving the distribution of the compound as shown in stage 3. Addition of a further 1 cm^3 of solvent leads to the distribution shown at stage 4, and a further 1 cm^3 aliquot to the situation at stage 5.

It is apparent that after five *equilibrations* the compound is distributed throughout the whole column, but is maximally concentrated at the centre of the column. If a compound had an effective distribution coefficient of less than 1, more than 50% of the compound would be left on the solid phase after each equilibration. Although after five equilibrations some of the compound would be present throughout the column, the concentration peak would be above the centre of the column. Alternatively, for a compound with an effective distribution of greater than 1, the concentration peak after five equilibrations would be below the centre of the column.

The greater the number of equilibrations that occur on a column, the

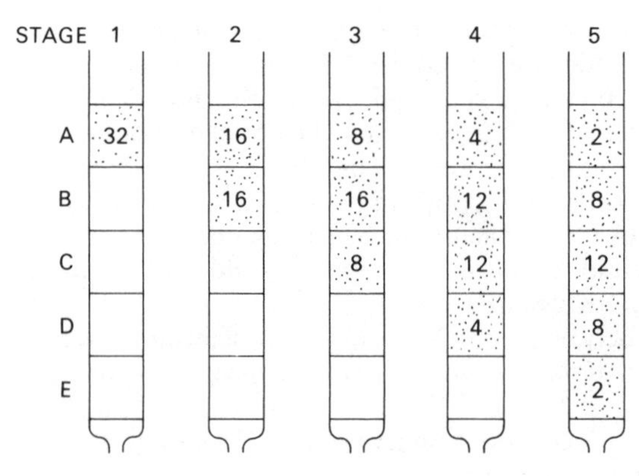

Fig. 6.1 Principle of column chromatographic separation.

greater becomes the concentration of the compound on a certain part of the column. There are, therefore, two important factors which influence the pattern of separation (*resolution*) of a mixture of compounds. The rate of progress of a compound through the column depends on its effective distribution coefficient, and the sharpness of the compound band on the column depends on the number of equilibrations that have taken place.

In a real situation, equilibration occurs continuously on a column since the solvent is being continuously added and, in normal working columns, thousands of equilibrations take place. Chromatography columns are considered to consist of a number of adjacent zones in each of which there is sufficient space for the solute to achieve complete equilibrium between the mobile and stationary phases. Each zone is called a *theoretical plate* and its length in the column is called the *plate height* (H) which has dimensions of length. The more efficient the column, the greater the number of theoretical plates that are involved. The way in which the number of theoretical plates (N) affects the distribution of a solute with an effective distribution coefficient of 1 is shown in Fig. 6.2.

In practice chromatographic separations may take one of three modes: *column chromatography* in which the stationary phase is packed into glass or metal columns; *thin layer chromatography* in which the stationary phase is thinly coated onto glass, plastic or foil plates; and *paper chromatography* in which the stationary phase is supported by the cellulose fibres of a paper sheet. Each of these three modes of chromatography have their specific advantages, applications and method of operation.

6.1.2 Column chromatography

All of the major types of chromatography are routinely carried out using the column mode. The apparatus and general techniques used for column adsorption, partition, ion-exchange, exclusion and affinity chromatography have much in common and are discussed below. Points specific to any of

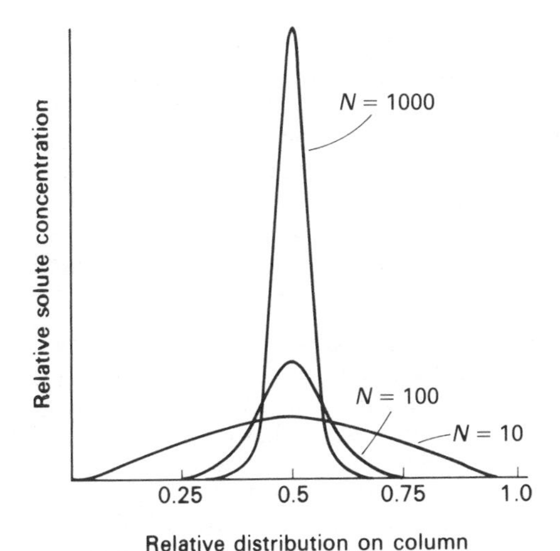

Fig. 6.2 Diagrammatic effect of the number of theoretical plates (N) on the shape of the solute band.

these forms are discussed in the appropriate section. Gas-liquid chromatography and high performance liquid chromatography each have their own unique apparatus, materials and practical procedures which are discussed in Sections 6.4 and 6.8 respectively.

Columns The glass column used should have a means of supporting the stationary phase as near to the base of the column as possible in order to minimise the *dead space* below the column support in which post-column mixing of separated compounds could occur. Commercial columns possess either a porous glass plate fused onto the base of the column, or a suitable device for supporting a replaceable nylon net which in turn supports the stationary phase. A cheaper alternative to these commercial column supports is to use a plug of glass wool together with a minimal amount of quartz sand or glass beads. A capillary tubing normally leads the effluent from the column to the monitor and/or collection system (Fig. 6.3). For some chromatographic separations, it is necessary to maintain the temperature of the column constant during the experiment. This is most simply achieved by jacketing the column so that liquid from a thermostatically-controlled bath, set at the required working temperature, which may be below ambience, may be pumped around the outside of the column. More sophisticated methods include placing the column in a heating block or in a thermostatically-controlled oven.

Stationary Phases The chemical nature of the stationary phase depends upon the particular form of chromatography to be carried out. Full details are given in later sections of this chapter. Most stationary phases are available in a range of sizes and shapes. Both properties are important since they influence the flow rate and resolution characteristics. The larger the particles, the faster the flow rate but opposing this, the smaller the particle the larger their surface area to volume ratio and potentially the greater their resolving power. In practice a balance has to be struck. The best packing characteristics are given by spherical particles and most stationary phases now have a spherical or approximately spherical shape. Particle size is commonly expressed by a *mesh size*, which is a measure of the openings per inch in a sieve, hence the larger the mesh size, the smaller the particle. 100 to 120 mesh is most common for routine use whilst 200 to 400 mesh is used for high resolution work

Packing of Columns This is one of the most critical factors in achieving a successful separation by any form of column chromatography. Packing a column is normally carried out by gently pouring a slurry of the stationary phase (adsorbent, resin or gel) into a column which has its outlet closed, whilst the upper part of the slurry in the column is stirred and/or the column is gently tapped to ensure that no air bubbles are trapped and that the packing settles evenly. Poor column packing gives rise to uneven flow (*channeling*) and reduced resolution. The slurry is added until the required height is obtained. The total volume of solid and liquid in the column is referred to as the *bed volume* and the volume of liquid phase outside the stationary phase as the *void volume*. (V_0). Once the required column height has been obtained, the flow of solvent through the packed column is started by opening the outlet, and continued until the packing has completely settled. This whole process generally requires considerable practice to achieve reproducible results. To prevent the surface of the column from being disturbed either by addition of solvent to the column or during the application of the sample to the column, it is normal to place a suitable protection device, such as a filter paper disc or nylon or rayon gauze, on the surface of the column. Some commercial columns possess an adaptor and plunger which serve the dual purpose of protecting the surface of the column and providing an inlet (often capillary tubing) to carry the solvent to the column surface. Once a column has been prepared, it is imperative that no part of it should ever be allowed to run dry, i.e. a layer of solvent should always be maintained above the column surface.

It is difficult to generalise about the ideal column height to diameter ratio and the total bed volume. They both influence the amount of material which can be separated on the column, and in practice will need to be determined by systematic trial and error. Experience frequently provides a guide; thus, for example, in exclusion chromatography, a height to diameter ratio of 10:1 to 20:1 is normally suitable.

Application of Sample Several methods are available for the actual application of the sample to the top of the prepared column. A simple way is to remove most of the solvent from above the column by suction and *just* to drain the remainder into the column bed. The sample is then carefully applied by pipette and it too allowed just to run into the column. A small volume of solvent is then applied in a similar manner to wash final traces of the sample into the bed. More solvent is then carefully added to the column to a height of 5 to 10 cm. The column is then connected to a suitable reservoir which contains more solvent, so that the height of the solvent in the column can be maintained at 5 to 10 cm. An alternative procedure, which avoids the necessity to drain the column to the surface of the bed, is to increase the density of the sample by addition of sucrose to a concentration of about 1%. When this solution is layered onto the solvent above the column bed, it will automatically sink to the surface of the column and hence be quickly passed into the column. This method of course assumes that the presence of sucrose in no way interferes with the separation and subsequent analysis of the sample. A third method involves the use of capillary tubing and/or syringe or peristaltic pump to pass the sample directly to the column surface. This latter method is probably the most satisfactory of the three.

In all cases, care must be taken to avoid overloading the column with sample, otherwise irregular separation will occur. It is also advantageous to apply the sample in as minimal a volume of solvent as possible since this ensures an initial tight band of material when the separation commences. In addition, to prevent anomalous adsorption effects, the sample should have been desalted before application to the column.

Column Development The components of the applied sample are separated by the continuous passage of a suitable eluant (mobile phase) through the column. This is known as *column development*. The volume of mobile phase required to elute a particular solute is known as the *elution volume*. (V_e) whilst the corresponding time for elution of the solute at a given flow rate is known as the *retention time* (t_r).

During the elution process it is essential that the eluant flow is maintained at a stable rate and this is most simply achieved by gravity feed. The flow rate may be regulated by adjusting the operating pressure which corresponds to the difference between the level of solvent in a reservoir situated above the column and the level at the outlet site of the column. An ordinary open reservoir is not satisfactory since the operating pressure will drop in the course of the experiment due to the drop in solvent level in the reservoir as the solvent runs through the column. This can be overcome by the use of a Mariotte flask which will keep the operating pressure constant (Fig. 6.3). An alternative and more effective method of maintaining stable flow rates is to use a peristaltic pump to pump the eluant at a predetermined rate on to the column. Care must be exercised, however, when using pumps, to ensure that the operating pressure used does not compact the column excessively and cause the column structure to change.

Column development using a single solvent as the eluant is known as an *isocratic separation*. However, in many cases in order to increase the resolving power of the eluant, it is necessary to continuously change its pH, ionic concentration or polarity. This is known as *gradient elution*. In order to produce a suitable gradient, two solvents have to be mixed in the correct proportions prior to entering the column. This may be achieved by use of commercially available *gradient mixers* or more simply as follows. The two solutions are placed in separate chambers, a recipient chamber linked to the column and the other, called the donor, linked to the recipient chamber by a

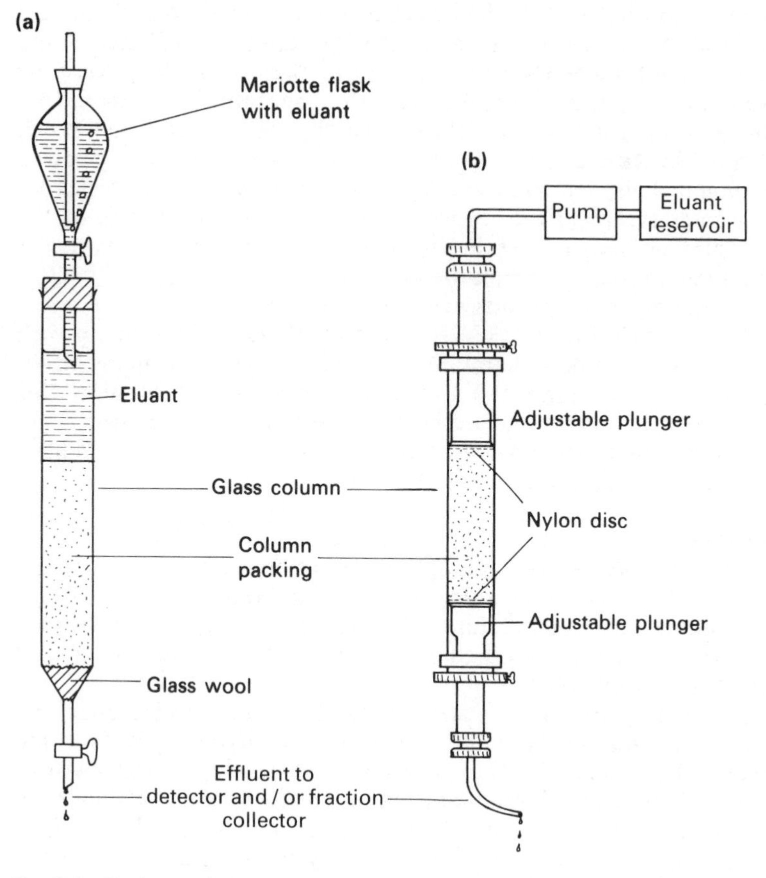

Fig. 6.3 Equipment for column chromatography: **(a)** simple version; **(b)** more sophisticated version.

siphon. As eluant enters the column from the recipient chamber, the solution in the donor replaces it and is mixed by a stirrer. The relative pH, ionic strength or polarity of the solution in the donor with respect to that in the recipient chamber will determine the direction in which the gradient will be formed. Moreover, the relative diameters of the two chambers will determine whether the gradient varies with time in a linear, convex or concave manner.

Fraction Collection and Analysis As the resolved compounds emerge in the effluent from the column, it is necessary to detect their presence in order to enable them to be isolated for further study. Two approaches are available to achieve this objective. Either the effluent can be continuously monitored and the part containing a particular compound collected, or the effluent can be divided into small (1 to 10 cm^3) fractions which are subsequently analysed and those containing a particular compound bulked together. Continuous monitoring can be carried out in several ways but in all cases the effluent from the column is passed through a flow cell with a small sample volume (typically 8 mm^3) located in the detector. The signal generated by the detector is recorded on a chart recorder each emerging compound giving a characteristic *peak* from which it is possible to calculate the retention time and/or elution volume (Fig. 6.4). Detection may be based on ultraviolet or visible absorption exploiting in particular the fact that the majority of unsaturated compounds including proteins and nucleic acids absorb at 254 nm, fluorescence, changes in the refractive index of the effluent, the presence of a radioactive label or on the ease of oxidation or reduction of the compounds as measured by an electrochemical detector (Section 10.7).

In the absence of a method for continuously monitoring the effluent, fractions of the order of 2 to 5% of the bed volume are collected. A particular compound may be distributed in several of these fractions, but if the separation has been successful, the number will be relatively small and they will be separated from fractions containing other compounds by intervening fractions which contain negligible amounts of any compound. A range of automatic fraction collectors is available commercially. They are designed to collect a certain amount of effluent in each tube before a new tube is placed in position automatically. The actual amount of effluent in each fraction may be determined in one of several ways. There may be a siphoning or similar system to deliver a predetermined volume into each tube, or there may be an electronic means of allowing a predetermined number of drops to enter each tube. This latter method has the slight disadvantage that if the composition of the effluent changes (e.g. during gradient elution), so too may its surface tension and hence droplet size, so that the actual volume collected also changes. A further possibility is that the effluent is allowed to enter each tube for a fixed interval of time. In this case, if the flow rate through the column varies, so too will the volume of each fraction. The fractions collected must subsequently be analysed by methods which specifically detect the compounds being separated. Such methods include colorimetry, ultraviolet absorption, fluorimetry, scintillation counting and radioimmunoassay.

When the column effluent has been monitored to produce a chart recording, then the area of each peak can be shown to be proportional to the amount of sample component present, and hence may be used to quantitatively determine the amount of a given component eluting from the column. The area of the peak may be determined by measuring the height of the peak (h_p) and its width at half the height (ω_{h_A}). The product of these dimensions is taken to be equal the area of the peak. Alternatively the peak may be cut out of the chart paper and weighed and the assumption made that area and weight are linearly related. When the peak area has been determined, the amount of the specific component present may be determined by use of a calibration curve obtained by chromatographing, under identical conditions, known amounts of the pure form of the compound being assayed. To aid this assay, by attempting to compensate for variations in chromatographic conditions and of any preliminary extraction procedure, use is made of an *internal standard*. The internal standard is a compound which has physical properties as similar as possible to the test compounds, and which chromatographs near to but distinct from them. A known amount of the standard is introduced into the test sample as early as possible in the extraction, and is therefore taken through any preliminary procedures with it. Any loss of standard during the analysis will thus be identical to the loss of the test compounds. The peak area associated with the fixed amount of internal standard is used to calculate the *relative peak area* for each peak in both the calibration data and the sample under analysis. A calibration curve therefore consists of a plot of relative peak area against the known amount of the compound thereby enabling the amount of the compound in the test sample to be calculated. An alternative procedure is to use an *external standard*. In this method the standard is added to the test sample immediately before the sample is chromatographed. It is therefore not taken through any preliminary extraction procedure and cannot compensate for variations in the efficiency of the extraction procedure. This method is only valid in those cases where the recovery of compound from the test sample is virtually quantitative.

The procedure for measuring peak areas and performing the necessary calculations can be very time consuming when complex and/or a large number of analyses are involved. The calculations are best performed by dedicated integrators or microcomputers. These can be programmed to compute retention time and peak area and to relate them to those of internal standards enabling relative retention times and relative peak areas to be calculated. These may be used to identify a particular solute and to quantify it using previously obtained and stored calibration data from internal or external standards. The data system can also be used to correct problems inherent in the chromatographic system. Such problems can arise either due to the characteristics of the detector or the efficiency of the separation process. Problems which are attributable to the detector are *baseline drift*, where the detector signal gradually changes with time, and *baseline noise*, a series of rapid minor fluctuations in detector signal, commonly the result of using too high a detector sensitivity.

Column Efficiency In practice, as the solute moves down the column, kinetic and flow phenomena cause the solute band to broaden to give a concentration profile within the band that has a Gaussian distribution about the mean (Fig. 6.4a). The width of the base of such a peak is taken to be four standard deviations (4σ) and the extent of peak broadening as the variance (σ^2). For symmetrical Gaussian peaks, the standard deviation, σ, is equal to the half-width of the peak at $0.607h_p$, which is the point of inflection (Fig. 6.4a). In some cases however, the peak is asymmetrical displaying either *fronting* (extended front portion of peak) or *tailing* (Fig. 6.4c). Peak asymmetry has many causes including the application of too much solute to the column, poor packing of the column, poor application of the sample to the column or solute-support interactions.

The success of any chromatographic procedure is measured by its ability to separate completely (*resolve*) one compound from a mixture of similar compounds. *Peak resolution* (R_s) is related to the properties of the peaks (Fig. 6.4) such that:

$$R_s = \frac{2\,(t_{R_B} - t_{R_A})}{\omega_A + \omega_B} \qquad 6.1$$

where t_{R_A} and t_{R_B} = the retention times of compounds A and B respectively
ω_A and ω_B = the base widths of the peaks for A and B respectively.

It can be shown that when $R_s = 1.5$ the separation of the two peaks is 99.7% complete. In most practical cases, R_s values of 1.0, corresponding to 98% separation, is satisfactory. According to the definition of resolution, R_s–1 peaks fit between any two peaks in question. Peak resolution is determined by three factors: *selectivity* which measures the discriminatory power of the system, *retention* which measures the retentive power of the system and which is related to the partition coefficients, K_d, of the compounds, and *efficiency* which measures the relative narrowness of the peaks in terms of the ratio of retention time to peak width.

Not uncommonly in chromatography it is not possible to completely resolve two peaks. As a result they produce *fused peaks* and for their analysis the assumption has to be made that the peak characteristics of the two unresolved compounds do not affect each other (Fig. 6.4b).

The number of theoretical plates (*plate number*) (N) involved in the elution of a particular compound is given by:

$$N = 16\left(\frac{t_R}{\omega}\right)^2 \qquad 6.2$$

$$\text{or } N = 5.54\left(\frac{t_R}{\omega_h}\right)^2 \qquad 6.3$$

where ω = the peak base width and is equal to 4σ
ω_h = the peak width at half the peak height and is equal to 2.355σ

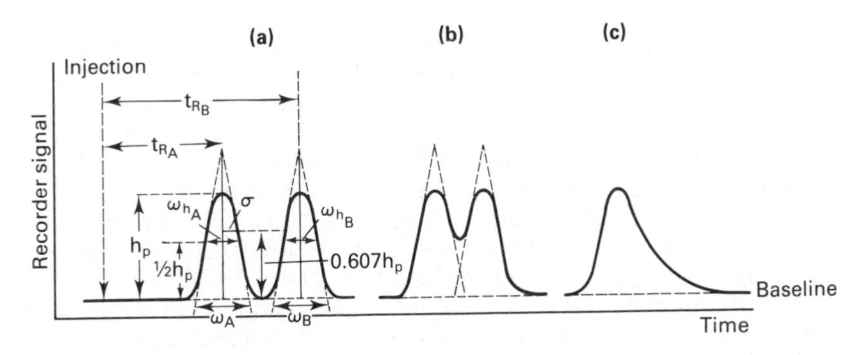

Fig. 6.4 **(a)** Chromatogram of two compounds showing complete resolution and the calculation of retention times; **(b)** two compounds giving incomplete resolution and the prodution of fused peaks; **(c)** a compound showing excessive tailing.

The value of N can therefore very easily be calculated from data on the chart recorder paper. The plate number can be increased by simply increasing the column length (L) but there is a limit to this since the retention time and peak width increase proportionally with L whilst the peak height decreases as the square root of N. Whilst N is a measure of the efficiency of the column, the plate height, which is also called the *height equivalent to a theoretical plate* (*HETP*), is useful for comparative purposes such as operating the column under different conditions. It can be shown that:

$$HETP = \frac{L}{N} = H \tag{6.4}$$

The maximum number of peaks that can be separated by a specific chromatographic system is called the *peak capacity* (n). It is related to the retention volumes of the first and last peaks (V_α and V_ω respectively) and to the plate number:

$$n = 1 + \sqrt{\frac{N}{16}}\left(\ln \frac{V_\omega}{V_\alpha}\right) \tag{6.5}$$

Peak capacity is determined by acceptable separation times and by detection sensitivity. In practice it can be increased either by the procedure of gradient elution as in liquid chromatography (Section 6.1.2) or by temperature programming as in gas-liquid chromatography (Section 6.4.1).

6.1.3 Thin layer chromatography (TLC)

Principle Partition, adsorption, exclusion and high performance liquid chromatography may all be carried out in the thin layer mode. The technique is simple, quick, and allows a large number of samples to be studied concurrently. It can be used for analytical and preparative purposes.

Thin Layer Preparation A slurry of the stationary phase, generally in water, is applied to a glass, plastic or foil plate as a uniform thin layer by means of a plate spreader starting at one end of the plate and moving progressively to the other. The thickness of the slurry layer used is dictated by the nature of the desired chromatographic separation. For analytical separations the layer is of the order of 0.25 mm thick and for preparative separations it may be up to 5 mm. Where the stationary phase is to be used for adsorption chromatography, a binding agent such as calcium sulphate is incorporated into the slurry in order to facilitate the adhesion of the adsorbent to the plate. With the exception of thin layer exclusion chromatography (Section 6.6.1), once the slurry layer has been prepared, the plates are dried to leave the coating of stationary phase. In the case of adsorbents, drying is carried out in an oven at 100 to 120°C. This also serves to activate the adsorbent. A range of pre-prepared plates is available commercially. So-called *polyamide layer sheets*, which consist of poly-ϵ-caprolactam coated onto *both* sides of a solvent-resistant polyester sheet, are unusual in that they are semi-transparent, allowing unknowns and standards run on opposite sides of the plate to be compared. They can also be re-used if immediately cleaned with ammonia-acetone, and are widely used in protein sequencing studies by the phenylthiohydantoin and dansyl methods (Section 5.5.3).

Sample Application The sample is applied to the plate by means of a micropipette or syringe. It is possible for this process to be automated. The spot of sample is generally placed 2.0 to 2.5 cm from the edge of the plate. The solvent may be removed from the spot by gentle heating or by use of an air blower, care being taken in the case of volatile or thermolabile compounds. It is then possible to apply more sample to the spot if necessary. In the case of adsorption chromatography, diffusion of the sample from the applied spot may be minimised by using a solvent in which components have a low R_F value (Section 6.1.4). For preparative thin layer chromatography, the sample is applied as a band across the plate rather than as a single spot.

Plate Development Separation takes place in a glass tank which contains the developing solvent to a depth of about 1.5 cm. This is allowed to stand for at least an hour with a lid over the top of the tank to ensure that the atmosphere within the tank becomes saturated with solvent vapour (*equilibration*). Unless this is done, irregular running of the solvent will occur as it ascends the plate by capillary action, resulting in poor separations being achieved. After equilibration, the lid is removed, and the thin layer plate is then placed vertically in the tank so that it stands in the solvent. The end of the plate bearing the sample should, of course, be the end in the solvent. The lid is replaced and separation of the compounds then occurs as the solvent travels up the plate. It is preferable to keep the system at a constant temperature whilst the development is occurring to avoid anomalous solvent-running effects. One of the biggest advantages of TLC is the speed at which separation is achieved. This is commonly 10 to 30 minutes and is hardly ever greater than 90 minutes. Thin layer exclusion chromatography is much slower and also requires special procedures which are discussed in Section 6.6.1.

In order to improve the resolution of partition and adsorption separations, the technique of *two-dimensional chromatography* may be used. The material to be chromatographed is placed towards one corner of the plate as a single spot and the plate developed in one direction and then removed from the tank and allowed to dry. It is then developed by another solvent system, in which the compounds to be separated have different K_d values, in a direction at right angles to the first development (Fig. 6.5). TLC can also be linked to thin layer electrophoresis (TLE) (Section 7.3.1).

Component Detection Several methods are available. Spraying the plate with 50% sulphuric acid or 25% sulphuric acid in ethanol and heating will result in most compounds becoming charred and showing up as brown spots. Examination of the plate under ultraviolet light will show the position of ultraviolet-absorbing or fluorescent compounds. Many commercially

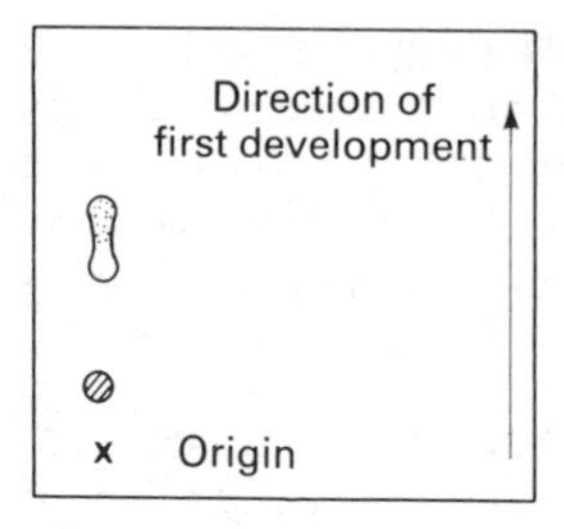

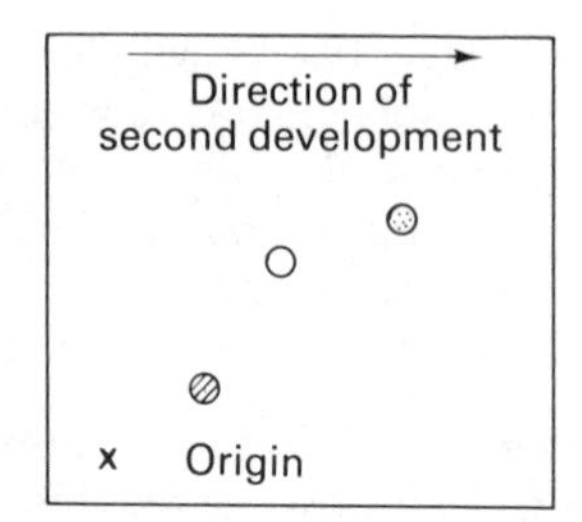

Fig. 6.5 A two-dimensional chromatogram.

available thin layer adsorbents contain a fluorescent dye so that when the plate is examined under ultraviolet light, the separated compounds show up as blue, green or black areas against a fluorescent background. Subjecting the plate to iodine vapour is useful if unsaturated compounds are being investigated. Spraying of plates with specific colour reagents will stain up certain compounds, for example ninhydrin for amino acids. Most of these colour reagents are based upon specific quantitative colour reactions which are listed in Section 8.2.3. If the compounds are radioactive, the plates may be subjected to autoradiography (Section 9.2.4) which will detect the spots as dark areas on X-ray film, or the plate may be scanned by a radiochromatogram scanner (Section 9.2.4).

Although the movement of compounds on TLC may be characterized by an R_F value (Section 6.1.4), such values are not so reliable as those in paper chromatography. Component identification is therefore most commonly made on the basis of a comparison of the movement of the components with those of reference compounds chromatographed alongside the sample on the TLC plate.

The amount of compound present in a given spot may be determined in a number of ways. On-plate quantification may be achieved by use of radiochromatogram scanning in the case of radio-labelled compounds or generally by means of densitometry. Precision densitometers are commercially available which measure the ultraviolet or visible absorption of the compound as well as simultaneously giving a complete absorption spectrum of the compound for identification purposes. Off-plate quantification may be carried out by scraping off the spot and the immediate surrounding stationary phase from the plate and eluting the compound with a suitable solvent. The amount of compound in solution can then be determined by standard methods, most commonly colorimetry or fluorimetry.

6.1.4 Paper chromatography

Principle The cellulose fibres of chromatography paper act as the supporting matrix for the stationary phase. The stationary phase may be water, a non-polar material such as liquid paraffin or impregnated particles of solid adsorbent. Papers are available which have different running

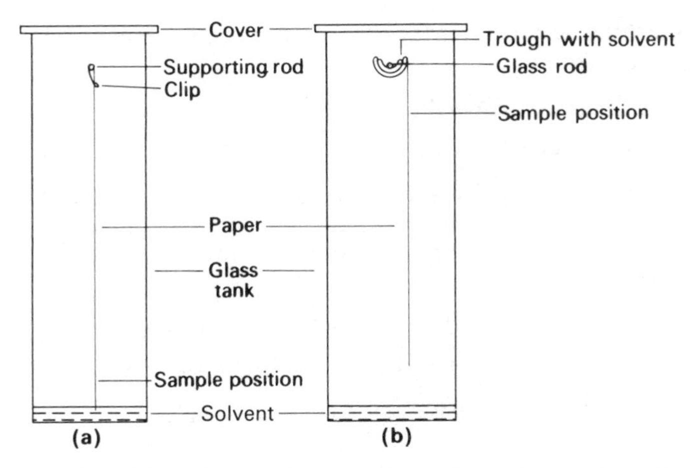

Fig. 6.6 **(a)** Ascending and **(b)** descending methods of paper chromatography.

characteristics, for example slow, medium and fast. Others have been acid washed to remove traces of impurities which may affect certain analyses. In one-dimensional chromatography, the paper should always be developed in its machined direction which is normally marked on the packet. For adsorption and normal phase partition paper chromatography, the paper in suitable form is commercially available. Paper for reverse phase chromatography must be prepared immediately before use.

Paper Development There are two techniques which may be employed for the development of paper chromatograms - *ascending* or *descending methods* (Fig. 6.6). In both cases the solvent is placed in the base of a sealed tank or glass jar to allow the chamber to become saturated with the solvent vapour. In the ascending method, the procedure is identical to that previously described for thin layer chromatography (Section 6.1.3). The sample spots should be in a position just above the surface of the solvent so that, as the solvent moves vertically up the paper by capillary action, separation of the sample is achieved. In the descending technique, the end of the paper near which the sample spots are located is held in a trough at the top of the tank and the rest of the paper allowed to hang vertically but not in contact with the solvent in the base of the tank. When the analysis is to be started, solvent is added to the trough. Separation of the sample then occurs as the solvent moves downwards under gravity. Although ascending chromatography is often preferred because of the simplicity of the set-up, the flow of solvent is faster in the descending technique. Two-dimensional chromatography may be used for paper systems in a manner similar to that described for TLC.

Component Detection The methods used for component detection are similar to those described for TLC, but spraying with sulphuric acid is not recommended as this causes the paper to disintegrate.

The identification of a given compound may be made on the basis of its R_F

value which is the distance moved during development relative to the distance moved by the solvent front:

$$R_F = \frac{\text{the distance moved by solute from origin}}{\text{the distance moved by solvent from origin}}$$

This value is a constant for a particular compound under standard conditions and closely reflects the distribution coefficient for that compound. In the case of carbohydrates the term R_G value is sometimes used for convenience. This is defined as:

$$R_G = \frac{\text{the distance moved by carbohydrate from origin}}{\text{the distance moved by glucose from origin}}$$

and necessitates the use of glucose as a reference compound.

6.2 Adsorption chromatography

6.2.1 Principle

An *adsorbent* may be described as a solid which has the property of holding molecules at its surface, particularly when it is porous and finely divided. It differs from an ion-exchanger (Section 6.5.2) in that the attraction of molecules to the surface of the adsorbent ideally does not involve electrostatic forces. Adsorption can be fairly specific so that one solute may be adsorbed selectively from a mixture. Separation of components by this method depends upon differences both in their degree of adsorption by the adsorbent and solubility in the solvent used for separation. These features are, of course, governed by the molecular structure of the compound. Adsorption chromatography can be carried out in both the column and thin layer modes.

6.2.2 Materials and applications

Adsorbents Compounds such as silicic acid (silica gel), aluminium oxide, calcium carbonate, magnesium carbonate, zinc carbonate, magnesium oxide and cellulose may be used as the stationary phase, the choice of any particular adsorbent and solvent elution system is dependent on the separation to be achieved. Hydroxyapatite (calcium phosphate) is widely used for the separation of proteins, nucleic acids and viruses. Unlike most other adsorbents, it has some ion-exchange properties which aids separation. In nucleic acid work it is particularly valuable because of its ability to bind double-stranded DNA but not single-stranded DNA (Section 5.5.4). Care is needed in the choice of adsorbent, since occasionally some adsorbents may degrade certain compounds during separation. Other adsorbents have a tendency to take up water from the atmosphere during storage and this may adversely affect their adsorption properties. In such cases it may be necessary to activate the adsorbent by keeping it for a period of time at 110°C to remove any water. Separations on adsorbents are often enhanced, however, by the presence of water in the adsorbent as this may result in both adsorption and partition effects being involved in the separation.

Adsorbents for thin layer chromatography are often impregnated with various ions during plate preparation and this may enhance the separation achieved. The incorporation of silver nitrate into the adsorbent enhances the separation of compounds differing in the number and position of double bonds. This is referred to as *argentation* TLC. Poly-ϵ-caprolactam TLC sheets (Section 6.1.3) are particularly versatile and do not require activation.

Solvents Virtually any organic solvent, provided it can be obtained in the necessary degree of purity, can be used as the mobile phase. The choice depends upon the polarity of the compounds to be resolved and upon their distribution coefficient. Commonly used low polarity solvents are hexane, heptane, toluene, acetonitrile, diethyl ether and chloroform. Intermediate polarity solvents include ethanoic acid, dichloromethane and pyridine. Highly polar solvents include propan-1-ol, butan-1-ol, acetone, ethanol, methanol and water. In the case of gradient elution, suitable mixtures of miscible solvents are used to give an eluant of gradually increasing polarity.

6.3 Partition chromatography

6.3.1 Liquid-liquid chromatography

In *normal phase* partition chromatography, the stationary phase is water supported by a matrix. In the case of column chromatography the matrix may be cellulose, starch or silicic acid. For thin layer chromatography it is most commonly cellulose, but this probably combines both adsorption and partition effects. Whereas for paper partition chromatography sufficient water is naturally present in the paper to act as the stationary phase, in column and thin layer chromatography the correct amount of water must be added to the support to achieve the correct partitioning. Some supports can be prepared to contain as much as 50% (w/v) water and yet remain free-flowing powders which can be packed into columns or coated onto plates in the normal way. The mobile solvent phase is generally an organic solvent immiscible with water or a mixture of aqueous organic solvents which have a dielectric constant less than that of water. Very often the solvent is miscible with water. In such cases partition of solute would not be expected to occur since there is apparently only a single phase. However, the stationary water phase is probably better regarded as an insoluble complex with the supporting matrix and is effectively immiscible with the mobile phase.

In *reverse phase* partition chromatography, the stationary phase is a non-polar compound such as liquid paraffin supported by a matrix similar to those employed in normal phase systems. The phase is prepared by treating the supporting matrix with a solution of the mobile phase in a suitable non-polar and volatile solvent. The solvent is subsequently removed by evaporation.

Liquid-liquid chromatography, especially in the high performance form (HPLC) is used for the separation of a very wide range of biological compounds. In its reverse phase form it is particularly successful for the separation of polar compounds.

6.3.2 Countercurrent chromatography (CCC)

This separation process is based upon the distribution of a compound between two immiscible liquid phases. These phases may be mixtures of solvents, buffers, salts and various complexing agents. This technique is atypical of normal partition chromatographic techniques in that neither of the phases is supported by an inert support. Nevertheless, the separation of compounds is based upon the different distribution coefficients between two immiscible phases and therefore the principle of the separation is the same as that of conventional partition chromatography.

The apparatus most commonly used is the *Craig Countercurrent Distribution Apparatus*. It consists of between 30 and 1000 interconnected vessels (the so-called *train*), each of which retains a fixed volume of the stationary liquid phase. The solute mixture is introduced to the first vessel in the train and equilibrated with the immiscible and less dense mobile phase by the repeated rocking of the vessel through 90°. After equilibration is complete (1 to 2 minutes), the mobile phase is transferred to the next vessel as a result of the complete tipping of the first vessel. When this returns to its original position, fresh upper phase is automatically introduced. The whole process is repeated so that the mobile phase is transferred progressively along the series of lower phases. The solutes are transferred at a rate determined by their partition coefficients and the relative volumes of the two solvents in each vessel. Each solute eventually accumulates in a specific group of vessels, the resolution being determined by the total number of transfers and the differences in partition coefficient.

Recently, several new versions of CCC have been developed. *Helix CCC* or *Toroidal CCC* uses a helically wound tube located circumferentially on a centrifuge rotor. The centrifugal field holds one of the liquid phases stationary in the coil so that when the second phase is pumped through the coil, mixing occurs without displacement of the stationary phase. An introduced sample therefore undergoes a series of partitions such that solutes more soluble in the stationary liquid phase elute more slowly than those more soluble in the pumped second phase. In *Droplet CCC*, droplets of the mobile phase are passed upwards through the stationary phase contained in a vertical column creating partioning of solutes between the two solvents. *Locular CCC* is a variant of Droplet CCC and uses a column divided into a series of small chambers (*locules*) by perforated partitions. The mobile phase is pumped upwards through the stationary phase, but this time droplet preformation is not necessary. The partitioning effect is enhanced by the rotation or gyration of the column.

CCC is the only form of chromatography which has been successfully used for cell organelle fractionation. It has also been used for cell fractionation and membrane receptor isolation. Its earlier application to nucleic acid isolation has been largely superceded by other techniques.

6.4 Gas-liquid chromatography (GLC)

6.4.1 Apparatus and materials

This technique, which is based upon the partitioning of compounds between a liquid and a gas phase, is a widely used method for the qualitative and quantitative analysis of a large number of compounds since it has high sensitivity, reproducibility and speed of resolution. It has proved to be most valuable for the separation of compounds of relatively low polarity. A stationary phase of liquid material such as a silicone grease is supported on an inert granular solid. This material is packed into a narrow coiled glass or steel column 1 to 3 m long and 2 to 4 mm internal diameter through which an inert carrier gas (the mobile phase) such as nitrogen, helium or argon is passed. The column is maintained in an oven at an elevated temperature which volatilises the compounds to be separated. The basis for the separation is the difference in the partition coefficients of the volatilised compounds between the liquid and gas phases as the compounds are carried through the column by the carrier gas. As the compounds leave the column they pass through a detector which is linked via an amplifier to a chart recorder which, in turn, records a peak as a compound passes through the detector (Fig. 6.7a).

GLC may also be performed using *capillary columns* which are made of glass or metal with diameters of between 0.03 to 1.0 mm and which may be up to 100 metres in length. There are two types of capillary column systems

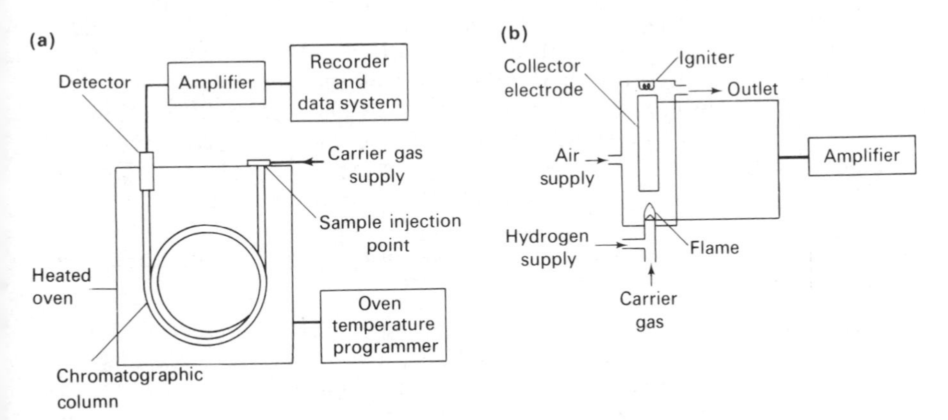

Fig. 6.7 Diagrammatic representation of **(a)** GLC system and **(b)** flame ionisation detector.

known as *wall coated open tubular* (WCOT) columns and *support coated open tubular* (SCOT), also known as *porous layer open tubular* (PLOT) colums. In WCOT columns the stationary phase is coated directly onto the walls of the capillary tubing. As there is only a small amount of stationary phase present, only very small amounts of sample may be chromatographed. Consequently a splitter system has to be used at the sample injection port so that only a small fraction of the sample injected reaches the column. The remainder of the sample is vented to waste. The design of the splitter is critical in quantitative analyses in order to ensure that the ratio of sample chromatographed/sample vented is always the same. As the length of these columns is very much greater than that of conventional columns, very high efficiencies are obtained (equation 6.4) and these systems are very useful for the analysis of complex mixtures.

In SCOT columns a support material is bonded to the walls of the capillary column, and the stationary phase is coated onto the support. The capacity of SCOT columns is considerably higher than that of WCOT columns and consequently small samples can be injected directly onto such columns without the need for a splitter system. SCOT systems are therefore considerably simpler to use for quantitative analyses than WCOT systems, their efficiency is less than WCOT systems but considerably better than conventional GLC columns.

The efficiency of a GLC column is determined by the principles outlined in Section 6.1.2. There is an optimum carrier gas flow rate for maximum column efficiency (i.e. minimum *HETP*). For capillary columns, the maximum number of theoretical plates that can be obtained is independent of the carrier gas used. In these cases, a decrease in column diameter should give a proportional increase in the number of plates per unit length i.e. *HETP*.

Solid Support Since this is used to provide a supporting surface on which is coated the film of stationary phase, it is important that the support should be inert to the sample. This is generally no problem when the support is holding a high percentage coating of stationary phase but, when the percentage coating is low, exposure of the support to the sample often hinders separation. The most commonly used support is Celite (diatomaceous silica) which because of the problem of support/sample interaction, is often treated so that hydroxyl groups which occur in the Celite are modified. This is normally achieved by silanisation of the support with such compounds as hexamethyldisilazane. As well as silanising the support, the glass column, the glass wool plug located at the base of the column and any other surface which may come into contact with the sample are also silanised. The support particles have an even size which, for the majority of practical applications, is 60 to 80, 80 to 100 or 100 to 120 mesh.

Stationary Phase The requirements for any stationary phase are that it must be involatile and thermally stable at the temperature used for analysis. Often the phases used are high boiling point organic compounds, and these are coated onto the support to give from 1% to 25% loading, depending upon the analysis. Such phases are of two types, either *selective*, where separation occurs by utilisation of different chemical characteristics of components, or *non-selective*, where separation is achieved on the basis of differences in boiling points of the sample components. The operating temperature for the analysis must be compatible with the phase chosen for use. Too high a temperature results in excessive *column bleed* due to the phase being volatilised off, contaminating the detector and giving an unstable baseline. The choice of phase for analysis depends on the compound under investigation, and is best chosen after reference to the literature. Commonly used stationary phases include the polyethylene glycols, methylphenyl- and methylvinylsilicone gums (so-called *OV phases*), Apiezon L, esters of adipic, succinic and phthalic acids, and squalene.

The colunms are dry-packed under a slight positive gaseous pressure and after packing must be *conditioned* for 24 to 48 hours by heating to near the upper working temperature limit whilst the carrier gas at normal flow rates is passed through the column. During this conditioning, the column should not be connected to the detector, to prevent its contamination.

Preparation and Application of Sample The majority of non-and low-polar compounds are directly amenable to GLC, but other compounds possessing such polar groups as —OH, $-NH_2$, —COOH are generally retained on the column for excessive periods of time if they are applied directly. This excessive retention is inevitably accompanied by poor resolution and peak tailing. This problem can be overcome by derivatisation of these polar groups. This increases the volatility and effective distribution coefficients of the compounds. Methylation, silanisation and trifluoromethylsilanisation are common derivatisation methods for fatty acids, carbohydrates and amino acids.

The sample for chromatography is dissolved in a suitable solvent such as ether, heptane or methanol. Chlorinated organic solvents are generally to be avoided as they may contaminate the detector. The sample is injected onto the column using a micro-syringe through a septum in the injection port which is attached to the top of the column. Normally between 0.1 and 10 mm^3 of solution is injected. It is common practice to maintain the injection region of the column at a slightly higher temperature than the column itself. This helps to ensure rapid and complete volatilisation of the sample. Sample injection is automated in many commercial instruments.

Separation Conditions Nitrogen, helium and argon are the three most commonly used carrier gases. They are passed through the column at a flow rate of 40 to 80 $cm^3\ min^{-1}$. The column temperature must be within the working range of the particular stationary phase and is chosen to give a balance between peak retention time and resolution. In *isothermal analysis* a constant temperature is employed. In the separation of compounds of widely differing polarity or molecular weight it may be advantageous to gradually increase the temperature. This is referred to as *temperature programming*. This, however, often results in excessive bleed of the stationary phase as the temperature is raised, resulting in baseline variation. Consequently many instruments have two identical columns and detectors, one set of which is used as a reference. The currents from the two detectors are opposed; hence, assuming equal bleed from both columns, the resulting current gives a steady baseline as the oven temperature is raised.

Detection Systems By far the most widely used detector is the *flame ionisation detector* (FID). It responds to almost all organic compounds, can detect as low as 1 ng and has a wide linear response range (10^6). A mixture of hydrogen and air is introduced into the detector to give a flame, the jet of which forms one electrode, whilst the other electrode is brass or platinum wire mounted near the tip of the flame (Fig. 6.7b). When the sample components emerge from the column they are ionised in the flame, resulting in an increased signal being passed to the recorder. The carrier gas passing through the column and the detector gives a small background signal, which can be offset electronically to give a baseline. A FID has a minimum detection quantity of the order of 5×10^{-12} g sec^{-1} and an upper temperature limit of 400°C.

The *nitrogen-phosphorus detector* (NPD), which is also called a *thermionic detector*, is similar in design to a FID but has a sodium salt fused onto the electrode system or a burner tip embedded in a ceramic tube containing a sodium salt or a rubidium chloride tip. The NPD has excellent selectivity towards nitrogen and phosphorus-containing compounds and shows a poor response to compounds possessing neither of these two elements. Its linearity (10^4), upper temperature limit (300°C) and detection limits (10^{-11} g sec^{-1}) are not quite as good as a FID. It is widely used in organophosphorus pesticide residue analysis.

The *electron capture detector* (ECD) responds only to substances which capture electrons, particularly halogen-containing compounds. This detector is, therefore, particularly used in the analysis of polychlorinated compounds, such as the pesticides DDT, dieldrin, aldrin. It has very high sensitivity (10^{-12} g sec^{-1}), an upper temperature limit of 300°C and can detect as little as one picogramme of a compound, but its linear range (10^2 to 10^4) is much lower than that of the FID. The detector works by means of a radioactive source (^{63}Ni) ionising the column gas, and the electrons so produced giving a current across the electrodes to which a suitable voltage is applied. When an electron capturing compound emerges from the column, the ionised electrons are captured, the current drops and this change in current is recorded. The carrier gas usually used in conjunction with an ECD is nitrogen or an argon +5% methane mixture.

The volatile solvent used to introduce the test sample gives rise to a *solvent peak* at the beginning of the chromatogram. The three main forms of detector respond to this solvent with varying sensitivity thereby affecting the detection and resolution of rapidly eluting solutes. In cases where authentic samples of the test compounds are not available for calibration purposes or in cases where the identity of the compounds is not known, the detector may be replaced by a mass spectrometer. Special separators are available for removing the bulk of the carrier gas from the sample emerging from the column and prior to its introduction in the mass spectrometer (Section 8.10.3). More recently, GLC has been linked to other types of detector including an infra-red spectrophotometer and to a nuclear magnetic resonance spectrometer, the resulting spectra aiding in the identification of unknown compounds.

6.4.2 Applications

Until the recent developments in HPLC (Section 6.8), GLC was probably the most commonly used form of chromatography. Its use nowadays is confined to volatile, non-polar compounds which do not need derivatisation. Compounds are characterised by their retention time or preferably by their relative retention time to a standard reference compound. In the analysis of compounds which form an homologous series, for example the methyl esters of the saturated fatty acids, there is a linear relationship between the logarithm of the retention time and the number of carbon atoms. This can be exploited, for example, to identify an unknown fatty acid ester in a fat hydrolysate. A widely used system for quantitative analysis is the *Retention Index* (RI) which is based on the retention of a compound relative to *n*-alkanes. The compound is chromatographed with a number of *n*-alkanes and a semi-logarithmic plot constructed of retention time against carbon number. Each *n*-alkane is assigned an RI of 100 times the number of carbon atoms it contains (pentane therefore has an RI of 500) allowing the RI for the compound to be calculated. Many commercially available GLC systems with data processing facilities have the capacity to calculate RI values automatically.

6.5 Ion-exchange chromatography

6.5.1 Principle

The principal feature underlying this form of chromatography is the attraction between oppositely charged particles. Many biological materials, for example amino acids and proteins, have ionisable groups and the fact that they may carry a net positive or negative charge can be utilised in separating mixtures of such compounds. The net charge exhibited by such compounds is dependent on their pK_a and on the pH of the solution in accordance with the Henderson-Hasselbalch equation (Section 1.2.1).

Ion-exchange separations are mainly carried out in columns packed with an ion-exchanger. There are two types of ion-exchanger, namely *cation* and *anion exchangers*. Cation exchangers possess negatively charged groups and these will attract positively charged molecules. These exchangers are also called *acidic ion-exchange materials* since their negative charges result from the protolysis of acidic groups. Anion exchangers have positively charged groups which will attract negatively charged molecules. The term *basic ion-exchange materials* is also used to describe these exchangers since positive charges generally result from the association of protons with basic groups.

The actual ion-exchange mechanism is thought to be composed of five distinct steps:

(i) diffusion of the ion to the exchanger surface. This occurs very quickly in homogeneous solutions;

(ii) diffusion of the ion through the matrix structure of the exchanger to the exchange site. This is dependent upon the degree of crosslinkage of the exchanger and the concentration of the solution. This process is thought to be the feature which controls the rate of the whole ion-exchange process;

(iii) exchange of ions at the exchange site. This is thought to occur instantaneously and is an equilibrium process:

Cation exchanger:

$$\underset{\text{exchanger}}{RSO_3^-} \ldots \underset{\text{counter ion}}{Na^+} + \underset{\text{charged molecule to be exchanged}}{\overset{+}{N}H_3R'} \rightleftharpoons RSO_3^- \ldots \underset{\text{bound molecular ion}}{\overset{+}{N}H_3R'} + \underset{\text{exchanged ion}}{Na^+}$$

Anion exchanger:

$$(R)_4\overset{+}{N} \ldots Cl^- + {}^-OOCR' \rightleftharpoons (R)_4\overset{+}{N} \ldots {}^-OOCR' + Cl^-$$

The more highly charged the molecule to be exchanged, the tighter it binds to the exchanger and the less readily it is displaced by other ions;

(iv) diffusion of the exchanged ion through the exchanger to the surface;

(v) selective desorption by the eluant and diffusion of the molecule into the external solution. The selective desorption of the bound molecule is achieved by changes in pH and/or ionic concentration or by affinity elution, in which case an ion which has greater affinity for the exchanger than has the bound molecule is introduced into the system.

6.5.2 Materials

Many commercial ion-exchangers which have been successfully employed for separation of biological materials are made by co-polymerising styrene with divinylbenzene. Polystyrene itself is a linear polymer which is soluble in several solvents. By condensing styrene with divinylbenzene, however, cross-linking of molecules occurs and this produces an insoluble resin. Various degrees of cross-linkage may be obtained by co-polymerising varying proportions of divinylbenzene and styrene. The higher the amount of divinylbenzene employed with respect to the styrene, the greater the degree of cross-linkage obtained. Resins with a low degree of cross-linking are more permeable to high molecular weight compounds than are highly cross-linked ones, but they are also less rigid and swell more when placed in a buffer. These swelling characteristics must be taken into account when a column is prepared. Sulphonation of cross-linked polystyrene results in a sulphonated polystyrene resin, such as Dowex 50, which is a strong acidic exchanger. The SO_3H groups are ionised at all except very low pH values. An analogous basic exchanger may be prepared by reacting cross-linked polystyrene with chlormethyl ether, and then reacting the chloro groups with tertiary amines. These attached amines exist as cations at all but very alkaline pH values. One of the main limitations of conventional ion-exchange resins is that they rely on the exchanging ion diffusing through the matrix. This is a slow process. Reducing the bead size of the resin helps but results in a reduced eluant flow rate due to the closer packing of the beads. This dilemma has been resolved by the development of so-called *pellicular resins* (Section 6.8.2).

Chemically modified celluloses have proved to be a particularly useful alternative to the polystyrene-based exchangers. Cellulose is a high molecular weight compound which can be obtained in a very pure state. Carboxymethylcellulose (CM-cellulose), where the $—CH_2OH$ group is converted to $—CH_2OCH_2COOH$, and diethylaminoethylcellulose (DEAE-cellulose, $—CH_2OCH_2CH_2N(CH_2CH_3)_2$) are examples of the main derivatives of practical value. Each form is commercially available in gel and bead forms which possess good flow and exchange properties. Closely related to the cellulose-based exchangers are those of the Sepharose type derived from cross-linked agarose (Section 6.6.2). Both the Sephadex and Sepharose types are particularly valuable for the separation of high molecular weight proteins and nucleic acids. Since all these exchangers are closely related to the materials used for exclusion chromatography and have a matrix structure, they all have exclusion limits so that it is probable that some molecular sieving accompanies the ion-exchange process which may help in the overall chromatographic resolution.

All exchangers are characterised by a *total exchange capacity* which is defined as the number of milliequivalents of exchangeable ions available, either per gramme of dried exchanger or per unit volume of hydrated resin. Thus the exchange capacity of Bio-Rad AGl-X4 is 1.2 meq cm^{-3}, Bio-Rex70 3.3 meq cm^{-3}, DEAE-Sephadex A-25 0.5 meq cm^{-3} and CM-Sepharose CL-6B 0.12 meq cm^{-3}. Sometimes *available capacity* is also used to express the available capacity for an arbitrarily chosen molecule such as haemoglobin. Thus the available capacity of DEAE-Sephadex A-25 for haemoglobin is 0.07 g cm^{-3}. These exchange capacities give an indication of the degree of substitution of the exchanger and are therefore a helpful guide in deciding on the scale of a particular application. Details of some commercially available resins are given in Table 6.1. The polystyrene exchangers are obtainable in a number of mesh sizes. All exchangers are generally supplied with an appropriate *counter ion*, normally sodium or chloride. In some cases, when the swollen exchanger has been packed into a column, it may need pre-treatment with acid or alkali to *generate* the desired salt form.

Chromatography on *ion-exchange papers* has also been employed successfully for separating compounds. Since chromatography papers consist primarily of cellulose, it is logical for workers to have modified the papers chemically to produce DEAE-cellulose (strongly basic) and

Table 6.1 Examples of ion-exchangers of biochemical importance.

Type	Polymer	Functional group	Examples of commercial products
Weakly acidic (cation exchanger)	Polyacrylic acid	$-COO^-$	Amberlite IRC 50 Bio-Rex 70 Zeocarb 226
	Cellulose or dextran	$-CH_2COO^-$	CM-Sephadex Cellex CM
	Agarose	$-CH_2COO^-$	CM-Sepharose
Strongly acidic (cation) exchanger)	Polystyrene	$-SO_3^-$	Amberlite IR 120 Bio-Rad AG 50 Dowex 50 Zeocarb 225
	Cellulose or dextran	$-CH_2CH_2CH_2SO_3^-$	SP-Sephadex
Weakly basic (anion exchanger)	Polystyrene	$-CH_2\overset{+}{N}HR_2$	Amberlite IR 45 Bio-Rad AG3 Dowex WGR
	Cellulose or dextran	$-CH_2CH_2\overset{+}{N}H(CH_2CH_3)_2$	DEAE-Sephadex Cellex D
	Agarose	$-CH_2CH_2\overset{+}{N}H(CH_2CH_3)_2$	DEAE-Sepharose
Strongly basic (anion exchanger)	Polystyrene	$-CH_2\overset{+}{N}(CH_3)_3$	Amberlite IRA 401 Bio-Rad AG 1 Dowex 1
		$-CH_2\overset{+}{N}(CH_3)_2$ with CH_2CH_2OH on N	Amberlite IRA 410 Bio-Rad AG2 Dowex 2
	Cellulose or dextran	$-CH_2CH_2\overset{+}{N}(CH_2CH_3)_2$ with $CH_2CH(OH)CH_3$ on N	QAE-Sephadex
		$-CH_2CH_2\overset{+}{N}(CH_2CH_3)_3$	Cellex T

CM-cellulose (weakly acidic) papers, as well as cellulose phosphate (strongly acidic), cellulose citrate (weakly acidic) and aminoethylcellulose (weakly basic) papers. Resin-impregnated papers are also commercially available, for example Amberlite SA-2 (strongly acidic) and Amberlite SB-2 (strongly basic).

6.5.3 Practical procedure and applications

The choice of the ion-exchanger depends upon the stability of the sample components, their molecular weight and the specific requirements of the separation. Many biological components, especially proteins, are only stable within a fairly narrow pH range so the exchanger selected must operate within this range. Generally, if the sample is most stable below its isoionic point giving it a net positive charge, a cation exchanger should be used, whereas if it is most stable above its isoionic point giving it a net negative charge, an anion exchanger should be used. Compounds which are stable over a wide range of pH may be separated by either type of exchanger. The choice between a strong and weak exchanger also depends on sample stability and the effect of pH on sample charge. Weak electrolytes requiring a very low or high pH for ionisation can only be separated on strong exchangers since only they operate over a wide pH range. In contrast, for strong electrolytes weak exchangers are advantageous for a number of reasons including a reduced tendency to cause sample denaturation, their inability to bind weakly charged impurities and their enhanced elution characteristics. Whilst the degree of cross-linking of an exchanger does not influence the ion-exchange mechanism, it does influence its capacity. The molecular weight of the sample component therefore determines the specific exchanger which should be used. The mesh size of polystyrene resins determines the flow rates that can be achieved.

The pH of the buffer used should be at least one pH unit above or below the isoionic point of the compounds being separated. In general, cationic buffers such as Tris, pyridine and alkylamines are used in conjunction with anion exchangers and anionic buffers such as acetate, barbiturate and phosphate are used with cation exchangers. The precise initial buffer pH and ionic strength should be such as to just allow the binding of the sample components to the exchanger. Equally, a buffer of the lowest ionic strength that effects elution should initially be used for the subsequent elution of the components. This ensures that, initially, the minimum number of undesired substances bind to the exchanger and that, subsequently, the maximum number of these impurities remain on the column. The amount of sample which can be applied to a column is dependent upon the size of the column and the capacity of the exchanger. Generally, if the starting buffer is to be used throughout the development of the column (isocratic elution), the sample volume should be 1 to 5% of the bed volume. If, however, gradient elution is to be used, the initial conditions chosen are such that the entire sample is bound by the exchanger at the top of the column. In this case the sample volume is not important and large volumes of dilute solution can be applied, thereby effectively introducing a concentration stage.

Gradient elution is far more common than isocratic elution. Continuous or stepwise pH and ionic strength gradients may be employed but continuous gradients tend to give better resolution with less peak tailing. Generally with an anion exchanger the pH gradient decreases and the ionic strength increases whilst for cation exchangers both the pH and ionic gradients increase.

The separation of amino acids (e.g. in a protein hydrolysate) is usually achieved using a strong acid cation exchanger. The sample is introduced onto the column at a pH of 1 to 2 thus ensuring complete binding of all of the various types of amino acid. Gradient elution using increasing pH and ionic concentration results in the sequential elution of the amino acids. The acidic amino acids, aspartic and glutamic, are eluted first, followed by the neutral amino acids such as glycine and valine. The basic amino acids such as lysine and arginine retain their net positive charge up to pH values of 9 to 11 and are

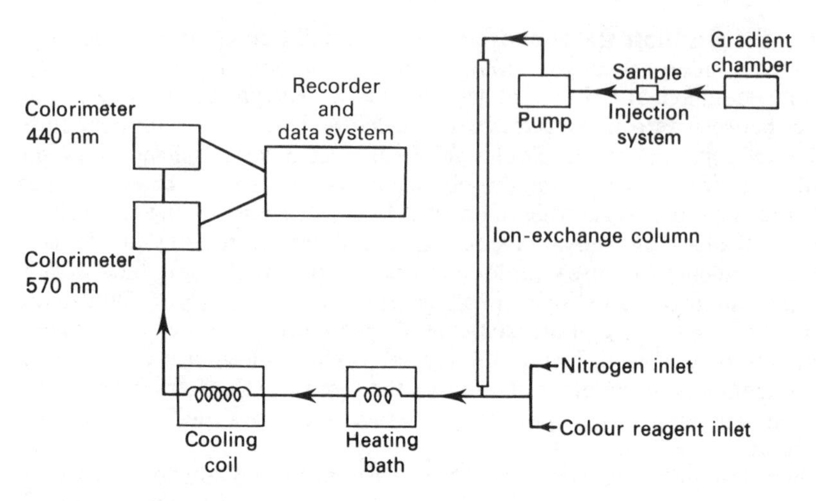

Fig. 6.8 Diagrammatic representation of an amino acid analyser.

eluted last. These principles are embodied in automatic *amino acid analysers*. A diagrammatic representation of such an analyser is shown in Fig. 6.8. The effluent from the column is mixed with ninhydrin colour reagent and nitrogen is also introduced to break the effluent stream into discrete bubbles. The mixture is heated to 105°C to develop the colour, the intensity of which is then determined by two colorimeters, one set at 570 nm to monitor the majority of the amino acids and a second set at 440 nm to specifically monitor the colour produced by proline and hydroxyproline. Alternatively, the amino acids may be detected by conversion to derivatives which fluoresce. Whilst this dispenses with the need for two detectors, the method is more tedious and generally less reproducible than the ninhydrin method. Many amino acid analysers use two separate columns, the second one containing an anion exchanger to separate the basic amino acids and ammonia faster and more effectively.

The ion-exchange chromatography of proteins is best carried out using the weakly acidic or basic exchangers derived from cellulose or agarose. Proteins with an isoionic point of less than 7 are best separated with DEAE-cellulose using a buffer of low ionic strength and pH 8 to 9, whilst proteins with an isoionic point greater than 7 can be resolved on CM-cellulose using a buffer pH 4 to 5. Proteins with an isoionic point in the region of 7 can be chromatographed on either type of exchanger and the choice would depend upon the relative stability of the protein in mildly acidic and mildly alkaline solution. The technique of *chromatofocusing*, the principle of which is similar to that of isoelectric focusing (Section 7.7), is particularly suitable for protein separations. A linear pH gradient is generated in the column by exploiting the buffering action of the exchanger and using amphoteric buffers which have even buffering capacity over a wide range of pH. Proteins elute from such gradients in order of their isoionic points. Focusing effects occur which result in sample concentration, band sharpening and very high resolution.

6.6 Exclusion (permeation) chromatography

6.6.1 Principle

The separation of molecules on the basis of their molecular size and shape utilises the *molecular sieve* properties of a variety of porous materials. Probably the most commonly used of such materials are a group of polymeric organic compounds which possess a three-dimensional network of pores which confer gel properties upon them. The term *gel filtration* is used to describe the separation of molecules of varying molecular size utilising these gel materials. Porous glass granules have been used as molecular sieves and the term *controlled-pore glass chromatography* introduced to describe this separation technique. The terms *exclusion or permeation chromatography* describe all molecular separation processes using molecular sieves. This section is mainly devoted to gel filtration since its principles and applications are best documented, but it must be appreciated that controlled-pore glass chromatography has much in common with it.

The general principle of exclusion chromatography is quite simple. A column of gel particles or porous glass granules is in equilibrium with a suitable solvent for the molecules to be separated. Large molecules which are completely excluded from the pores will pass through the interstitial spaces, while smaller molecules will be distributed between the solvent inside and outside the molecular sieve and will then pass through the column at a slower rate. Three stages in such a column are represented diagrammatically in Fig. 6.9.

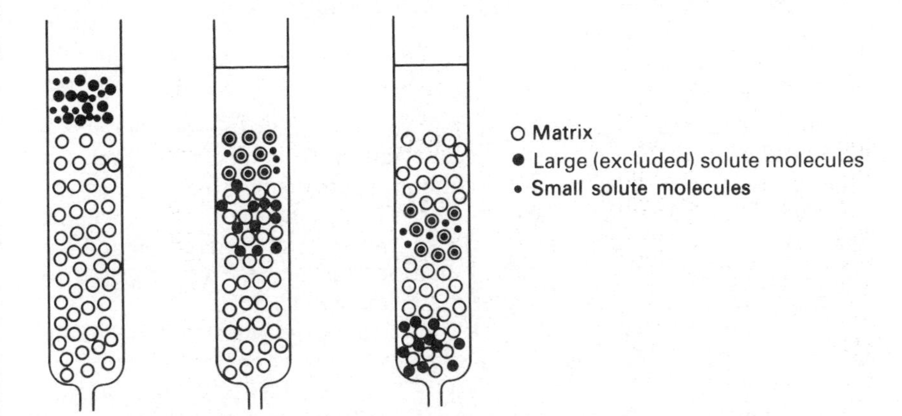

Fig. 6.9 Diagrammatic representation of separation by exclusion chromatography.

The solvent absorbed by a swollen gel is available to a solute to an extent which is dependent upon the porosity of the gel particle and the size of the solute molecules. Thus the distribution of a solute in a column of a swollen gel is determined solely by the total volume of solvent, both inside and outside the gel particles, which is available to it.

For a given type of gel, the distribution coefficient, K_d, of a particular solute between the inner and outer solvent is a function of its molecular size. If the solute is large and completely excluded from the solvent within the gel, $K_d = 0$, whereas if the solute is sufficiently small to gain complete accessibility to the inner solvent, $K_d = 1$. Due to variation in pore size for a given gel, there is some inner solvent which will be available and some which will not be available to solutes of intermediate size, hence K_d values vary between 0 and 1. It is this complete variation of K_d between these two limits which makes possible separation of solutes within a narrow molecular size range on a given gel.

The elution volume, V_e of a given solute depends on the void volume, V_o, the distribution coefficient, and the volume inside the gel matrix itself, V_i,

Thus:

$$V_e = V_o + K_d V_i \qquad 6.6$$

The *inner volume,* V_i, can be calculated from the known dry weight of the gel, a, and the *water regain* value, W_r since:

$$V_i = aW_r \qquad 6.7$$

The numerical value of V_e for a given solute will vary with the size of the column, whereas K_d is a characteristic value for the solute and is independent of the geometry of the gel bed.

For two substances of different molecular weight and K_d values, $K_{d'}$ and $K_{d''}$, the difference in their effluent volumes, V_s, is given by:

$$V_s = V_{e'} - V_{e''} = (V_o + K_{d'} V_i) - (V_o + K_{d''} V_i)$$

therefore:

$$V_s = (K_{d'} - K_{d''}) V_i \qquad 6.8$$

Thus, for complete separation of the two substances, the sample volume must not be larger than V_s. In practice, deviations from ideal behaviour, for example due to poor packing of the column, make it advisable to reduce the sample volume below the value of V_s since the ratio between sample volume and inside gel volume affects both the sharpness of the separation and the degree of dilution of the sample. Equations 6.6 and 6.7 can be used to calculate the optimum bed volume for a given purification.

It is also possible to undertake gel filtration using the thin layer mode. *Thin layer gel filtration* (TLG) and TLC have much in common, but there are some important distinctions. In TLG a layer of swollen gel is spread onto a glass plate. The gel beads adhere to the plate without the addition of a fixative, and form the stationary phase; the interstitial fluid forms the mobile phase. In contrast to TLC, the layer must not be dried, hence in TLG there is no solvent front. The TLG plate is placed in an airtight container and connected to reservoirs at either end by means of filter paper bridges. The plate is inclined at an angle of 20° to the horizontal, thus facilitating transport of the mobile phase through the layer. Equilibration must be carried out for a minimum of

12 hours. The main function of this equilibration is to normalise the ratio between the stationary and mobile phase volumes. It is possible to use a horizontal plate and produce solvent flow by having the two reservoirs at different levels. The sample is applied as a spot or band and the plate developed for a suitable period. The separated spots are detected by an appropriate method.

Whereas TLC is mainly used for the separation of amino acids, sugars and oligosaccharides, alkaloids, steroids, and lipophilic substances in general, TLG is used for the separation of hydrophilic substances requiring mild conditions, such as proteins, peptides, nucleic acids, i.e. high molecular weight biological material.

The great advantage of TLG over column gel filtration is that a number of samples can be chromatographed at the same time, under identical conditions. In addition, very small amounts of the sample can be used; thus it is ideal for clinical samples.

6.6.2 Materials

Gels which are commonly used include cross-linked dextrans (trade name Sephadex), agarose (Sepharose, Bio-Gel A, Sagavac), polyacrylamide (Bio-Gel P), polyacryloylmorphine (Enzocryl Gel) and polystyrenes (Bio-Beads S).

The dextran gels are obtained by cross-linking the polysaccharide dextran with epichlorhydrin. In this way the water soluble dextran is made water insoluble, but it retains its hydrophilic character and swells rapidly in aqueous media, forming gel particles suitable for gel filtration. By varying the degree of cross-linking several types of Sephadex have been obtained. They differ in porosity and consequently are useful over different molecular size ranges. Due to the random distribution of cross-linking there is also a wide distribution of pore sizes in each gel type. This means that molecules of a size below the limit where complete exclusion occurs are either partly or fully able to enter the gel. Each type of Sephadex is characterised by its water regain, i.e. the amount of water taken up in the completely swollen gel granules by one gramme of Sephadex.

Agarose gels, which are produced from agar, are linear polysaccharides consisting of alternating residues of D-galactose and 3,6-anhydro L-galactose units. Their gelling properties are attributed to hydrogen bonding of both inter- and intra-molecular type. Due to their hydrophilic nature and the nearly complete absence of charged groups, agarose gels, like dextran gels, cause very little denaturation and adsorption of sensitive biochemical substances. By virtue of their greater porosity they complement the dextran gels. Whereas the latter allow fractionation of spherical molecules such as globular proteins, of dimensions corresponding to molecular weights of up to 800 000 daltons, or randomly coiled polymers like dextran of molecular weights up to 200 000 daltons, the agarose gels may be used to separate molecules and particles up to a molecular weight of several million

Table 6.2 Some commonly used gels for exclusion chromatography.

Polymer	Trade name		Fractionation range* (Daltons)	Bed volume ($cm^3\,g^{-1}$ dry gel)
Dextran	† Sephadex	G10	<700	2–3
		G25	1×10^3 to 5×10^3	2–3
		G50	1.5×10^3 to 3×10^4	4–6
		G100	4×10^3 to 1.5×10^5	15–20
		G200	5×10^3 to 6×10^5	30–40
	† Sephacryl	S200	5×10^3 to 2.5×10^5	**
		S300	1×10^4 to 1.5×10^6	**
		S400	2×10^4 to 8×10^6	**
Agarose	† Sepharose	2B	1×10^4 to 4×10^6	**
		4B	6×10^4 to 2×10^7	**
		6B	7×10^4 to 4×10^7	**
	†† Bio-Gel	A5m	1×10^4 to 5×10^6	**
		A15m	4×10^4 to 1.5×10^7	**
		A50m	1×10^5 to 5×10^7	**
		A150m	1×10^6 to 1.5×10^8	**
Polyacrylamide	†† Bio-Gel	P2	1×10^2 to 1.8×10^3	3–4
		P6	1×10^3 to 6×10^3	7
		P30	2.5×10^3 to 4×10^4	11
		P100	5×10^3 to 1×10^5	15
		P300	6×10^4 to 4×10^5	30

*Determined for globular proteins. The range is approximately the same for single-stranded nucleic acids and smaller for fibrous proteins and double-stranded DNA.
†Manufactured by Pharmacia Biotechnology, Uppsala, Sweden.
††Manufactured by Bio-Rad, Richmond, California, USA.
**Supplied fully hydrated.

daltons. They have, therefore, been widely used in the study of viruses, nucleic acids and polysaccharides.

Polyacrylamide gels are prepared by the polymerisation of acrylamide and methylene bisacrylamide. By varying the relative proportions of the two monomers, a range of gels with differing porosities may be obtained. They have characteristics very similar to the dextran and agarose gels. They have a molecular weight exclusion limit ranging from 1800 to 400 000 daltons.

Some commonly used gels are listed in Table 6.2. The Sephadex and polyacrylamide gels must be converted to their swollen form before use whereas the Sephacryl and agarose gels are supplied in the pre-swollen form. Many gels are available in a range of sizes: superfine, fine, medium and coarse. The coarser the bead, the better column flow rate but poorer resolution it gives. Thus the fine and superfine beads are preferred for analytical work and the coarse ones for preparative purposes. The *capacity* of a particular gel is a measure of the weight of a solute that can penetrate a particular weight of gel. It therefore gives an indication of the amount of solute that can be separated by a particular column of gel.

6.6.3 Applications

Purification The main application of exclusion chromatography is in the purification of biological macromolecules. Viruses, proteins, enzymes, hormones, antibodies, nucleic acids and polysaccharides have all been separated and purified by use of appropriate gels or glass granules. Mixtures of lower molecular weight compounds may also by separated. Thus amino acids can be separated from peptides, and peptides obtained from the partial hydrolyses of a protein can be fractionated, as can oligonucleotides from a nucleic acid hydrolysate. Low molecular weight dextrans, in such mixtures as corn syrup oil, can also be separated.

Molecular Weight Determination The effluent volumes of globular proteins are largely determined by their molecular weight. It has been shown, that over a considerable molecular weight range, the effluent volume is approximately a linear function of the logarithm of the molecular weight. Hence the construction of a calibration curve, with proteins of a similar shape and known molecular weight, enables the molecular weight of other proteins, even in crude preparations, to be estimated.

Solution Concentration Solutions of high molecular weight substances can be concentrated by the addition of dry Sephadex G-25 (coarse). Water and low molecular weight substances are absorbed by the swelling gel while the high molecular weight substances remain in solution. After ten minutes the gel is removed by centrifugation, leaving the high molecular weight material in a solution whose concentration has increased but whose pH and ionic strength are unaltered.

Desalting By use of a column of Sephadex G-25, solutions of high molecular weight compounds may be desalted. The high molecular weight substances move with the void volume while the low molecular weight components are distributed between the mobile and stationary phases and hence move slowly. This method of desalting is faster and more efficient than dialysis. Applications include removal of phenol from nucleic acid preparations, ammonium sulphate from protein preparations as well as monosaccharides from polysaccharides and amino acids from proteins.

Protein-Binding Studies Exclusion chromatography is one of a number of methods commonly used to study the reversible binding of a ligand to a macromolecule such as a protein including receptor proteins (Section 3.7). A sample of the protein/ligand mixture is applied to a column of a suitable gel (e.g. G-25) which has previously been equilibrated with a solution of the ligand of the same concentration as that in the mixture. The sample is eluted with buffer in the standard way and the concentration of ligand and protein in the effluent determined. The early fractions will contain unbound ligand, but the subsequent appearance of the protein will result in an increase in the total amount of ligand (bound plus unbound). By repeating the experiment at a series of ligand concentrations the appropriate binding constants can be calculated.

6.7 Affinity chromatography

6.7.1 Principle

Purification by affinity chromatography is unlike all other forms of chromatography and such techniques as electrophoresis and centrifugation, in that it does not rely on differences in the physical properties of the molecules to be separated. Instead, it exploits the unique property of extremely specific biological interactions to achieve separation and purification. As a consequence, affinity chromatography is theoretically capable of giving absolute purification, even from complex mixtures, in a single process. The technique was originally developed for the purification of enzymes, but it has since been extended to nucleotides, nucleic acids, immunoglobulins, membrane receptors and even whole cells and cell fragments.

The technique requires that the material to be isolated is capable of reversibly binding to a specific ligand which is attached to an insoluble matrix.

$$\underset{\text{macro-molecule}}{\mathrm{M}} + \underset{\text{ligand (attached to matrix)}}{\mathrm{L}} \underset{k_{-l}}{\overset{k_{+l}}{\rightleftharpoons}} \underset{\text{complex}}{\mathrm{ML}}$$

Under the correct experimental conditions when a complex mixture containing the specific compound to be purified is added to the insolubilised ligand, generally contained in a conventional chromatography column, only that compound will bind to the ligand. All other compounds can therefore be washed away and the compound subsequently recovered by displacement from the ligand (Fig. 6.10).

The method therefore requires a detailed preliminary knowledge of the structure and biological specificity of the compound to be purified, so that the separation conditions which are most likely to be successful may be carefully planned. In the case of an enzyme, the ligand may be a substrate or a reversible inhibitor or activator. The conditions chosen would normally be those that are optimum for enzyme-ligand binding. Since the success of the method relies on the reversible formation of the complex and on the numerical values of the first order rate constants k_{+l} and k_{-l}, as the enzyme is progressively added to the insolubilised ligand in a column, the enzyme molecules will be stimulated to bind and a dynamic situation develops in which the complex concentration and the strength of binding increases. It is because of this progressive increase in effectiveness during the addition of the sample to the column, that column procedures are invariably more successful than batch type methods. Nevertheless, alternative forms have been developed and are particularly suitable for large-scale work. They include *affinity precipitation* in which the ligand is attached to a soluble carrier which subsequently can be precipitated by, for example, a pH change, and *affinity partitioning* in which the ligand is attached to a water-soluble polymer such as polyethylene glycol and which, with the ligand bound, preferentially

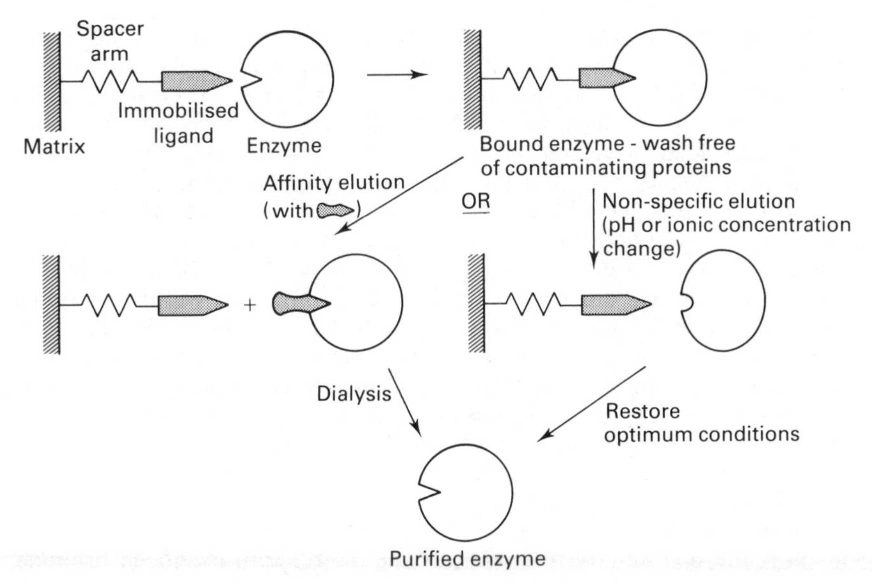

Fig. 6.10 Diagrammatic representation of purification of an enzyme by affinity chromatography.

partitions into an aqueous polymer phase in equilibrium with a pure aqueous phase.

6.7.2 Materials

Matrix An ideal insoluble matrix for affinity chromatography must possess the following characteristics:

(i) it must contain suitable and sufficient chemical groups to which the ligand may be covalently coupled and it must be stable under the conditions of the attachment;
(ii) it must be stable during binding of the macromolecule and its subsequent elution;
(iii) it must at the most interact only weakly with other macromolecules to minimise non-specific adsorption;
(iv) it should exhibit good flow properties.

In practice, particles which are uniform, spherical and rigid are used. The most common ones are the cross-linked dextrans (e.g. Sephacryl S), agarose (e.g. Sepharose, Bio-Gel A), polyacrylamide gels (e.g. Bio-Gel P) the characteristics of which were described in Section 6.6.2, polystyrene (e.g. Bio-Beads S), cellulose and porous glass and silica.

Selection and Attachment of Ligand The chemical nature of the ligand is determined by the prior knowledge of the biological specificity of the compound to be purified. In practice it is frequently possible to select a ligand which displays *absolute specificity* in that it will bind exclusively to one particular compound. Alternatively, it is possible to select a ligand which displays *group selectivity* in that it will bind to a closely related group of compounds which possess a similar in-built chemical specificity. An example of the latter type of ligand is 5′AMP which can reversibly bind to many NAD^+-dependent dehydrogenases since 5′AMP is structurally similar to part of the NAD^+ structure. It is essential that the ligand possesses a suitable chemical group which will not be involved in the reversible binding of the ligand to the macromolecule, but which can be used to attach the ligand to the matrix. The most common such groups are $—NH_2$, —COOH, —SH and —OH (phenolic and alcoholic). To prevent the attachment of the ligand to the matrix interfering with its ability to bind the macromolecule, it is generally advantageous in interpose a *spacer arm* between the ligand and the matrix. The optimum length of this spacer arm is six to ten carbon atoms or their equivalent. In some cases, the chemical nature of this spacer is critical to the success of the separation. Some spacers are purely hydrophobic, most commonly consisting of methylene groups, others are hydrophilic, possessing carbonyl or imido groups.

The most common method of attachment of the ligand to the matrix involves the preliminary treatment of the matrix with cyanogen bromide (CNBr) at pH 11. The reaction conditions and the relative proportion of the reagents will determine the number of ligand molecules which can be attached to each matrix particle. CNBr-activated polysaccharides such as Sepharose 4B and 6B are commercially available. Other activated matrices commercially available include thiopropyldextran or thiopropylagarose agarose, tresyl (2,2,2-trifluoroethanesulphonyl)dextran and epoxy-1, 4-bis (2,3-epoxypropoxy) butane derivatives of agarose or dextran. Full details of the procedure for the attachment of the ligand are provided by the manufacturer.

Many different spacer arms are used. Examples include 1,6-diaminohexane, 6-aminohexanoic acid and 1,4-bis-(2,3-epoxypropoxy)butane. They must possess a second functional group to which the ligand may be attached by conventional organo-synthetic procedures which frequently involve the use of succinic anhydride and a water soluble carbodiimide. A number of supports of the agarose, dextran and polyacrylamide type are commercially available with a variety of spacer arms and ligands pre-attached ready for immediate use. Examples of ligands are given in Table 6.3.

Practical Procedure The procedure for affinity chromatography is similar to that used in other forms of liquid chromatography. The ligand-treated matrix is packed into a column in the normal way for the particular type of support (Section 6.1.2). A buffer is used which will encourage the binding macromolecule to be strongly bound to the ligand. The buffer generally has a high ionic strength to minimise non-specific adsorption of polyelectrolytes onto any charged groups in the ligand. The buffer must also contain any cofactors such as metal ions necessary for ligand-macromolecule interaction. Once the sample has been applied and the macromolecule bound, the column is eluted with more buffer to remove non-specifically bound contaminants.

Table 6.3 Group specific ligands commonly used in affinity chromatography.

Ligand	Affinity
1. 5′ AMP	NAD^+-dependent dehydrogenases and certain kinases
2. 2′,5′ADP	$NADP^+$-dependent dehydrogenases
3. $-C_6H_4-B(OH)_2$	Compounds with coplanar cis-diol groups, e.g. sugars, nucleosides, nucleotides, catecholamines
4. $-C_6H_4-HgCl$	Proteins containing SH-groups
5. Poly(U)	m-RNA which contain a poly(A) tail
6. Poly(A)	Ribonucleic acids which contain a poly(U) sequence; RNA-specific proteins such as nucleic acid polymerases
7. Lysine	rRNA; plasminogen
8. Concanavalin A	Glycoproteins and glycopeptides; glycolipids; membrane fragments containing α-D-mannopyranosyl and α-D-glucopyranosyl residues
9. Calmodulin	Proteins regulated by calmodulin
10. Heparin	A wide range of proteins including lipoproteins, lipases, coagulation proteins and steroid receptors
11. Protein A (a protein isolated from cell walls of *Staphylococcus aureus*)	IgG and molecules which contain the Fc region of IgG
12. Cibacron Blue F3GA (a reactive anthraquinone-type dye)	Nucleotide-requiring enzymes; blood coagulation factors; albumin
13. Lectin (from *Triticum vulgare*)	Cells and macromolecules containing *N*-acetyl-β-glucosamine residues
14. Lectin (from *Helix pomatia*)	Cells and macromolecules containing *N*-acetyl-α-galactosamine residues

The purified compound is finally recovered by either *specific or non-specific elution*. Non-specific elution may be achieved by a change in either pH or ionic strength. pH shift elution using dilute acetic acid or ammonium hydroxide results from a change in the state of ionisation of groups in the ligand and/or the macromolecule which are critical to ligand-macromolecule binding. A change in ionic strength, not necessarily with a concomitant change in pH, also causes elution due to a disruption of the ligand-macromolecule interaction; 1M NaCl is frequently used for this purpose. *Affinity elution* involves the addition of substrates or reversible inhibitors of the macromolecule if it is an enzyme, otherwise the addition of compounds for which the ligand has a higher affinity than it has for the bound material. The purified material is eventually recovered in a buffered solution which may be contaminated with specific eluting agents or high concentrations of salt and these must finally be removed before the isolation is complete (Fig. 6.10).

6.7.3 Applications

A wide range of enzymes and other proteins, including receptor proteins and immunogobulins, has been purified by affinity chromatography. The application of the technique is limited only by the availability of immobilised ligands. The principles have been extended to nucleic acids and have made a considerable contribution to recent developments in molecular biology. Messenger RNA is routinely isolated by selective hybridisation on Poly(U)-Sepharose 4B and viral RNA on the corresponding Poly(A). Immobilised single-stranded DNA can be used to isolate complementary RNA and DNA. Whilst this separation can be achieved on columns, it is usually performed using single-stranded DNA immobilised on nitrocellulose filters (Section 5.9.1). Immobilised nucleotides are useful for the isolation of proteins involved in nucleic acid metabolism. Affinity chromatography can be exploited to concentrate dilute solutions and to separate native and denatured macromolecules.

Cibacron-Blue is one of a number of triazine dyes which can be immobilised and used as a ligand to purify proteins. The dyes contain ionic groups and a conjugated ring system which have the ability to bind to the catalytic or effector site of some proteins. The interaction is not genuinely specific so the term *dye-ligand chromatography* is sometimes preferred for this technique. Proteins purified by the procedure include interferon, plasminogen and restriction endonucleases. *Metal chelate chromatography* or *immobilised metal affinity chromatography* is another logical extension of the basic technique. It enables proteins with similar molecular weights and isoelectric points to be separated on the basis of their differential binding to metal ions which have been immobilised by chelation. Binding to such metal ions as Zn^{2+}, Cu^{2+}, Cd^{2+}, Hg^{+}, Co^{2+} and Ni^{2+} is pH dependent. The sample is applied at neutral pH and the compound eluted by reducing the pH and ionic strength of the buffer or by including EDTA in the buffer. The two most commonly used chelating ligands for the metal ions are iminodiacetic acid and tris(carboxymethyl)ethylenediamine. The binding of the metal atom to the protein invariably involves a histidine residue. Proteins purified by the technique include fibrinogen, superoxide dismutase and non-histone nuclear proteins.

A valuable development of affinity chromatography is its use for the separation of a mixture of cells into homogeneous populations. The technique relies on the antigenic properties of the cell surface or the chemical nature of exposed carbohydrate residues on the cell surface or on a specific membrane receptor-ligand interaction. The immobilised ligands used include Protein A, which is capable of binding to the Fc region of IgG, a lectin or the specific ligand for the membrane receptor.

Covalent chromatography differs from other forms of affinity chromatography in that it involves covalent bond formation between the bound ligand and the compound to be separated (most commonly proteins). The most common form involves the formation of a disulphide bond between thiol groups in the compound and ligand. Commercially available ligands include thiopropyl-Sepharose and thiol-Sepharose. Elution is carried out with dithiothreitol or cysteine, the success of the technique depending on the number of thiol groups in the protein and the ease with which the disulphide bonds are broken by the eluant. Papain and urease which have many thiol groups are both readily purified by the technique and newly synthesised mRNA can be separated from other RNA and DNA by a similar procedure. The method has also been used to isolate whole genes by hybridising partially single stranded DNA with the complementary mercurated mRNA which is then reacted with a thiolated matrix.

6.8 High performance (pressure) liquid chromatography (HPLC)

6.8.1 Principle

As can be seen from equations 6.1 to 6.5, the resolving power of a chromatographic column increases with column length and the number of theoretical plates per unit length, although there are limits to the length of a column due to the problem of peak broadening. As the number of theoretical plates is related to the surface area of the stationary phase it follows that the smaller the particle size of the stationary phase, the better the resolution. Unfortunately, the smaller the particle size, the greater the resistance to eluant flow. All of the forms of column chromatography so far discussed rely on gravity or low pressure pumping systems for the supply of eluant to the column. The consequences of this is that the flow rates achieved are relatively low and this gives greater time for band broadening by simple diffusion phenomena. The use of faster flow rates is not possible because it creates a back-pressure which is sufficient to damage the matrix structure of the stationary phase, thereby actually reducing eluant flow and impairing resolution. In the past decade there has been a dramatic development in column chromatography technology which has resulted in the availability of new smaller particle size stationary phases which can withstand these pressures and of pumping systems which can give reliable flow rates. These developments, which have occurred in adsorption, partition, ion-exchange, exclusion and affinity chromatography, have resulted in faster and better resolution and explain why HPLC has emerged as the most popular, powerful and versatile form of chromatography.

Originally, HPLC was referred to as high pressure liquid chromatography but nowadays the term high performance liquid chromatography is preferred since it better describes the characteristics of the chromatography and avoids creating the impression that high pressures are an inevitable pre-requisite for high performance. This is now known not to be the case and the term *medium pressure liquid chromatography* (MPLC) has been coined for some separations. The principal components of a HPLC system are shown in Fig. 6.11.

The new technology in stationary phases has been applied to thin-layer chromatography giving rise to *high performance thin-layer chromatography* (HPTLC). In general, however, the impact of this new technology has not been quite so great as it has been in column chromatography.

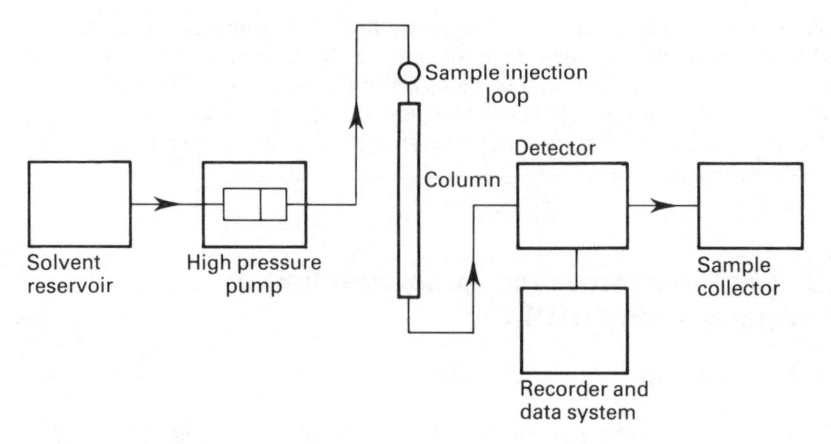

Fig. 6.11 Diagrammatic representation of the components of an isocratic HPLC system.

6.8.2 Apparatus and materials

The Column The columns used for HPLC are generally made of stainless steel and are manufactured so that they can withstand pressures of up to 5.5×10^7 Pa (8000 p.s.i). Straight columns of 20 to 50 cm in length and 1 to 4 mm in diameter are generally used though smaller capilliary columns are available. The best columns are precision bored with an internal mirror finish which allows efficient packing of the column. Porous plugs of stainless steel or teflon are used in the ends of the columns to retain the packing material. The plugs must be homogeneous to ensure uniform flow of solvent through the column. It is important in some separations involving liquid partition and ion-exchange that the column temperature is thermostatically controlled during the analysis.

Column Packing Three forms of column packing material are available based on a rigid solid (as opposed to gel) structure. These are:

(i) *microporous supports* where micropores ramify through the particles which are generally 5 to 10 μm in diameter;

(ii) *pellicular* (superficially porous) supports where porous particles are coated onto an inert solid core such as a glass bead of about 40 μm in diameter:

(iii) *bonded phases* where the stationary phase is chemically bonded onto an inert support.

For adsorption chromatography, adsorbents such as silica or alumina are available as microporous or pellicular forms with a range of particle sizes. Pellicular systems generally have a high efficiency but low sample capacity, and therefore microporous supports are preferred where applicble. All forms of HPLC column packing are characterised by their regular spherical shape which distinguishes them from conventional materials. These small spheres pack most efficiently and give good flow properties.

In liquid-liquid partition systems, the stationary phase may be coated onto the inert support. Both microporous and pellicular supports are used for supporting the liquid phase. One disadvantage of supports coated with liquid phases is that the developing solvent may gradually wash off the liquid phase with repeated use. To overcome this problem bonded phases have been developed where the liquid phase has been covalently bonded to the supporting material which may be silica or a silicone polymer. The silicone polymer bonded phases have the particular advantage that as well as not being eluted by the developing solvent, they are chemically, hydrolytically and thermally stable. In normal phase liquid-liquid chromatography, the stationary phase is a polar compound such as alkyl nitrile or alkylamine derivatives and the mobile phase a non-polar solvent such as hexane. For reverse-phase chromatography, the stationary phase is a non-polar compound such as a C_8 or C_{18} hydrocarbon and the mobile phase a polar solvent such as water/acetonitrile or water/methanol mixtures.

Many different types of ion-exchangers are available of which the cross-linked microporous polystyrene resins are widely used. Pellicular resin forms are also available, as are bonded phase exchangers covalently bonded to a cross-linked silicone network. These resins are classed as *hard gels* and readily withstand the pressures required during analysis.

The stationary phases for exclusion separations are generally porous silica, glass, polystyrene or polyvinylacetate beads. These are generally used where the eluting solvent is an organic system, and the beads are available in a range of pore sizes. Semirigid gels such as Sephadex or Bio-Gel P and non-rigid gels such as Sepharose and Bio-Gel A are only of limited use in HPLC since they can withstand only low pressures. The supports for affinity separations are similar to those for exclusion separations. The spacer arm and ligand are attached to these supports by similar chemical means to those used in conventional low pressure affinity chromatography (Section 6.7.2). Table 6.4 lists some examples of commonly used stationary phases and their applications.

Column Packing Procedure Columns may be purchased already packed from commercial companies with specified packing material structure and dimensions. Many workers, however, prefer to pack their own columns since this is cheaper than purchasing pre-packed columns. Several methods are available for packing columns and the method used will depend on the nature of the packing material and the dimensions of the particles. The major priority in the packing of a column is to obtain a uniform bed of material with no cracks or channels. Rigid solids and hard gels should be packed as densely as possible, but without fracturing the particles during the packing process. The most widely used technique for column packing is the high pressure slurrying technique. A suspension of the packing is made in a solvent of equal density to the packing material. The slurry is then rapidly pumped at high pressure into a column with a porous plug at its outlet. The resulting bed of packed material within the column can then be prepared for use by running the developing solvent through the column, hence equilibrating the packing with the developing solvent. When hard gels are packed, it is necessary for them to be allowed to swell first in the solvent to be used in the chromato-

Table 6.4 Some examples of HPLC stationary phases and their applications.

Chromatographic separation principle	Commercial name	Nature of stationary phase	Type of support	Applications
Adsorption	Corasil	Silica	Pellicular	Steroids; vitamins; chlorinated pesticides; polar herbicides; plant pigments; triglycerides; alkaloids
	Pellumina	Alumina	Pellicular	
	Partisil	Silica	Microporous	
	MicroPak A1	Alumina	Microporous	
Partition	Bondapak-C_{18}/Corasil	Octadecylsilane	Pellicular	Dansylated amino acids; drugs; pesticides; aflatoxins; saccharides; fatty acids
	μ Bondapak-C_{18}	Octadecylsilane	Porous	
	UltroPac TSK ODS	Octadecylsilane	Porous	
	μ Bondapak-NH_2	Alkylamine	Porous	
	UltroPac TSK-NH_2	Alkylamine	Porous	
Ion-exchange	Partisil-SAX	Strong base	Porous	Amino acids; peptides; proteins; nucleotides; adrenaline; drugs and their polar metabolites
	MicroPak-NH_2	Weak base	Porous	
	Partisil-SCX	Strong acid	Porous	
	AS Pellionex-SAX	Strong base	Pellicular	
	Zipak-WAX	Weak base	Pellicular	
	Perisorb-KAT	Strong acid	Pellicular	
Exclusion	Bio-Glas	Glass	Rigid solid	Proteins; peptides; nucleic acids; nucleotides; polysaccharides; oligosaccharides
	Styragel	Polystyrene-divinylbenzene	Semi-rigid gel	
	Sephadex	Agarose	Soft gel	
	Fractogel TSK	Polyvinyl	Semi-rigid gel	

graphic process before packing under pressure. Soft gels cannot be packed under pressure and have to be allowed to pack from a slurry in the column under gravitational sedimentation only, in a similar way to the packing of columns for conventional column chromatography (Section 6.1.2).

Chromatographic Solvent (mobile phase) The choice of mobile phase to be used in any separation will depend on the type of separation to be achieved. Isocratic separations may be made with a single solvent, or two or more solvents mixed in fixed proportions. Alternatively a gradient elution system may be used where the composition of the developing solvent is continuously changed by use of a suitable gradient programmer. In the majority of cases this involves the use of two pumps. All solvents for use in HPLC systems must be specially purified since traces of impurities can affect the column and interfere with the detection system. This is particularly the case if the detection system is measuring absorbance at below 200 nm. Purified solvents for use in HPLC systems are available commerically, but even with these solvents a 1 to 5 μm microfilter is generally introduced into the system prior to the pump. It is also essential that all solvents are degassed before use otherwise *gassing* tends to occur in most pumps. It tends to be particularly bad for aqueous methanol and ethanol solvents. Gassing (the presence of air bubbles in the solvent) can alter column resolution and interfere with the continuous monitoring of the column effluent. Degassing may be carried out in several ways; by warming the solvent, by stirring it vigorously with a magnetic stirrer, subjecting it to a vacuum, ultrasonic vibration, or by bubbling helium gas through the solvent reservoir.

Pumping Systems The pumping system is one of the most important features of an HPLC system. There is a high resistance to solvent flow due to the narrow columns packed with small particles, and high pressures are therefore required to achieve satisfactory flow rates. The main feature of a good pumping system is that it is capable of outputs of at least 3.4×10^7 Pa (5000 p.s.i) and ideally there must be no pulses of flow through the system. There must be a flow delivery of at least 10 cm^3 min^{-1} for normal analysis, and up to 30 cm^3 min^{-1} for preparative analysis. All materials in the pump should be chemically resistant to all solvents. Various pumping systems are available which operate on the principle of constant pressure or constant displacement.

Constant pressure pumps produce a pulseless flow through the column, but any decrease in the permeability of the column will result in lower flow rates for which the pump will not compensate. These pumps operate by the introduction of high pressure gas into the pump, and the gas in turn forces the solvent from the pump chamber into the column. The use of an intermediate solvent between the gas and the eluting solvent reduces the chances of dissolved gas directly entering the eluting solvent and causing problems during the analysis.

Constant displacement pumps maintain a constant flow rate through the column irrespective of changing conditions within the column. One form of constant displacement pump is a motor-driven syringe type pump where a fixed volume of solvent is forced from the pump to the column by a piston driven by a motor. Such pumps, as well as providing uniform solvent flow rates, also yield a pulseless solvent flow which is important as certain detectors are sensitive to changes in solvent flow rate. The *reciprocating pump* is the most commonly used form of constant displacement pump. The piston is moved by a motorised crank, and entry of solvent from the reservoir to the pump chamber and exit of solvent to the column is regulated by check valves. On the compression stroke solvent is forced from the pump chamber into the column. During the return stroke the exit check valve closes and solvent is drawn in via the entry valve to the pump chamber, ready to be pumped onto the column on the next compression stroke. Such pumps produce pulses of flow and pulse dampeners are usually incorporated into the system to minimise this pulsing effect. All constant displacement pumps have in-built safety cut-out mechanisms so that if the pressure within the chromatographic systems changes from pre-set limits the pump is inactivated.

Detector Systems Since the quantity of material applied to the column is frequently very small, it is imperative that the sensitivity of the detector system is sufficiently high and stable. Most commonly the detector is a variable wavelength ultraviolet-visible spectrophotometer, a fluorimeter, a refractive index monitor or an electrochemical detector. A recent development has been the interfacing of HPLC to a mass spectrometer.

Practical Procedure The correct application of a sample onto a HPLC column is another particularly important factor in achieving successful separations. Ideally the sample ought to be introduced as an infinitely narrow band onto the column. There are two methods which are generally used. The first method makes use of a microsyringe designed to withstand high pressures. The sample is injected through a septum in an injection port, either directly onto the column packing or onto a small plug of inert material immediately above the column packing. This can be done while the system is under pressure, or the pump may be turned off before injection, and when the pressure has dropped to near atmospheric, the injection is made and the pump switched on again. This is termed a *stop flow injection*. The second method of sample introduction is by use of a *loop injector*. This consists of a metal loop of small volume which can be filled with the sample. By means of an appropriate valve, the eluant from the pump is channelled hrough the loop, the outlet of which leads directly onto the column. The sample is thus flushed onto the column by the eluant, without interruption of solvent flow to the column. Automatic versions of loop injectors are commercially available.

Repeated application of highly impure samples such as sera, urine, plasma or whole blood, which have preferably been deproteinated, may eventually cause the column to lose its resolving power. To prevent this occurrence a *guard column* is installed between the injector and the analytical column. This guard column is a short (2 to 10 cm) column of the same internal diameter, and packed with similar material to that present in the analytical column. The packing of the guard column can be replaced at regular intervals.

6.8.3 Applications

The wide applicability, speed and sensitivity of HPLC have resulted in it becoming the most popular form of chromatography and virtually all types of biological molecules have been purified using the approach. *Reverse phase partition HPLC* is particularly useful for the separation of polar compounds such as drugs and their metabolites, peptides, vitamins, polyphenols and steroids. Prior to the advent of this form of chromatography, the separation of such polar compounds was not easily accomplished and often required pre-derivatisation to less polar compounds. The technique is particularly widely used in clinical and pharmaceutical work as it is possible to apply biological fluids such as serum and urine directly to the column, preferably using a guard column. The separation of some highly polar compounds, such as amino acids, organic acids and the catecholamines which are difficult to resolve, adequately, by reverse phase chromatography, can often be improved by one of two possible approaches. The first is *ion-suppression* in which the ionisation of the compound is suppressed by chromatographing at an appropriately high or low pH. Weak acids, for example, can be chromatographed using an acidified mobile phase. The second is *ion-pairing* in which a *counter ion* with opposite charge to that to be separated is added to the mobile phase so that the resulting ion-pair has sufficient lipophilic character to be retained by the non-polar stationary phase of a reverse phase system. Thus to aid the separation of acidic compounds which would be present as their conjugate anions, a quaternary alkylamine ion such as tetrabutylammonium would be used as the counter ion whereas for the separation of bases which would be present as a cation, an alkyl sulphonate such as sodium heptanesulphonate would be used. The mechanism by which ion-pairing results in better separation is not clear but two theories have been proposed. The first suggests that the ion-pair behaves as a single neutral species, whilst the second suggests that an active ion-exchange surface is produced in which the counter ion, which has considerable lipophilic properties, and the ions to be separated are adsorbed by the hydrophobic, non-polar stationary phase. In practice, the success of the ion-pairing approach is variable and somewhat empirical. The size of the counter ion, its concentration and the pH of the solution are all factors which may profoundly influence the outcome of the separation.

HPLC has probably had the biggest impact on the separation of oligopeptides and proteins. Instruments dedicated to the separation of proteins have given rise to the technique of *fast protein liquid chromatography* (FPLC). There are no unique principles associated with FPLC, it is simply based on reverse phase and ion-exchange chromatography and on chromatofocusing (Section 6.5.3). Microbore glass-lined stainless steel columns 1 mm diameter and 2.5 cm long have recently been developed which enable very small amounts of sample to be used with separation taking as little as 10 minutes. The technique enables such complex mixtures as tryptic digests of proteins and the culture supernatant of microorganisms to be applied directly to the column which most commonly contains an ion-exchange system. Protein mixtures from cell extracts still need some form of preliminary fractionation (Section 3.2.3) prior to study. Although high performance exclusion and ion-exchange chromatography are so successful for protein separations, not all proteins can be completely purified using them. In these cases, the technique of *hydrophobic interaction chromatography* which exploits hydrophobic regions on the surfaces of proteins, may be successful. The stationary phase is strongly hydrophobic and most commonly is octyl- or phenylagarose. The hydrophobic regions of the proteins' surface interacts with this phase by π–π bond interaction. This minimises interaction of the protein with the aqueous environment. Binding is accomplished in dilute (0.01M) buffer and elution carried out either with aqueous ethylene glycol or ethanol or by the addition of so-called *chaotropic compounds* (perchlorate, trifluroacetate or thiocyanate ions or urea) which disrupt water structure and thus discourage hydrophobic interactions. Proteins purified by this technique include aldolase, transferrin, cytochrome *c* and thyroglobulin.

6.9 Selection of a chromatographic system

It is possible to rationalise to some extent the type of system most likely to be applicable to the separation of compounds for which the physical characteristics are known. Figure 6.12 summarises the approach.

The majority of chromatographic procedures exploit differences in physical properties of compounds. The exception is affinity chromatography which is based upon the specific ligand-binding properties of biological macromolecules. If this form of chromatography can be applied it is the most likely to be successful. Otherwise, volatile compounds are best separated by gas-liquid chromatography whilst non-volatile compounds which are soluble in organic solvents are generally best separated either by adsorption or normal phase partition chromatography. If the compounds have different functional groups, adsorption chromatography on silica with a non-polar solvent is probably the better method. To separate compounds in a homologous series, normal partition systems are preferred, where a polar stationary phase is used with a non-polar mobile phase such as hexane. If water soluble compounds are non-ionic or weakly ionic, reverse phase partition chromatography is preferable where a non-polar stationary phase such as a hydrocarbon is used together with a polar mobile phase such as water/acetonitrile or water/methanol mixtures. Water soluble compounds which are strongly ionic are best chromatographed by an ion-exchange system, using either an anionic or cationic resin, together with a suitable buffer system for elution. Strongly ionic compounds can, however, be chromatographed by reverse phase partition systems by the technique of

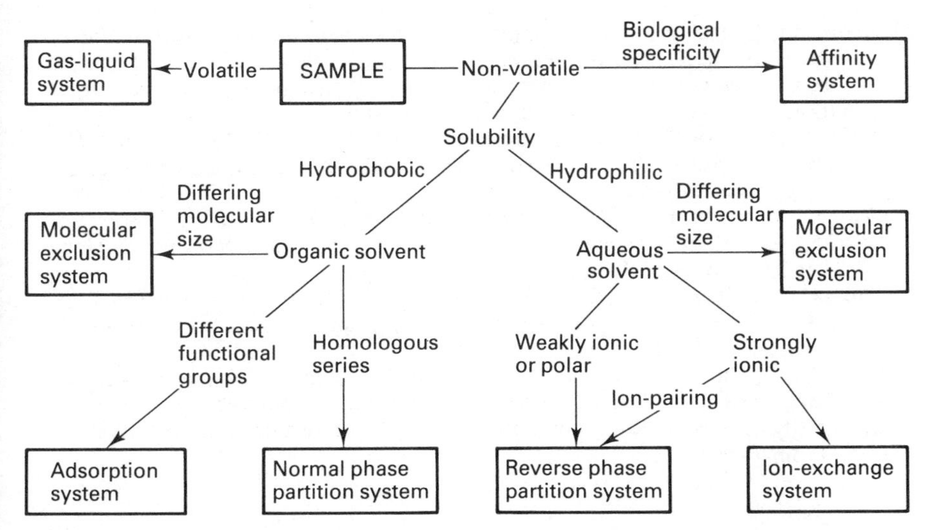

Fig. 6.12 Rationale for the choice of a chromatographic system.

ion-pairing. Compounds differing in molecular size are best separated by exclusion chromatography.

Whatever form of chromatography is chosen for a particular biochemical study, the decision to use conventional low pressure liquid chromatography or HPLC depends on many factors including the availability of apparatus, cost, preparative or analytical separation, qualitative or quantitative assay and the procedure adopted in successful separations recorded in the literature. The modern trend is to select HPLC which is certainly capable of giving fast, accurate and precise data. In part, however, it is fashion and to some extent overlooks the advantages of GLC and TLC. HPLC apparatus and solvents can be expensive and are not always trouble-free. The simplicity of TLC, especially for qualitative work and its facility for concurrent investigation of many samples including standards, remains a considerable attraction. Equally, the recent developments in capillary gas chromatography make it a fast and sensitive system for volatile compounds.

6.10 Suggestions for further reading

Bertsch, W., Jennings, W.G. and Kaiser, R.E. (Eds) (1982). *Recent Advances in Capillary Chromatography*. Huthig, Amsterdam. (A good coverage of recent development in this important aspect of column chromatography.)

Fritz, J.S., Gjerde, D.T. and Pohlandt, C. (1982). *Ion Chromatography*. Huthig, Amsterdam. (Comprehensive account of all aspects of this form of chromatography.)

Grob, R.L. (Ed.) (1985). *Modern Practice of Gas Chromatography*, 2nd edition. Wiley-Interscience, New York. (An excellent up-to-date review of the subject.)

Heftmann, E. (Ed.) (1983). *Chromatography-Fundamentals and Applications of Chromatographic and Electrophoretic Methods*. Vols 22A and 22B of *Journal of Chromatography Library*. Elsevier, New York. (A full account of the theoretical aspects of chromatography and some of its applications.)

Scott, R.P.W. (1977). *Liquid Chromatography Detectors*. Vol. 11 of *Journal of Chromatography Library*. Elsevier, New York. (A detailed discussion of the principles of chromatography detectors.)

Scouten, W.H. (1981). *Affinity Chromatography-Bioselective Adsorption on Inert Matrices*. Vol. 59 *Chemical Analysis*. Wiley-Interscience, New York. (Good detailed account of the subject.)

Sulkowski, E. (1985). Purification of proteins by IMAC. *Trends in Biotechnology*, **3(1)**, 1–7. (An excellent review of this important new procedure for protein purification.)

Yau, W.W., Kirkland, J.J. and Bly, D.D. (1979). *Modern Size-Exclusion Liquid Chromatography*. Wiley-Interscience, New York. (A comprehensive discussion of the principles and applications of gel permeation and gel filtration chromatography.)

7
Electrophoretic techniques

7.1 General principles

Many important biological molecules such as amino acids, peptides, proteins, nucleotides and nucleic acids possess ionisable groups and can therefore be made to exist in solution as electrically charged species, either as cations (+) or anions (−). Even typically non-polar substances such as carbohydrates can be given weak charges by derivatisation, for example as borates or phosphates. Moreover, molecules which have a similar charge will have different charge/mass ratios when they have inherent differences in molecular weight. In combination these differences form a sufficient basis for a differential migration when the ions in solution are subjected to an electric field. This is the principle of *electrophoresis*.

The equipment required for electrophoresis consists basically of two items, a *power pack* and an *electrophoresis unit* (Section 7.3.2). The power pack supplies a *direct current* between the electrodes in the electrophoresis unit. Cations move to the cathode (−) and anions move to the anode (+) at rates which depend on the balance between the impelling force of the electric field on the charged ion and the frictional and electrostatic retarding effects between the sample and the surrounding medium. The sample must be dissolved or suspended in buffer for electrophoresis to take place and any supporting medium must also be saturated with buffer to conduct the current (Section 7.2.1). A buffer is also important to maintain a constant state of ionisation since changes in pH would alter the charge on molecules being separated, particularly when sample ions are zwitterions.

The current is maintained throughout the circuit by electrolysis taking place at the electrodes, both of which dip into large *buffer reservoirs*. During electrolysis, hydroxyl ions and hydrogen are produced at the cathode while oxygen and hydrogen ions are produced at the anode.

$$2e^- + 2H_2O \xrightarrow{\text{Cathode}} 2OH^- + H_2\uparrow \qquad H_2O \xrightarrow{\text{Anode}} 2H^+ + \tfrac{1}{2}O_2\uparrow + 2e^-$$

The hydroxyl ions produced at the cathode cause increased dissociation of the weak acid component (HA) of the buffer mixture (Section 1.2.3) resulting in the formation of more A^- to conduct the current to the anode. At the anode, A^- ions combine with H^+ ions to reform HA and the electrons are fed into the electric circuit. Thus most of the current between the electrodes is conducted by the buffer ions in solution with only a small component being provided by the sample ions. Provided the electric field is removed before the ions in the sample mixture reach the electrodes, the components may be separated according to their *electrophoretic mobility*. Electrophoresis is thus an incomplete form of electrolysis.

Electrophoresis may be conducted in free solution without a supporting medium, in which case there is minimal frictional resistance between the sample ions and solution and rapid migration takes place. This occurs in *continuous flow electrophoresis* (Section 7.9) which is now used for large scale preparative separations. Since the charge properties of similar molecules are close to one another they tend to move together as a band, with boundaries formed between substances having slightly different electrophoretic mobilities. This technique, suitably termed *moving boundary electrophoresis*, was pioneered by Tiselius and co-workers in Sweden and it is fundamentally a specialised analytical tool requiring sophisticated Schlieren scanning optics to detect changes in refractive indices at separated boundary interfaces. This principle is currently exploited in the analytical technique of *isotachophoresis* (Section 7.8). The principle of moving boundary electrophoresis also operates in some forms of electrophoresis in which separation is achieved on a *supporting medium*, for example *isoelectric focusing* (Section 7.7).

The use of an inert and relatively homogeneous supporting medium instead of free solution has given rise to the versatility of electrophoresis in the separation of charged substances, ranging from small inorganic ions to large macromolecules. Separation of a sample on a supporting medium causes its components to migrate as distinct zones, which can subsequently be detected by suitable analytical techniques (Section 7.10). The term *zone electrophoresis* has been applied to this method which has widespread applications in both preparative and analytical work.

There are many different types of supporting media available such as sheets of absorbent paper or cellulose acetate, a thin layer of silica or alumina or a gel of starch, agar or polyacrylamide. Each may offer some advantage over the others for a particular separation. All supporting media have a capillary structure which has good anticonvectional properties. Sometimes the medium used may be designed specifically to interact with the sample ions being separated, i.e. to exploit differences in the charge/mass ratios and to introduce special retardation forces to suit the analysis.

7.2 Factors affecting electrophoresis

7.2.1 The electric field

Voltage If the separation of the electrodes is d metres and the potential difference between them is V volts, the *potential gradient* is V/d volts m^{-1}. The force on an ion bearing a charge q coulombs is then Vq/d newtons. This force causes migration and the rate of migration is proportional to Vq/d. The rate of migration under unit potential gradient is called the *mobility* of the ion. An increase in the potential gradient will therefore increase the rate of migration proportionally.

Current When a potential difference is applied between the electrodes, a current is generated, measured in coulombs sec^{-1} or amperes. The size of this current is determined by the resistance of the medium and is proportional to the voltage. The current in the solution between the electrodes is conducted mainly by the buffer ions with a small proportion being conducted by the sample ions. An increase in voltage will therefore increase the total charge per second conveyed towards the electrode. The distance migrated by the ions will be proportional to both current and time.

Resistance Ohm's law expresses the relationship between current I (measured in amperes, A), voltage V (measured in volts, V) and resistance R (measured in ohms, Ω) in which:

$$\frac{V}{I} = R \qquad 7.1$$

The current, and hence the rate of migration are thus inversely proportional to the resistance, which in turn is a function of the medium, the buffer and its concentration (Section 7.2.3). Resistance will increase with the length of the supporting medium but will decrease with its cross-sectional area and with increasing buffer ion concentration.

During electrophoresis the power dissipated in the supporting medium (W, measured in watts) is such that:

$$W = I^2R \qquad 7.2$$

An increase in temperature will cause the resistance to fall. Part of this effect is due to an increase in the mobility of the ions as a result of a decrease in the viscous resistance offered by the liquid to the motion of the ions through it as the temperature rises. The heating will produce evaporation of the solvent from the supporting medium causing a decrease in resistance. Although the rate of migration and the total charge per second conveyed towards the electrode will increase, the increase in buffer ion concentration will result in slower migration of the sample (Section 7.2.3).

In order that results should be as reproducible as possible, stabilised power packs are used which can automatically maintain either a constant voltage or constant current despite unavoidable changes in resistance due to temperature fluctuation. When a constant voltage is applied, the current will increase during electrophoresis due to a decrease in resistance of the medium with the rise in temperature. Consequently, more heat will be produced resulting in more evaporation of solvent and a decrease in resistance. A constant current avoids these problems but may lead to a drop in voltage due to decreased resistance, resulting in reduced rate of migration.

If a number of supporting media are run in parallel from one power supply, the total resistance will decrease such that:

$$1/R = 1/r_1 + 1/r_2 + 1/r_3 \ldots + 1/r_n \qquad 7.3$$

where R is the total resistance and r_1, r_2, r_3, etc., are the resistances of each supporting medium. Although it might be possible to obtain the same degree of separation, as in a single medium, in all of the media run in parallel by applying a constant voltage, the associated heating effects make this impracticable. Running at constant current will overcome these problems, but in this case the total current supplied must be increased in proportion to the number of media used, assuming that they all have the same resistance.

The voltages used can be low (100 to 500 V) or high (500 to 10 000 V) with potential gradients up to 20 and 2000 V cm^{-1} respectively. High voltages are used mainly for the separation of low molecular weight compounds for reasons to be explained later (Section 7.4). When a low voltage is used with paper electrophoresis, constant voltage or constant current may be applied as the heat generated is small and is easily dissipated. However, with all gels and cellulose acetate, heat dissipation is a problem and constant current tends to be used to reduce heat production. Direct current must of course be used. Evaporation is minimised by enclosing the apparatus under an air-tight cover. Additional cooling for high voltage may be achieved by incorporating a cooling system into the apparatus.

7.2.2 The sample

The nature of charged compounds being separated affects their migration rates in several ways. These include:

Charge The rate of migration increases with an increase in the net charge. The magnitude of the charge is generally pH dependent in accordance with the Henderson-Hasselbalch equation (Section 1.2.2).

Size The rate of migration decreases for larger molecules, due to the increased frictional and electrostatic forces which are exerted by the surrounding medium.

Shape Molecules of similar size but different shapes such as fibrous and globular proteins exhibit different migration characteristics because of the differential effect of frictional and electrostatic forces.

7.2.3 The buffer

This determines and stabilises the pH of the supporting medium. The buffer can also affect the migration rate of compounds in a number of ways. These include:

Composition The buffers in common use are formate, acetate, citrate, barbitone, phosphate, Tris, EDTA and pyridine. The buffer should be such that it does not bind with the compounds being separated as this may alter the rates of migration. In some cases, however, binding can be advantageous, for example borate buffers are used to separate carbohydrates since they produce charged complexes with carbohydrates. Since the buffer acts as a solvent for the sample, some diffusion of the sample is inevitable, being particularly noticeable for small molecules such as amino acids and sugars. The extent of diffusion can be minimised by avoiding overloading of the sample, by applying samples as narrow bands, using a high voltage for as short a time as possible and by rapid removal of the supporting medium after the separation has been completed.

Concentration As the ionic strength of the buffer increases, the proportion of current carried by the buffer will increase and the share of the current carried by the sample will decrease, thus slowing its rate of migration. High ionic strength of the buffer will also increase the overall current and hence heat production. At low ionic strengths the proportion of current carried by the buffer will decrease and the share of the current carried by the sample will increase, thus increasing its rate of migration. A low ionic buffer strength reduces the overall current and results in less heat production, but diffusion and the resulting loss of resolution are higher. Therefore, the choice of ionic strength must be a compromise and this is generally selected within a range of ionic strength of 0.05 to 0.10 M.

$$\text{Ionic strength} = \tfrac{1}{2}\Sigma cz^2, \qquad 7.4$$

where c = the molar concentration of the ion
z = charge of the ion

pH This has little effect on fully ionised compounds such as inorganic salts, but for organic compounds pH determines the extent of ionisation (Section 1.2.2). The ionisation of organic acids increases as pH increases whereas the reverse applies for organic bases; therefore their degree of migration will be pH dependent. Both effects can apply to compounds such as amino acids that have basic and acidic properties (*ampholytes*).

$$H_3N^+{-}\underset{}{\overset{R}{\overset{|}{C}}}H{-}COOH \underset{H^+}{\overset{OH^-}{\rightleftharpoons}} H_3N^+{-}\overset{R}{\overset{|}{C}}H{-}COO^- \underset{H^+}{\overset{OH^-}{\rightleftharpoons}} H_2N{-}\overset{R}{\overset{|}{C}}H{-}COO^-$$

pH:	Acidic	Isoionic point	Alkaline
Ionic form:	Cation	Zwitterion	Anion
Migration:	Towards cathode	Stationary	Towards anode

The direction and also the extent of migration of ampholytes are thus pH dependent and buffers ranging from pH 1 to 11 can be used to produce the required separations. The buffer present in both reservoirs is normally the same buffer which is used to saturate the supporting medium. This is known as a *continuous buffer system*. However, in some forms of gel electrophoresis such as SDS polyacrylamide gel (Section 7.6.2) where the buffer acts as part of the supporting medium, a different buffer may be used in the gel to that in the reservoirs (a *discontinuous buffer system*).

7.2.4 The supporting medium

Although the supporting medium is relatively inert, its precise composition may cause *adsorption, electro-osmosis* and *molecular sieving*, each of which may influence the migration rate of compounds. A consideration of these properties and how sample ions will be influenced by them will determine the choice of supporting medium for a particular separation (Sections 7.3.1 and 7.5.1).

Adsorption This is the retention of sample molecules by the supporting medium, as shown in adsorption chromatography. Adsorption causes *tailing* of the sample so that it moves in the shape of a comet rather than as a distinct, compact band, thus reducing both the rate and the resolution of the separation.

Electro-osmosis (Electro-endosmosis) This phenomenon results from a relative charge being produced between water molecules in the buffer and the surface of the supporting medium. The charge may be caused by surface adsorption of ions from the buffer and the presence of stationary carboxyl groups on paper or sulphonate groups on agar. This generates a motive force for the movement of fixed anions to the anode and results in the movement of hydroxonium ions (H_3O^+) in the buffer to the cathode carrying along neutral substances by solvent flow. This *electro-osmosis* will accelerate the movement of cations but retard anion transference. The effects of electro-osmosis can normally be ignored, but if the isoelectric points of compounds are required it must be allowed for. This is usually achieved by measuring the extent of migration of neutral substances such as urea or glucose in the same system.

Molecular Sieving This feature is shown by gels in which the randomly intertwined molecular chains distributed throughout the gel result in a sieve-like structure. The principle of molecular sieving in agar, starch and polyacrylamide gels (Section 7.5.1) is that the movement of large molecules is hindered increasingly by decreasing the pore size since all molecules have to traverse through the pores. The situation is different if a Sephadex-type gel is used since, in this case, small pores exclude larger molecules from access to the stationary phase inside the particle, thereby causing movement outside the pores whereas small molecules are tightly held within the pores (Section 6.6.2).

7.3 Low voltage thin sheet electrophoresis

7.3.1 Materials

This is the simplest and cheapest form of electrophoresis, and has been used extensively for routine analytical separation of a wide range of charged compounds. The supporting medium may be paper, cellulose acetate, or a thin layer of material such as silica. The equipment and operational procedures are very similar for each of these supporting media.

Paper Chromatography paper is suitable for electrophoresis and needs no preparation other than to be cut to size. When using paper there is always some adsorption which can be diminished by employing a buffer more alkaline than the isoelectric point of the sample, although some degree of electro-osmosis will always take place whenever paper is used. Low voltage paper electrophoresis has been extensively used in the past for the separation of a range of charged compounds, for example amino acids, peptides, proteins, nucleotides, nucleic acids and charged carbohydrate derivatives. Considerable diffusion of small molecules occurs on paper during low voltage electrophoresis and better resolution may be obtained by applying high voltage, where the time required for separation is reduced and less diffusion of the molecules occurs. The use of paper media for particular separations has nowadays been superseded by other forms of support which afford improved resolution.

Cellulose Acetate High purity cellulose acetate is available commercially in thin uniform strips which have a homogeneous micropore structure. Very little adsorption occurs even with macromolecules. Cellulose acetate is therefore a suitable medium for the separation of radio-labelled substances and for such microtechniques as immunodiffusion and immunoelectrophoresis (Section 4.4). In contrast, it is not suitable for preparative work. Cellulose acetate is less hydrophilic than paper, less buffer is held by it, and therefore better resolution is obtained in a shorter time (Section 7.2.3). Another feature of the reduced buffer content of cellulose acetate membranes is that, under a given constant current or constant voltage, there will be greater heat production so that, particularly when using high voltages, care must be taken to prevent the strip from drying out by evaporation. Cellulose acetate has the additional advantage that the background of the strips may be rendered translucent after separation and staining by treatment with clearing agents such as Whitemor-oil 120, thus facilitating quantification. Cellulose acetate will, in general, separate the same range of compounds as paper but has found particular application in clinical investigations for the separation of blood proteins, including glycoproteins, lipoproteins and haemoglobins.

Thin Layer Electrophoresis (TLE) Thin layers of silica, kieselguhr, alumina or cellulose can be prepared on glass plates as for thin layer chromatography (TLC) (Section 6.1.3). The plates are placed horizontally into the electrophoresis unit and the thin layer is allowed to saturate with buffer by diffusion from the reservoir via the connecting wicks. Thin layer electrophoresis, like TLC, is rapid and gives good resolution and high sensitivity. When used in conjunction with chromatography, TLE is a very convenient method for two-dimensional separations of protein and nucleic acid hydrolysates. Thin layer media such as those described above may also be used for separations at high voltages (Section 7.4).

7.3.2 Apparatus and methods

Equipment A power pack provides a stabilised direct current and has controls for both voltage and current output. Power packs are available for low voltage use which have an output of 0 to 500 V and 0 to 150 mA, and can give either constant voltage or constant current. The electrophoresis unit contains the electrodes, buffer reservoirs, a support for the electrophoresis medium and a transparent insulating cover (Fig. 7.1). Stainless steel electrodes can be used, but some buffers cause corrosion and platinum electrodes are more satisfactory.

The two buffer reservoirs are normally each partitioned into two interconnected sections, one contains the electrode, the other has the supporting medium in contact with the buffer. Separate compartments are necessary so that any change in pH occurring at the electrodes does not affect the buffer supporting the saturating medium. Contact between the supporting medium (always saturated with buffer prior to electrophoresis) and the buffer in the reservoirs is normally maintained by wicks, consisting of several layers of filter paper or gauze. Wicks can be dispensed with for low voltage paper electrophoresis and contact maintained by having the paper dipping directly into the buffer. The supporting medium loaded with sample is usually arranged horizontally on a flat surface of insulating material such as Perspex (Fig. 7.1). Horizontal electrophoresis units are available for use with paper, cellulose acetate, and thin layers, though simple vertical units are also available for low voltage paper electrophoresis.

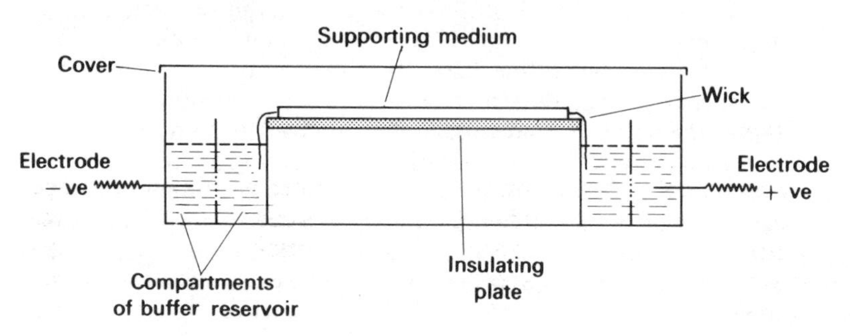

Fig. 7.1 Horizontal electrophoresis unit.

Medium Saturation The supporting medium must be saturated with buffer before electrophoresis is started since the buffer conducts the majority of the current. Saturation of the medium is completed before the sample is applied to prevent the sample spreading at the origin. Paper is dipped in buffer and excess moisture is removed by blotting. Cellulose acetate strips are impregnated with buffer by floating them on the surface in a shallow tank; rapid immersion traps air bubbles which are difficult to remove. Thin layer media are best saturated by capillary action as described in Section 7.3.1.

Sample Application The sample solution is applied, usually with a micropipette, as a small spot or narrow streak, on or along a suitable origin. If the individual components of a mixture have opposite charges they would be expected to migrate towards both electrodes and in this case the origin should be near the middle of the supporting medium. When all components are either positively or negatively charged, movement would be expected towards only one of the electrodes and the origin should be near the end of the medium furthest away from this electrode to allow for a greater distance for separation. It is important to avoid overloading of the sample and excessive diffusion into the supporting medium. For analytical electrophoresis sample concentrations of 1 to 5 mg cm^{-3} are normally used, with volumes applied within the range of 1 to 5 mm^3.

Running of Samples After the sample has been applied the power is switched on at the required voltage for the period necessary for separation. Electrophoresis units should always be enclosed by a cover during use to minimise evaporation and to provide electrical insulation. The equipment must be watched carefully throughout electrophoresis, even when using stabilised voltage and current supplies, since overheating (and charring of paper) may occur if the supporting medium has not been prepared properly. When heating is expected under low voltage it is useful to place the entire apparatus inside a cold room (0 to 4°C). Low voltage separation of proteins on paper may take longer.

When electrophoresis is complete the power must be switched off before the supporting medium is removed. Paper, cellulose acetate strips and thin layer plates can be removed and air dried directly, usually in an oven at 110°C, unless thermolabile compounds are present. Methods of detecting the samples after electrophoresis are described in Section 7.10.

7.4 High voltage electrophoresis (HVE)

When low molecular weight compounds are separated by low voltage paper electrophoresis, considerable diffusion occurs. This can largely be overcome by using much higher voltages, resulting in better resolution and very rapid separations (10 to 60 min). High voltage power packs are available which will supply up to 10 000 V and 500 mA, producing potential gradients of up to 200 V cm^{-1}.

High voltage electrophoresis generates so much heat that a direct cooling system is required. This is normally achieved by using cooling plates (Fig. 7.2). The two plates, normally aluminium, are insulated from the supporting medium by polythene and the plates are pressed against the insulated supporting medium by an inflatable pressure pad. Equipment is available with plates of up to 50 × 50 cm, to enable large sheets of paper (generally Whatman 3 MM) to be used. Cold water is circulated through channels in the cooling plates, and for large plates, a flow rate of between 10 and 15 dm^3 min^{-1} is required to disperse the heat produced. In high ambient temperature conditions the cooling water should first be refrigerated; in areas

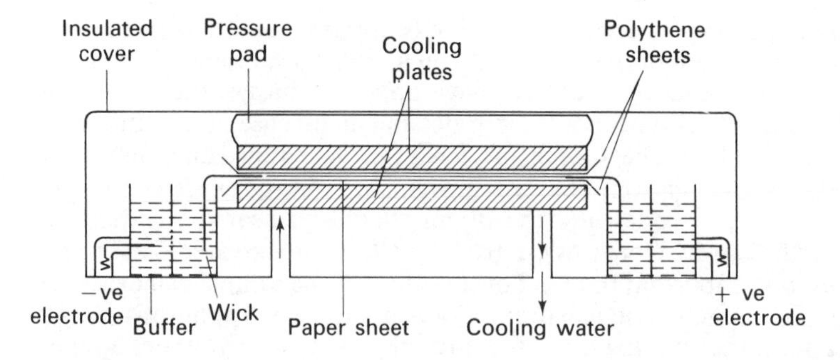

Fig. 7.2 High voltage electrophoresis (HVE) unit.

where water is very hard, the use of a water softener is recommended in order to avoid furring up the channels in the plates. Temperature gradients across the plates must be avoided, since a difference of 1 °C causes a 3% change in migrational velocity, and this affects reproducibility. Complete electrical insulation of the equipment is vital because lethal voltages are used. Normal safety precautions should be strictly observed even though commercial systems have a number of cut-out (fail safe) devices to avoid accidents.

For the separation of protein hydrolysates, HVE in one dimension followed by chromatography in a second dimension is frequently used. This two-dimensional system allows characteristic fingerprints of different proteins to be obtained, as with two-dimensional thin layer electrophoresis (Section 7.3.1). HVE has proved to be a very effective method for the separation of small peptides and amino acids.

7.5 Gel electrophoresis

7.5.1 Materials

The use of gels as supporting media has largely superseded low voltage thin sheet electrophoresis systems (Section 7.3) for the separation of high molecular weight substances such as proteins and nucleic acids, because of the improved resolution obtained. This is due to the physical properties of gels which are water insoluble, hydrophilic, semi-solid colloids. Suitable gels may be prepared shortly before use from a variety of powdered solids, for example starch, agar and polyacrylamide. The molecular sieving property of the semi-rigid gel helps to separate large ionic compounds such as proteins which have similar charge properties but which differ in size and shape. Smaller molecules can be separated only in Sephadex-type gels (Section 7.2.4).

Starch gels are prepared by heating and cooling a mixture of partially hydrolysed starch in an appropriate buffer. This causes the branched chains of the amylopectin component of starch to intertwine and form a semi-rigid gel.

The choice of buffer suitable for starch gels is often empirical and a wide variety has been successfully used. Weak *high porosity gels* may be prepared by incorporating less than 2% (w/v) starch to buffer and strong *low porosity gels* by adding 8 to 15% (w/v) starch. However, there is no way of determining the exact pore size of starch gels and different batches of starch are inconsistent with respect to pore size for the same percentage of starch.

Starch gels may be used preparatively or analytically with continuous or discontinuous buffer systems. The molecular sieving properties of starch makes it a good choice for the separation of complex mixtures of structural molecules and physiologically active proteins. An important application of starch gel electrophoresis is the analysis of isoenzyme patterns (*zymograms*) because of the ease of applying histochemical tests after electrophoresis, although isoelectric focusing has now become the preferred technique for this (Section 7.7). Thick blocks of starch are used for preparative electrophoresis (Section 7.9).

Agar/Agarose Agar is a cheap, non-toxic, chemically ill-defined, complex, powdery mixture containing two galactose-based polymers, agarose and agaropectin. Agar dissolves in boiling aqueous buffers and sets to form a gel at about 38°C. As a 1% (w/v) gel in buffer, agar has a high water content, good fibre structure (and therefore good anticonvectional properties), a large pore size and low frictional resistance. Consequently, movement of ions is very rapid during electrophoresis, thereby assisting the separation of macromolecules in particular. Agar has the disadvantage that electro-osmosis is severe unless its sulphate content is removed by prior purification. Agar gel is very suitable for chemical staining after separation and its low diffusion resistance makes it excellent for separation and detection of antigenic proteins by isotachophoresis and immunoelectrophoresis (Sections 7.8 and 4.4.4).

Purified agarose gel is now being used extensively for the separation of nucleic acids and DNA restriction fragments because of its lack of molecular sieving and electro-osmosis.

Polyacrylamide gels are prepared immediately before use from a number of highly toxic synthetic chemicals. Acrylamide monomer ($CH_2{=}CHCONH_2$) is copolymerised with a cross linking agent, usually *N,N'*-methylenebisacrylamide ($CH_2(NHCOCH{=}CH_2)_2$), in the presence of a catalyst accelerator-chain initiator mixture. This mixture may consist of freshly prepared ammonium persulphate as catalyst (0.1 to 0.3% w/v) together with about the same concentration of a suitable base, for example dimethylaminopropionitrile (DMAP) or *N,N,N',N'*-tetramethylenediamine (TEMED) as initiator. TEMED is most frequently used and proportional increases in its concentration speeds up the rate of gel polymerisation. Prior degassing of solutions is required since molecular oxygen inhibits chemical polymerisation. Gel photopolymerisation may also be achieved using riboflavin and TEMED in which case traces of oxygen are required. Other chemicals such as detergents, enzyme substrates or enzymes may be added in buffers used to make up the gels. Gelation is due to vinyl polymerisation, as shown in Fig. 7.3.

The porosity of a gel is determined by the relative proportion of acrylamide

$CH_2{=}CHCONH_2$ acrylamide + $CH_2(NH\ COCH{=}CH_2)_2$ *N,N'* methylenebisacrylamide

free radical catalyst

```
—CH2—CH—[—CH2—CH—]x—CH2—CH—
      |          |           |
      CO         CO          CO
      |          |           |
      NH         NH2         NH
      |                      |
      CH2                    CH2
      |                      |
      NH         NH2         NH
      |          |           |
      CO         CO          CO
      |          |           |
—CH2—CH—[—CH2—CH—]x—CH2—CH—
```

Fig. 7.3 Polyacrylamide gel formation.

monomer to cross linking agent. Gels may be defined in terms of the total percentage of acrylamide present. Gels may be prepared containing from 3% to 30% acrylamide, corresponding to pore sizes of 0.5 nm and 0.2 nm diameter respectively. Lower percentage acrylamide gels have therefore larger pore sizes, thereby offering less resistance to the passage of larger molecules. In general terms a 30% gel is suitable for the separation of compounds having molecular weights around 10^4 daltons whereas a 3% gel will separate compounds having molecular weights around 10^6 daltons. Most protein separations are carried out using gels ranging from 5 to 15% acrylamide.

Polyacrylamide gels may be prepared with a high degree of reproducibility and the degree of porosity may be selected to enhance the separation of molecules having similar charges but different shape and size. This feature makes the method particularly suitable for resolving mixtures of proteins. Other features of polyacrylamide gels which extend their usefulness for macromolecule separation include their minimal adsorption capacity, their lack of electro-osmosis and their general suitability for *in situ* quantitative analysis (they do not absorb in the ultraviolet) and for various types of histochemical analysis. Polyacrylamide gels are also used in some specialised electrophoresis systems described later. These include SDS polyacrylamide gel electrophoresis (Section 7.6) and isoelectric focusing (Section 7.7).

7.5.2 Apparatus and methods

Equipment The power packs used are the same as were described for low voltage electrophoresis (Section 7.3.2). Gels can be run as horizontal slabs

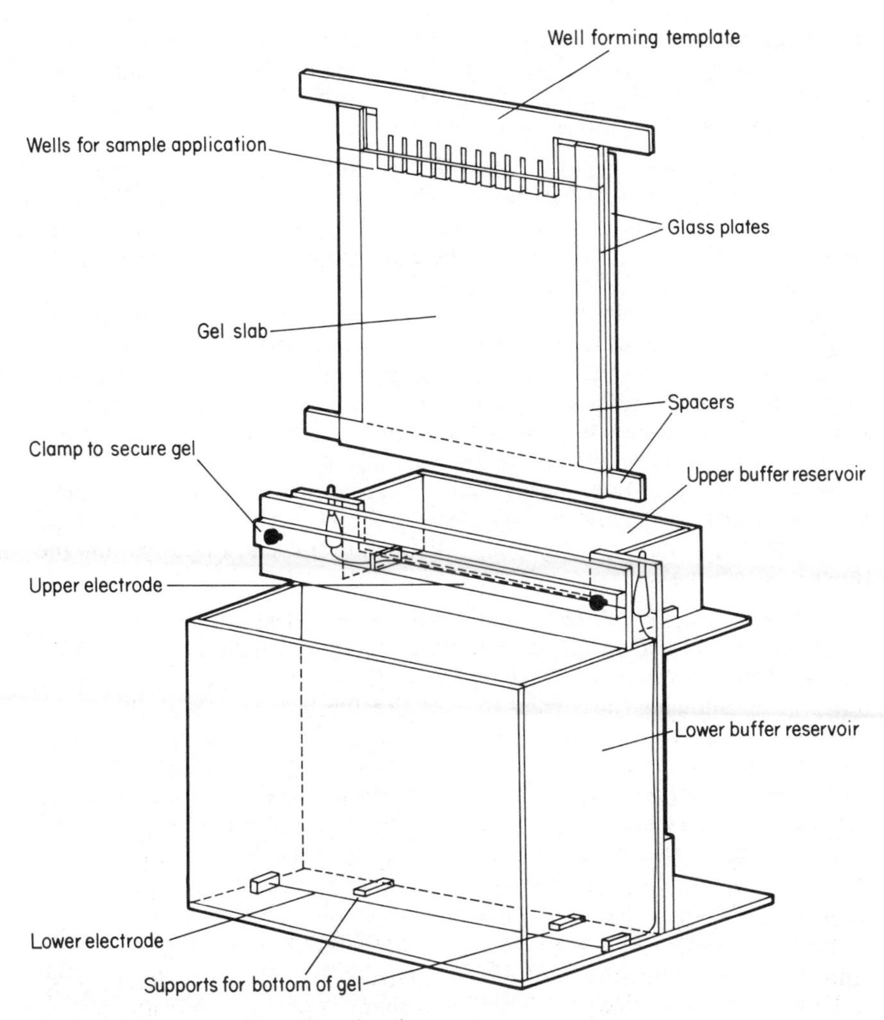

Fig. 7.4 Vertical gel electrophoresis unit.

using equipment similar to that shown in Fig. 7.1. Vertical slabs are more frequently used, and a variety of Perspex units of the type shown in Fig. 7.4 are commercially available. For some analytical procedures vertical cylinders of gel are used, supported in precision-bored glass tubes contained in specially designed apparatus. Slabs of gel carry more samples than cylinders (up to 25 samples can be run on a single gel plate) making them more economical to use, and enabling more samples to be compared with each other when run under identical conditions.

Preparation Gels are prepared in the glass (or Perspex) containers in which they are to be used. In the case of slabs, the gels are cast between two clean glass plates which are clamped together but held apart by plastic spacers. Gel dimensions of up to 12 × 25 cm are used for starch and agarose gels, with thicknesses up to 6 and 3 mm respectively. For polyacrylamide gels typical dimensions are 12 × 14 cm, with thicknesses ranging from 1 to 3 mm. Vertical slabs are run with the glass plates left on both sides of the gel. For horizontal slabs, the plate above the gel is removed before the run.

Sample Application Dissolved samples can be applied to the surface of horizontal gel slabs via filter paper strips, but more commonly for both horizontal and vertical slabs the sample solutions are injected from a microsyringe into slots or wells in the gel. These are prepared by inserting a comb-like template into the gel before it sets (Fig. 7.4).

The buffer in which the sample is dissolved usually contains sucrose or glycerol (10 to 15%) to increase its density and ensure that the solution sinks into the well. A marker dye such as bromophenol blue is often added to aid observation of loading and monitoring the migration. Urea or sodium dodecylsulphate may be added to protein samples to facilitate their solubilisation, and also disulphide reducing agents such as dithiothreitol or 2-mercaptoethanol. Only microgramme quantities of proteins and nucleic acids are used for analytical gel electrophoresis, so sample volumes of approximately 1 mm^3 are generally used, with concentrations within the range of 1 to 3 mg cm^{-3}.

Running the Samples For horizontal systems, electrical contact between the gel and the buffer in the electrical compartments can be maintained by wicks, as described in Section 7.3.2. Alternatively the gel can be submersed in the buffer, thus allowing the current to pass directly through it, with the added benefit of the buffer dispersing heat from the gel. For vertical systems of the type shown in Fig. 7.4, the gel slab, sandwiched between glass plates, is placed in the lower reservoir, with the top of the gel in contact with the buffer in the upper reservoir. The gel thus completes the electrical circuit between the electrodes in the upper and lower compartments. Although the buffer surrounding the gel helps to disperse heat generated by the current, additional cooling may be required during long runs. This can be achieved by carrying out the run in a cold room or by circulating the buffer through a cooling system. The precise voltage and time required to obtain optimal separations will depend on the nature of the samples and the type of gel used, but several hours at a few hundred volts are generally required. Marker dyes such as bromophenol blue for proteins and ethidium bromide for nucleic acids enable the progress of the run to be monitored. After the run sample bands can be visualised by the methods described in Section 7.10.

7.6 Sodium dodecylsulphate (SDS) polyacrylamide gel electrophoresis

7.6.1 Principle

This form of polyacrylamide gel electrophoresis is one of the most widely used methods for separating protein mixtures and for determining their molecular weights. Sodium dodecylsulphate (SDS) is an anionic detergent which binds strongly to proteins, causing their denaturation. In the presence of excess SDS about 1.4 g of the detergent binds to each gramme of protein, giving the protein a constant negative charge per unit mass. Protein–SDS complexes will therefore all move towards the anode during electrophoresis, and owing to the molecular-sieving properties of the gel, their mobilities (and therefore the distances they migrate in a given time) are inversely proportional to the $\log_{10}$ of their molecular weights. If standard proteins of known molecular weight are also run, the molecular weights of the sample proteins can thus be determined.

7.6.2 Apparatus and methods

Standard SDS Polyacrylamide Gels are run vertically in the type of apparatus shown in Fig. 7.4, with the anode as the lower electrode. The protein samples are usually dissolved in a Tris buffer of pH 6 to 8, containing SDS, 2-mercaptoethanol (to reduce disulphide bonds), sucrose or glycerol (to increase density) and bromophenol blue (as a marker dye). Resolution of the protein bands is greatly increased by applying the samples onto a short *stacking gel* set on top of the main *separating gel.* Differences in pH and composition between these two gels cause the samples to be concentrated into narrow bands before separation occurs during migration through the main gel. This increased resolution is due to the involvement of a phenomenon called *isotachophoresis* (Section 7.8). Adequate separations are usually achieved in 3 to 4 hours using a current of 20 to 30 mA.

Gradient Gels contain a *concentration gradient* of acrylamide increasing from around 5 to 25%, with a corresponding decrease in pore size, causing improved resolution of protein bands. The gradients are formed by running high and low concentrations of acrylamide solutions between the gel plates via a gradient mixer. The migration of the proteins will be impeded once the pores become too small, producing narrower bands and thus increased resolution. Separation is therefore primarily based on differences in the size of the proteins. This means that samples can be run in which there are wide ranges of molecular weights. Where molecular weights are similar, better separation is achieved than in a uniform (non-gradient) gel. Although gradient gels can be used without SDS and stacking gels, these should be utilised if optimal separations are required.

Two-Dimensional Gels enable the protein components of a complex mixture to be separated with still greater resolution. The samples are first partially separated by isoelectric focusing (Section 7.7) on the basis of differences in isoelectric points, using a cylindrical column of gel. This gel is then applied along the top of the stacking gel for SDS gel electrophoresis in the second dimension for separation to be completed on the basis of differences in molecular size. Either uniform or gradient SDS gels may be used, depending on requirements.

7.7 Isoelectric focusing (IEF)

7.7.1 Principle

This technique, sometimes called *electrofocusing* is based on moving boundary rather than zone electrophoresis (Section 7.1). *Amphoteric* substances such as amino acids and peptides are separated in an electric field across which there are both *voltage* and *pH gradients.* The anode region is at a lower pH than the cathode region and a stable pH gradient is maintained between the electrodes. A pH range is chosen such that the samples being separated will have their isoelectric points within this range. Substances which are initially at pH regions below their isoelectric point will be positively charged and will migrate towards the cathode, but as they do so the surrounding pH will be steadily increasing until it corresponds to their isoelectric

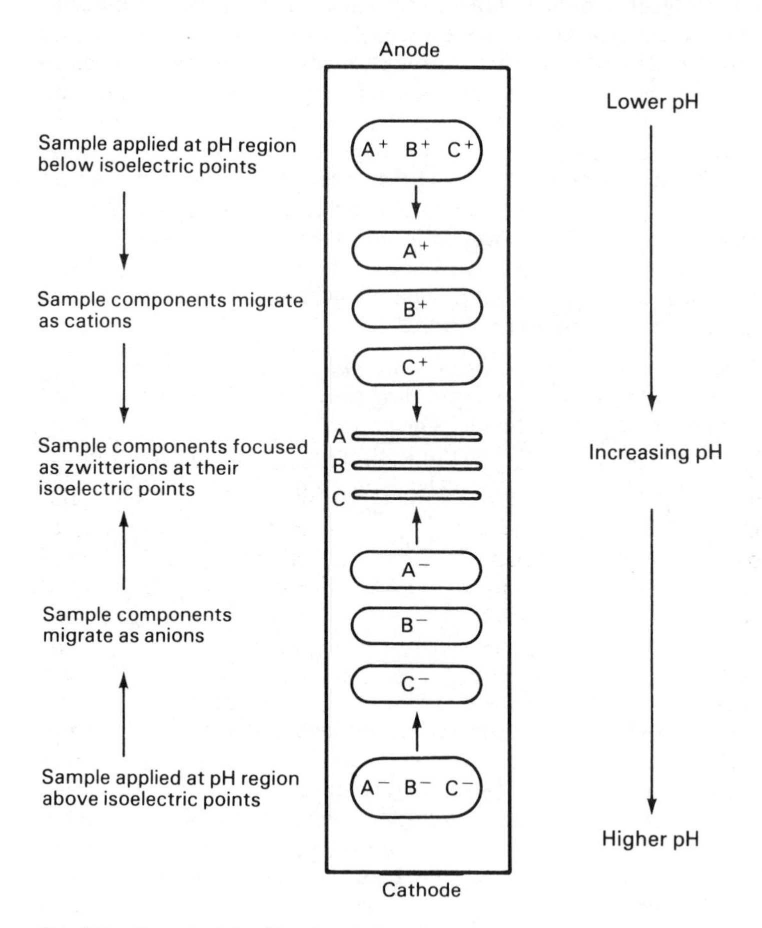

Fig. 7.5 The principle of isoelectric focusing.

points. They will then be in the zwitterion form with no net charge so further movement will cease. Likewise, substances which are initially at pH regions above their isoelectric points will be negatively charged and will migrate towards the anode until they reach their isoelectric points and become stationary. Amphoteric substances thus become focused into narrow stationary bands. This is illustrated in Fig. 7.5. As the samples will always move towards their isoelectric points it is not crucial where they are applied. Samples can thus be separated with very high resolution, making the technique particularly suitable for separating isoenzymes, as differences between isoelectric points of only 0.01 pH units are sufficient for separation.

The stable pH gradient between the electrodes is maintained by using a mixture of low molecular weight carrier ampholytes whose individual isoelectric points cover a preselected pH range. These carrier ampholytes are usually synthetic aliphatic polyamino-polycarboxylic acids and are available commercially in mixtures covering a wide pH band (e.g. 3 to 10) or various narrow bands (e.g. pH 4 to 5). Commercial ampholytes include Ampholine (LKB), Bio-Lyte (Bio Rad), and Pharmalyte (Pharmacia). Separation can be carried out in a vertical column or on a horizontal gel plate, but in both cases purpose-made equipment is required.

7.7.2 Apparatus and methods

Vertical Column IEF was developed first but has now been largely superseded by the horizontal plate system, although the column system is still a valuable technique for preparative IEF. In the column system, a water-cooled vertical glass column is filled with a mixture of carrier ampholytes suspended in a sucrose solution containing a density gradient to prevent diffusion. The upper (anode) end of the column is connected to a reservoir containing an acidic solution (e.g. phosphoric acid) and the lower (cathode) end is connected to a reservoir containing an alkaline solution (e.g. sodium hydroxide). On opening the two reservoir valves the two solutions are allowed to diffuse into the column from their respective ends, setting up a pH gradient between the acidic anode and the alkaline cathode. The valves are then closed and the current switched on, causing the carrier ampholytes to migrate until they reach the pH regions where they have no net charge. They will then remain stationary at these points, thus stabilising the pH gradient. The sample is then allowed to enter the upper end of the column by opening a valve, and the charged sample components migrate until they reach the pH regions of the tube at which they have no net charge, where they remain focused. When the separation is complete (after 1 to 3 days) the current is switched off and the sample components run out through a valve in the base of the column, which can be linked to a fraction collector for subsequent analysis. The power supplied should be kept constant to minimise thermal fluctuations. Vertical columns containing polyacrylamide gel impregnated with carrier ampholytes can also be used for IEF, reducing the time required for separation to only 1.5 to 3 hours. These can be used for two-dimensional electrophoresis by combining with SDS gel electrophoresis (Section 7.6.2).

Horizontal Gel IEF is the most widely used system for analytical IEF. Thin slabs of gel impregnated with carrier ampholytes are mounted on glass plates or plastic sheets. Polyacrylamide gel is commonly used, but high grade agarose gel is also used, particularly for very high molecular weight proteins, as this avoids the molecular sieving effect of polyacrylamide. A solution of carrier ampholytes covering a suitable pH range is mixed with a low percentage acrylamide solution before the gel sets on its backing plate. Photopolymerisation is carried out with riboflavin as catalyst, since ammonium persulphate interferes with isoelectric focusing (Section 7.5.1). Gel thicknesses of 1 to 2 mm have normally been used, but *ultra-thin gels* of only 0.15 to 0.25 mm thickness are now being increasingly used. These have the advantages of lower cost (ampholytes are very expensive), decreased sample loading, decreased running time (since higher voltages can be used), improved resolution and decreased staining and de-staining times. However, ultra-thin gels are very fragile and extra care is needed in their preparation and handling. Ready prepared polyacrylamide gel plates on sheets of plastic are available commercially, containing ampholytes in various pH ranges, and up to 24 samples can be run on one of these plates.

To achieve rapid separations, relatively high voltages are used (up to 2000 volts). Power packs should be used which stabilise the power output, as well as voltage and current to minimise thermal fluctuations. As considerable heat is produced, cooling plates are used, similar to those described for high voltage electrophoresis (Fig. 7.2). Wicks consisting of thick strips of filter paper are soaked in appropriate acid or alkali (e.g. phosphoric acid or sodium hydroxide solutions) and then laid along the anode and cathode sides of the plate. Platinum wire electrodes fixed to an insulated cover make contact with the full length of the wicks. Sample solutions containing a high concentration of inorganic ions should be desalted before use to avoid disturbing the pH gradient. Samples can be applied to the gel surface by dipping small pieces of filter paper into dilute solutions of the samples (1 to 5 mg cm^{-3}) and laying these on the gel surface to allow absorption of the sample by the gel. Only a few microgrammes of protein are required for analysis. The insulating cover is then placed over the gel, the cooling system turned on and the current applied for 30 minutes. The samples will by then have diffused into the gel so the current is switched off and the filter paper pieces carefully removed to avoid later interference with the gel. The run is then continued for a further 1 to 2 hours.

If the precise isoelectric points of the sample components need to be known, they can be determined after the run by measuring the pH gradient across the gel using a surface electrode to measure the pH at 1 cm intervals. As diffusion of the samples will occur while this is being done, it is advisable to refocus them by applying the current for a further 10 minutes. As surface electrodes can be both slow to respond and low in sensitivity, a more satisfactory method of determining the pH gradient is to run a mixture of marker proteins of known isoelectric points. Suitable mixtures are available commercially covering both broad and narrow pH ranges, and usually include coloured marker proteins. Fixing, staining and destaining are carried out as described in Section 7.10.

IEF is suitable only for amphoteric substances and is mainly used for separating proteins and peptides, for which it is now recognised as the most effective method of separation. The sensitivity and high resolution of the technique have led to it being used very extensively, despite the high cost of the equipment and materials. It is used in clinical, forensic and human genetics laboratories for the separation and identification of serum proteins, by the food and agricultural industries, and for research in enzymology, membrane biochemistry, microbiology, immunology, cytology and taxonomy. Although gel IEF is mainly used for analytical separations, preparative IEF can be carried out using flat beds of granulated gel, giving a separation in 14 to 16 hours.

7.8 Isotachophoresis

7.8.1 Principle

Like isoelectric focusing (Section 7.7) this technique is a form of moving boundary electrophoresis (Section 7.1). The principle underlying isotachophoresis is utilised in the stacking gel system in SDS polyacrylamide electrophoresis (Section 7.6.2). The name of this technique, derived from the Greek, refers to the fact that the ions being separated all travel (*phoresis*) at the same (*iso*) speed (*tacho*). Any charged substances can be separated by isotachophoresis, including inorganic ions. Separation of the ionic components of the sample is achieved through stacking them into discrete zones in order of their mobilities, producing very high resolution.

For the separation of a mixture of anions, a leading anion (e.g. chloride) is chosen which has a higher mobility than the sample ions, and a trailing (or terminating) anion (e.g. glutamate) which has a lower mobility than the sample ions. All the anions must have a common cation (e.g. Tris). Likewise for the separation of cations, such as metal ions, there are requirements for leading and trailing cations and a common anion. When the current is switched on, the leading ions will move towards the appropriate electrode, the sample ions will follow in order of their mobilities and the trailing ion will follow behind the sample ions. Once equilibrium is achieved the ions will all move at the same speed in discrete bands in the order of their mobilities (Fig. 7.6).

7.8.2 Apparatus and methods

Commercial equipment is available for analytical isotachophoresis which gives complete separations in 10 to 30 minutes. As voltages up to 30 kV are used, a thermostatically controlled cooling bath is used. Separation is achieved in a capillary tube, into which the samples are injected between the

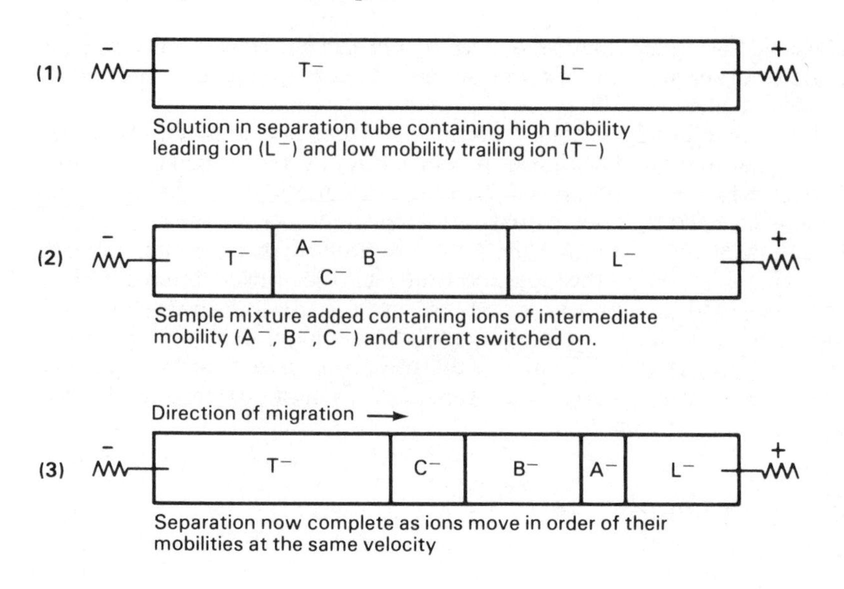

Fig. 7.6 The principle of isotachophoresis.

leading and trailing ions. Analytical isotachophoresis is carried out in an aqueous environment, the only solutions being the leading and trailing electrolytes and the samples. As the length of each separated band is proportional to the amount of ion present, the separation can be made quantitative (e.g. by measuring ultraviolet absorption).

Where the mobilities of the sample ions are very similar, their resolution may be enhanced by including, with the sample, synthetic ampholytes called *spacer ions*. These have mobilities intermediate to those of the sample ions and hence help to separate them by taking up positions between the sample ions. The spacer ions are similar to the ampholytes used in isoelectric focusing (Section 7.7.1).

Isotachophoresis is used for separating charged substances ranging from inorganic ions and organic acids to proteins and nucleic acids. Samples as small as a few microgrammes can be separated quantitatively, and large quantities of samples can be separated preparatively, using columns containing polyacrylamide gel. As well as being used extensively in research laboratories, isotachophoresis is now finding industrial applications in pollution control (detecting detergents and inorganic ions in effluent water) and in quality control in the food, brewing and pharmaceutical industries.

7.9 Preparative electrophoresis

The electrophoretic techniques described so far are used primarily for analytical separations, although the gel systems can also be used preparatively. The techniques described briefly in this section have been developed primarily for preparative applications.

Block Electrophoresis is a technique mainly used for preparative procedures, as large amounts of sample can be applied (up to 1 g). Inferior resolution however limits its analytical applications. The supporting medium, in the form of a dry powder, which can bind water on its surface, is made up into a slurry with buffer and then used in the form of blocks. The material used must be able to conduct heat adequately, stabilise against convection, not absorb the sample, and be free of fixed charges on its surface. The most commonly used material is granulated potato starch, but cellulose, agarose, Sephadex, powdered plastic (e.g. polyvinyl chloride) and ground fibre glass are also used. A thick slurry is poured into a plastic or glass rectangular container (e.g. $20 \times 10 \times 1.5$ cm) and surplus moisture is removed by applying thick filter paper to the surface. Samples are applied into slots cut into the block and filter paper wicks make contact between the ends of the block and the electrode compartments. The block remains in the container for the separation, and is covered by a lid. Separations are best carried out in a cold room overnight at a current of approximately 25 mA. After the zones have been located (Section 7.10) they can be cut out and the components eluted by filtration or centrifugation.

Block electrophoresis has proved useful for preparative separations of a variety of macromolecules including enzymes, blood proteins, nucleoproteins, and nucleic acids.

Continuous Flow Electrophoresis is used for separations in free solution on a large production scale. Like isoelectric focusing and isotachophoresis it is a form of moving boundary electrophoresis (Section 7.1). Electrophoresis takes place continuously as the separating material is carried upwards by a flow of carrier buffer through an annular space between two vertical concentric cylinders (Fig. 7.7). The outer cylinder is rotated to maintain a stable laminar flow of the buffer solution. An electric field is applied between the two cylinders, causing the sample material to separate radially as it is carried upwards by the buffer flow. At the top of the inner cylinder a series of radial slits enables the buffer stream to be separated into up to 30 individual fractions.

Equipment suitable for large-scale separations is now available commercially. It can be used not only for protein separations, e.g. blood plasma fractionation and enzyme purification, but also for separation of particles such as cells, cell organelles and viruses. Separation is very rapid (particles take about one minute to pass through the annulus), large quantities can be processed (up to 100 g protein per hour) and as the process operates under mild conditions, high product yields are obtained with a high retention of biological activity. Continuous flow electrophoresis has the distinction of having been used on a space flight, to test the effect of zero gravity on the separation process. It is a technique likely to find increasing applications in biotechnology.

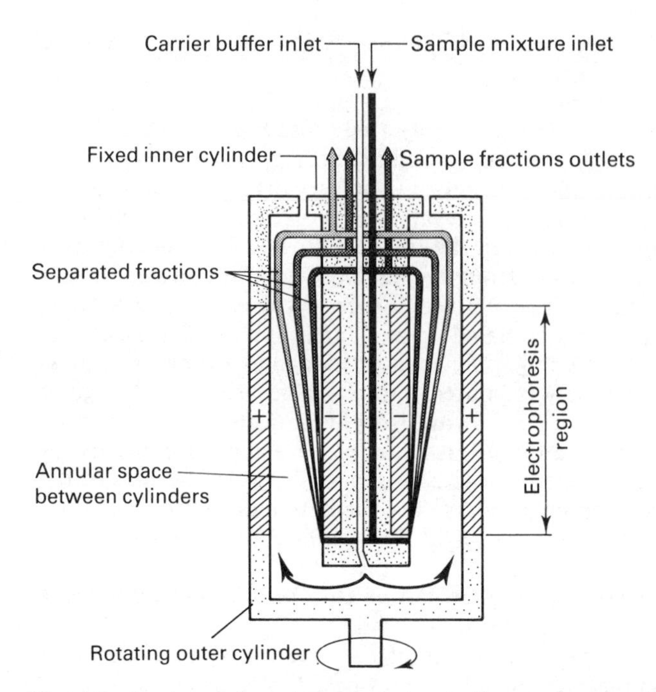

Fig. 7.7 Continuous flow electrophoresis unit. The sample enters the annular space via the sample inlet and is subject to upward movement due to electrophoresis. The result is radial separation into a series of bands (fractions) which may be collected via a series of sample outlets. (Reproduced by permission of AERE Harwell.)

7.10 Detection, recovery and estimation

The identification of the components of unknown mixtures on the *electrophoretogram* after electrophoresis is achieved by comparing the migration characteristics of the resolved mixture with pure samples of known compounds, separated under identical conditions to the test sample. In this respect electrophoresis is identical to chromatography. Individual compounds are usually detected and identified *in situ*. If substances naturally absorb or fluoresce under ultraviolet light and the supporting medium does not, this may be used as a means of detection. For example nucleic acids and proteins can be identified by ultraviolet absorbance within the 260 to 280 nm range in transparent polyacrylamide gel containing a low percentage of cross linking agent. Ultraviolet absorbance can also be exploited if compounds form complexes with agents which then induce fluorescence under ultraviolet light, for example amino acids, peptides and proteins with fluorescamine or dansyl chloride.

Most biological molecules are colourless and have to be treated with reagents which specifically produce stable coloured compounds (Table 7.1). Excess dye or colour-inducing reagent must be removed from the supporting

Table 7.1 Staining methods for visualising compounds on electrophoresis strips.

Compound	Reagent	Comments
Proteins	Nigrosine in acetic acid or trichloroacetic acid.	Very sensitive
	Bromophenol blue-$ZnSO_4$-acetic acid	Quantitative
	Lissamine green in aqueous acetic acid	Quantitative
	Coomassie Brilliant Blue R-250 (PAGE Blue)	Quantitative and very sensitive
	Silver stains	Ultra-sensitive but not quantitative
Glycoproteins	Periodic acid oxidation followed by treatment with Schiff's reagent	Quantitative
Lipoproteins	Sudan black in 60% ethanol	
Peptides	ClO_2 or NaOCl chlorination followed by KI-starch or benzidine-acetic acid	Reacts with all NH_2 compounds
Nucleic acids	Methyl green-pyronine	RNA-red, DNA-blue
	Ethidium bromide	Fluoresces under UV when bound to DNA
	Silver stains	Very sensitive for DNA and RNA
Polysaccharides	Iodine	
Acid mucopoly-saccharides	Toluidine Blue in methanol-water	

medium by elution in a suitable solvent or by further electrophoresis. Before staining, electrophoretograms may be treated with a fixative to reduce zone spreading. This is particularly important for isoelectric focusing gels (Section 7.7.3), which are treated as follows. After the run is complete the samples are immediately fixed, to avoid diffusion, by soaking the gel in a fixing solution (e.g. 10% trichloroacetic acid) for 30 minutes. The samples can then be stained by immersing the gel in a suitable staining solution (Coomassie Brilliant Blue R-250 is particularly suitable for proteins) for 10 minutes. Several rinses in a destaining solution (an aqueous solution containing 25% ethanol and 8% acetic acid) are required to remove excess stain. After partial drying, the stained gel can then be preserved between two layers of plastic film. Ultra-thin gels can be dried out completely on their backing plates.

Detection of enzymes *in situ* on unfixed electrophoretograms largely exploits principles of histochemistry whereby specific substrates are normally converted into insoluble coloured products. Non-ionic substrates may be immobilised in gels during gel preparation and enzyme activity detected after electrophoresis by incubation in a more appropriate buffer. Alternatively, after electrophoresis, a support medium containing substrate may be placed in contact with the electrophoretogram to localise activity when appropriately incubated. In cases where separated compounds are radioactive, supporting media may be subjected to autoradiography (Section 9.2.4) or a radiochromatogram scanner (Section 9.2.4).

Qualitative analysis is also possible for electrophoretograms in which the separated compounds are removed from the supporting medium. The most common method of recovery is to divide up the supporting medium into standard uniform lengths and to treat the supporting medium in a way that releases the test compounds. Compounds are poorly eluted with buffers from paper due to its strongly adsorptive properties. Cellulose acetate can be dissolved in acetone, leaving the stained material in solution. Starch gel is readily eluted following either mechanical maceration (an amylase treatment is sometimes included) or maceration by alternate freezing and thawing. Macromolecules may be recovered from starch gels by electrodialysis (enclosing the gel inside a dialysis membrane which, in turn, is placed inside an electrophoresis unit). Polyacrylamide gels can be sliced up only when semi-frozen, they will not break down on freezing and thawing but can be eluted by electrodialysis. Radioactive compounds can be eluted into vials containing a suitable scintillation cocktail for counting. Quenching (Section 9.2.3) in gels may be minimised by prior treatment with suitable reagents, for example H_2O_2 at 60°C. Large volumes of eluent should be avoided, particularly when eluting enzymes, due to the decrease in specific activity on dilution.

Recovery of compounds by electrophoresis, continued until the sample migrates off the end of the supporting medium, can also be achieved for some forms of electrophoretogram, for example preparative polyacrylamide rods, although this requires a special attachment at the base of the column to bleed off the sample as it emerges.

The basic procedures used for *in situ* and elution qualitative analyses after electrophoresis can also be used quantitatively. Different extinction coefficients of separated compounds must, of course, be accounted for in quantification. Technically, staining for quantitative work is complicated by differences in uptake and elution of dyes, but even so, the method has widespread application (Table 7.1). If the uptake of a dye is quantitative, *in situ* analysis may be performed by densitometry. *Densitograms* are plots of absorbance versus distance and for quantification, peak area should be proportional to concentration. This condition prevails for narrow concentration ranges where the Beer-Lambert law holds. Calibration of the instrument is required and is achieved by analysis of standard amounts of known compounds. Completed densitograms may be conveniently photographed for permanent recording.

7.11 Suggestions for further reading

Gaal, O., Medgyesi, G.A. and Vereczkey, L. (1980). *Electrophoresis in the Separation of Biological Macromolecules.* Wiley, Chichester. (A detailed study of the principles and applications of all electrophoretic techniques.)

Righetti, P.G. (Ed.) (1983). *Isoelectric Focusing: Theory, Methodology and Applications.* In *Laboratory Techniques in Biochemistry and Molecular Biology*, Vol. II (Work, T.S. and Burdon, R.H., Eds). Elsevier, Amsterdam. (A detailed and updated text, primarily for research workers.)

Simpson, C.F. and Whittaker, M. (Eds) (1983). *Electrophoretic Techniques*. Academic Press, London. (An up to date collection of articles on all the current techniques.)

Smith, I. (Ed.) (1976). *Zone Electrophoresis. Chromatographic and Electrophoretic Techniques*, Vol 2, 4th edition. Heinemann Medical, London. (A standard text on the older techniques.)

Walker, J.M. (Ed.) (1984). *Methods in Molecular Biology*, Vol. 1 *Proteins* and Vol. 2 *Nucleic Acids*. Humana Press, Clifton, New Jersey. (Contains much useful information for laboratory workers.)

Walker, J.M. and Gaastra, W. (Eds) (1983). *Techniques in Molecular Biology*. Croom Helm, London. (Short chapters with lots of practical tips.)

Detailed information about specific electrophoresis equipment and its applications is available from the manufacturers.

8
Spectroscopic techniques

8.1 General principles

8.1.1 Radiation, energy and atomic structure

Light, heat and other so called *electromagnetic radiations* consist of electromagnetic waves travelling at 3×10^8 m s^{-1}. However, for convenience such radiations may also be considered as consisting of a stream of packets of energy called either *quanta* or *photons*. The amount of energy in each quantum determines the wavelength of the radiation.

In the *ground state* of an atom, the electrons occupy the lowest energy levels commensurate with the laws of quantum mechanics. The ground state of sodium, which contains 11 electrons, is illustrated in Fig. 8.1a. In order for an electron to move from its energy level in the ground state to an *excited state*, its energy must be raised by the absorption of a discrete quantum of energy, exactly equivalent to that involved in the transition. When an excited electron returns to its ground state, it emits radiation of a specific wavelength giving rise to the spectroscopic phenomena of fluorescence and emission flame photometry (Sections 8.5 and 8.7). When bonding occurs between atoms to form a molecule, the electrons occupy new energy levels. In addition, the atoms in a molecule can vibrate and rotate about a bond axis, giving rise to vibrational and rotational energy sub-levels, as shown in Fig. 8.1b. As with atoms, each of the electrons in a molecule usually occupies the lowest energy level that is available to it, *i.e.* the ground state. Similarly, under the right conditions, an electron may acquire energy which elevates it to a higher energy level, *i.e.* to an excited state. Since, for molecules, each ground and excited state is really subdivided into a number of energy sublevels, molecular spectra are usually seen as *band spectra*. Due to the lack of vibrational energy sublevels in elemental systems, atomic spectra are relatively simple line spectra.

As described for atoms, electrons in a molecule may change their energy level only when distinct quanta of radiation are absorbed or emitted by the molecule: hence the terms *absorption* and *emission spectra*, respectively. The frequency of the absorbed or emitted radiation is a direct function of the change in energy of the electron:

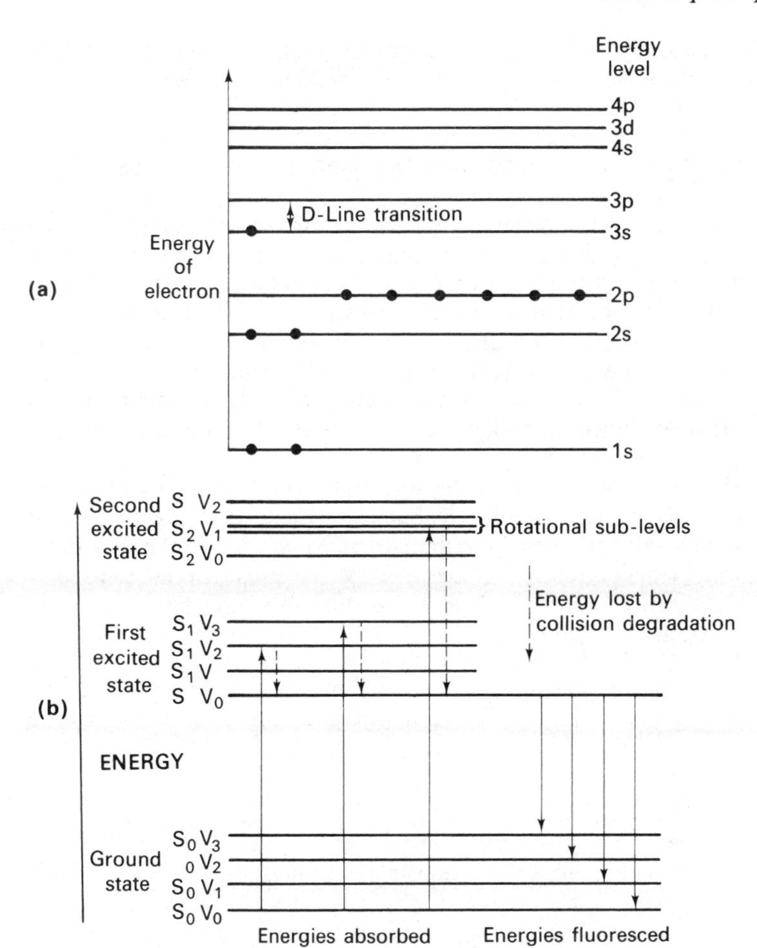

Fig. 8.1 Energy levels and transitions of electrons: **(a)** in the sodium atom and **(b)** in a fluorescent organic molecule. (For clarity, rotational sublevels have been indicated only for vibrational sublevel S_2V_1.)

$$E = E_1 - E_2 = h\nu \qquad 8.1$$

where
E = the energy of radiation absorbed or emitted by the molecule
E_1 = the energy of the electron in the original level
E_2 = the energy of the electron in the final level
h = Planck constant = 6.63×10^{-34} J s
ν = frequency of the radiation in hertz = c/λ
c = speed of light = 3×10^8 m s^{-1}
λ = wavelength of radiation = $1/\bar{\nu}$
$\bar{\nu}$ = wave number of radiation in waves cm^{-1} (kaysers)

Wavelength is usually measured in centimetres (cm), micrometres (μm), or nanometres (nm); angströms (Å), mμ and μ should not be used.

8.1.2 Types of spectra and their biochemical usefulness

Molecules interact with radiation having a vast range of wavelengths, giving rise to spectra in a number of distinct regions as illustrated in Fig. 8.2. A spectrum may be represented as a graph of the amount of energy absorbed or emitted by a system against wavelength or a similar electromagnetic parameter. Different types of instrument are required for the study of different regions. Some of these regions readily yield information of great use to the biochemist and are used routinely, whereas other regions requiring more sophisticated apparatus and expertise are used only in detailed research of biological macromolecules.

Electronic spectra arise due to the outer electrons of atoms changing between major electronic energy levels. Such spectra occur in the visible and ultraviolet regions and are usually accompanied by changes in the rotational and vibrational energy levels. These spectra are used routinely in biochemistry (Sections 8.2 and 8.7). *Fluorescence spectra* may also arise owing to these transitions (Section 8.5).

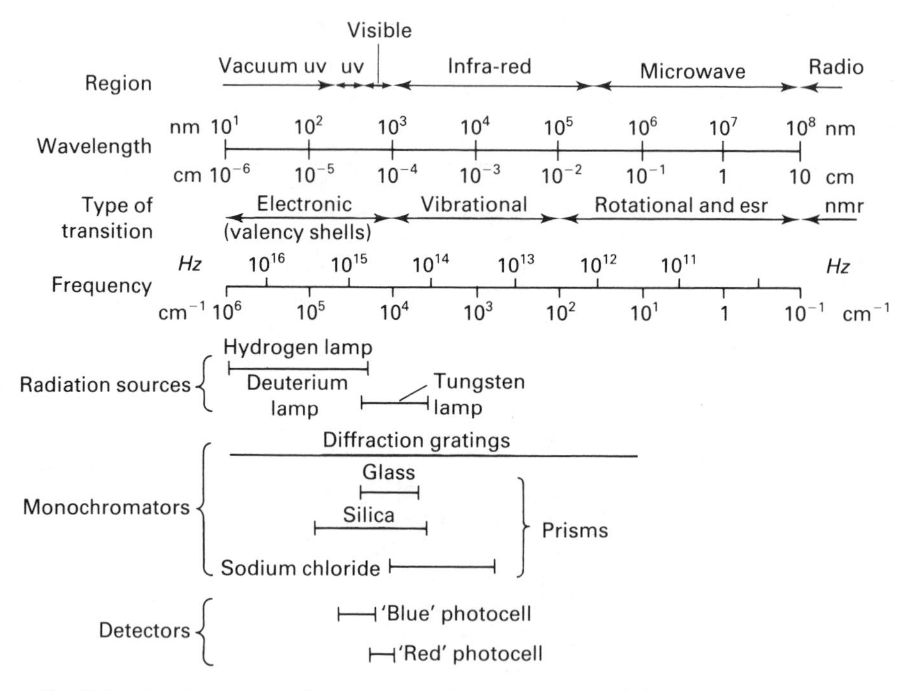

Fig. 8.2 The electromagnetic spectrum and spectral characteristics of spectrophotometer components.

Vibration-rotation spectra are caused by changes in the vibration energy levels. They occur in the near infra-red region and may be accompanied by changes in the rotational energy levels. Such spectra are sometimes used in studies of the detailed structure of biological macromolecules in non-aqueous environments (Section 8.3).

Electron spin resonance spectra and *nuclear magnetic resonance spectra* arise due to changes in the direction of the spins of electrons and nuclei respectively in a magnetic field. These two types of spectra are valuable for studying the structure of biological macromolecules (Sections 8.8 and 8.9).

Molecular band spectra may be resolved into a number of very close line spectra, corresponding to the vibrational and rotational energies of the electrons, only at extremely high resolution. (Resolution is the ability of the instrument to distinguish between two closely spaced absorption lines.)

8.1.3 Basic laws of light absorption

For a uniform absorbing medium the proportion of the radiation passing through it is called the *transmittance*, T, where

$$T = \frac{I}{I_0} \qquad 8.2$$

where I_0 = intensity of the incident radiation
I = intensity of the transmitted radiation

The extent of radiation absorption is more commonly referred to as the *absorbance (A)* or *extinction (E)* which are equal to the logarithm of the reciprocal of the transmittance, i.e.

$$A = E = \log 1/T = \log I_0/I$$

The historic term *optical density* (OD) should not be used.

Transmittance is usually expressed on a range 0 to 100% but is rarely used. Extinction has no units and varies from 0 to ∞. *The Beer-Lambert law* states that the extinction is proportional to concentration of the absorbing substance and to the thickness of the layer:

$$E = \epsilon_\lambda cd \qquad 8.3$$

where ϵ_λ = *molar extinction coefficient* for the absorbing material at wavelength λ (in units of $dm^3\ mol^{-1}\ cm^{-1}$)
c = concentration of the absorbing solution (molar)
d = light path in the absorbing material in (cm)

Since the molar extinction coefficent, ϵ_λ, of a compound may be extremely large, an alternative is to quote the extinction given by a 1 cm thick sample of a 1% solution of the compound, i.e. $E^{1\%}_{1cm}$.

Often the Beer-Lambert law may not be applicable to a system for a number of reasons. Firstly, the specimen may ionise or polymerise at higher concentrations, or it may coagulate to give a turbid suspension, which may increase or decrease the apparent extinction. Furthermore, the instrument may be susceptible to stray radiation as well as being capable of producing beams of radiation of finite waveband only.

8.2 Visible and ultraviolet (UV) spectrophotometry

8.2.1 Principles

The *absorption spectrum* (plural, spectra), or more correctly the absolute absorption spectrum, of a compound may be shown as a plot of the light absorbed (extinction) by that compound against wavelength. Such a plot for a coloured compound will have one or more absorption (extinction) maxima in the visible region of the spectrum (400 to 700 nm) as illustrated in Fig. 8.3 for the reduced form of cytochrome *c* which normally appears cherry pink. Absorption spectra in the ultraviolet (200 to 400 nm) and visible regions are due to energy transitions of both bonding and non-bonding outer electrons of the molecule. Usually delocalised electrons are involved such as the π-bonding electrons of carbon–carbon double bonds and the lone pairs of nitrogen and oxygen. Since most of the electrons in a molecule are in the ground state at room temperature, spectra in this region give information about this state and the next higher one. As the wavelengths of light absorbed are determined by the actual transitions occurring, specific absorption peaks may be recorded and related to known molecular sub-structures. The term *chromophore* is given to a small part of a molecule which gives rise independently to distinct parts of an absorption spectrum, for example the carbonyl group, C = O. Conjugation of double bonds lowers the energy required for electronic transitions and hence causes an increase in the wavelength at which

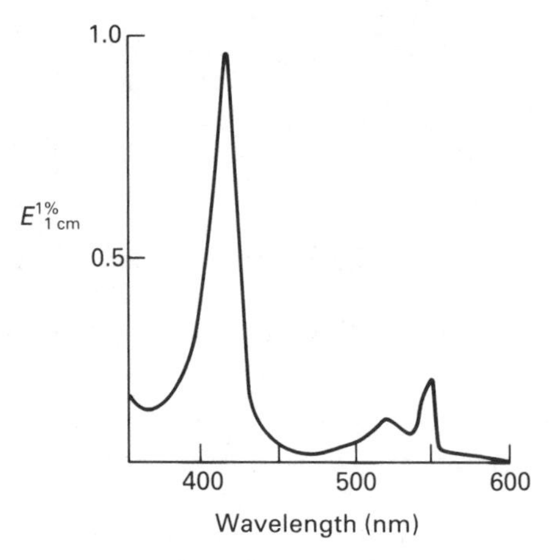

Fig. 8.3 Absolute absorption spectrum of reduced cytochrome *c*.

a chromophore absorbs. This is referred to as a *bathochromic shift*, whereas a decrease in conjugation, caused for example by protonating a ring nitrogen atom, causes a *hypsochromic shift* which leads to a decrease in wavelength. *Hyperchromic* and *hypochromic effects* refer to an increase and a decrease in extinction respectively. Sometimes absorption bands in a spectrum overlap. This may be reduced by making measurements at temperatures approaching zero due to decreased thermal vibrations.

8.2.2 Instrumentation

To obtain an absorption spectrum, the extinction of a substance must be measured at a series of wavelengths. Absorption in the visible and ultraviolet regions may be detected by eye or by a photographic emulsion. (The instruments used are called spectroscopes and spectrographs respectively.) However, the technique of *spectrophotometry* utilises the fact that the potential developed in a photoelectric cell is proportional to the intensity of radiation impinging upon that photocell. The main components of a simple spectrophotometer are illustrated in Fig. 8.4.

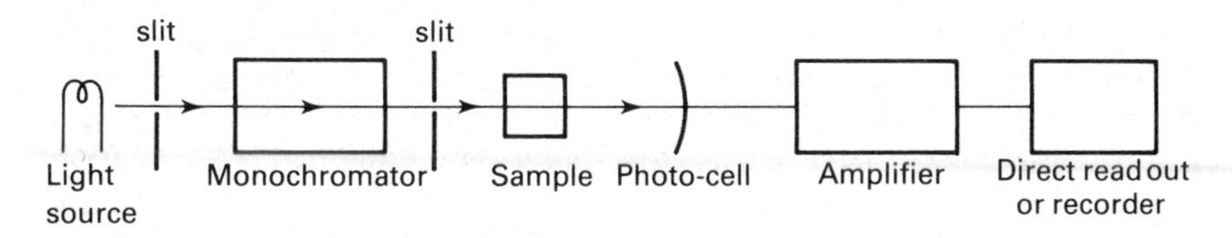

Fig. 8.4 The main components of a simple spectrophotometer.

Modern systems tend to use mirrors instead of lenses wherever possible, both for cheapness and because reflectance systems tend to lose less radiation due to chromatic aberration than refractive systems. The light source is usually a tungsten lamp for the visible region, and either a hydrogen or deuterium lamp for the ultraviolet.

Monochromators are optical systems which produce, from a multiwavelength source of radiation, a parallel beam of monochromatic radiation, i.e., theoretically, radiation of a single wavelength. Usually it is based upon *refraction* by a prism or *diffraction* by a grating. Prisms are best made of glass for the visible region, but must be of quartz or silica for the ultraviolet region since glass absorbs at wavelengths less than 400 nm. The light emerging from any monochromator does not consist of a single wavelength, but of a group of wavelengths known as the *spectral slit width, band width* or *wave band.* Band width is important because it is indicative of the actual wavelengths involved in a given extinction measurement. It is usually defined as twice the half intensity band width, which is the range of wavelengths for which the transmitted intensity is greater than half the intensity of the chosen wavelength and it is a function of the slit width. (Band widths for simple spectrophotometers range from 5 to 35 nm depending on the quality of the instrument, but more sophisticated spectrophotometers give band widths of less than 3 nm/mm slit width.)

Diffraction Grating Monochromators consist of a series of ruled lines on a transparent or reflecting base. Diffraction of white light gives rise to a series of overlapping spectra. Usually in a monochromator a fore-prism preselects a portion of the spectrum of a light source which is then diffracted to obtain monochromatic light. The major advantage of diffraction grating monochromators is that their resolving power is directly proportional to the closeness of the lines, and hence they can be made superior to prisms. In addition, they yield a linear resolution of the spectrum, unlike prisms, which resolve lower wavelengths more than higher ones.

Photocells convert quanta of radiation to electrical energy which may be amplified, detected and recorded. In the photo-emission type, photons impinging on a metal surface in a vacuum cause the emission of electrons in proportion to the intensity of radiation. A positive electrode attracts these emitted electrons and hence a current flows, which causes a potential difference across a resistor incorporated in the system. By electronically amplifying this potential and balancing it against a potentiometer calibrated directly in extinction units, the light absorbed can be measured. Typical photocells are optimally sensitive to radiation with a wavelength of about 400 nm, and are relatively insensitive at wavelengths greater than 550 nm, so it is usual to change to a red sensitive photocell at about this latter wavelength. The theoretical accuracy of these photocells is 1 $\pm$ 0.003 extinction units, i.e. $\pm 0.3\%$. *Photomultiplier tubes* are more sensitive than simple photocells because the electrons emitted by the photoemissive surface are accelerated by a high potential and generate secondary electrons by collision with the gas phase, resulting in a larger current.

Slit widths are important because they affect the observed extinction due to their effects both on the band width and the variations in sensitivity of the photocell with wavelength. To obtain very reliable data, the narrowest possible slit width should be used. Increasing or decreasing the slit width by a factor of two should not cause changes in the apparent extinction if a narrow enough slit width is being used. Zero extinction is usually obtained with the best instruments by amplification of the photocell current or potential allowing the instrument to be operated with predetermined band widths.

Cuvettes (optically transparent cells) are usually used to contain the material under study which is normally dissolved in a suitable solvent. *A reference cuvette* optically identical to, and containing the same solvent (and impurities) as, the test cuvette is always required for setting the spectrophotometer to read zero extinction. For accurate work, the optical matching of the two cuvettes should always be checked. Since glass and plastic absorb strongly below 310 nm, quartz or silica cells, transparent to wavelengths greater than 180 nm, must be used when studying the ultraviolet region. If the solvent used is volatile or corrosive, stoppered cuvettes should be used.

When an absorption cuvette is positioned in the light path it becomes an integral part of the instrument's optical system, so it should be treated with the same care given to other optical components. Scratched or contaminated cuvette optical windows reflect and absorb radiation, resulting in inaccurate measurements. Since all organic molecules absorb in some part of the region, touching the optical surfaces of a cuvette should be avoided. It is obvious that the contents of a cuvette must be homogeneous if valid data are to be obtained, but it is often overlooked that both the formation of bubbles or turbidity inside, or of condensation on the outside, of the cuvettes will lead to completely erroneous results.

The cuvettes commonly used for accurate work have an optical path length of 1 cm and require 2.5 to 3.0 cm^3 of sample for an accurate reading. However, optical microcells, which may require only 0.3 to 0.5 cm^3 of sample, are particularly useful when valuable reagents such as enzymes are being used. When temperature is likely to affect the results, a thermostatically controlled temperature housing for the cuvettes should be used. Flow-through cells are also available enabling continuous monitoring of, for instance, the effluent from chromatography columns.

Specialised Types of Spectrophotometer *Recording spectrophotometers* are generally capable of both scanning a predetermined spectrum and measuring the change in extinction at a predetermined wavelength with time. The most significant difference between these instruments and that illustrated in Fig. 8.4 is that they contain a beam splitting arrangement which enables a simultaneous comparison of the absorption by both the sample and reference cuvettes. The data are commonly recorded on a paper chart but more recent instruments present them on a *visible display unit* (VDU) and store them on floppy discs. The better instruments enable the chart to be run at varying speeds, the spectrum to be scanned at varying speeds, and any desired part of the extinction scale to be expanded; thus for example, the full height of the trace may be employed for extinctions of 0 to 0.1, 0 to 1, 1 to 2, 2 to 2.1. Some instruments also have automatic cell changers, which allow the determination of the extinction at any fixed wavelength of a number of samples at predetermined time intervals. *Low temperature spectrophotometers* allow spectra to be obtained down to $-196°C$ by cooling the cell compartment with liquid nitrogen. This results in greatly increased resolution due to decreased thermal motion and an apparent increase in extinction coefficient due to the greatly increased path length as much of the radiation is internally reflected within the frozen sample. *Reflectance spectrophotometers*, which measure the radiation absorbed when a light beam is reflected by the sample, allow the determination of absorption spectra of pastes and suspensions of microorganisms which are too opaque to transmit radiation. Since these instruments measure the absorption during internal reflection and refraction, and hence the true optical path length is unknown, quantification of the data is complicated. The reference reflecting surface is usually magnesium oxide. *Multi-beam recording spectrophotometers*, capable of recording extinction changes at two predetermined wavelengths at the same time, are now commercially available, and instruments working at more than two wavelengths have been constructed and used. *Microscale spectrophotometers* use a very narrow beam of monochromatic radiation coupled to a microscope system. This instrument enables the extinction at different points within a single living cell to be determined.

8.2.3 Applications

Colorimetry Many substances which do not posses significant extinction coefficients in the visible region will react quantitatively with some other reagent to give a coloured product. This property is used as a basis for assaying such substances. The colour (*chromophore*) is produced under standard conditions from known quantities of the substance, and the extinctions of these samples are measured, using a reagent blank in the reference cuvette which contains all of the reagents except the substance being estimated. The extinction reading is set to zero using this blank by amplification of the photocell current or potential. The measured extinctions are plotted against the quantity or concentration of the test substance producing the colour. This graph is known as the *calibration curve.* Unknown quantities of the substance may then be assayed by producing the colour under the same standard conditions, measuring the extinctions and using the calibration curve to determine the quantity of the substance which produces that extinction. Since it is often possible to produce comparatively high extinctions with relatively small amounts of material, colorimetry is widely used in biochemistry to assay a wide range of biologically important molecules. Table 8.1 summarises some of the most commonly used assays. Some important points to bear in mind when carrying out colorimetry are as follows:

(i) unlike straightforward spectrophotometry, colorimetry is a destructive technique i.e. that portion of the sample analysed cannot be recovered;
(ii) a chromophore exhibits the complementary colour to that which it absorbs, i.e. a yellow compound appears yellow because it absorbs blue light and therefore it must be estimated in the blue region of the spectrum;
(iii) colorimetric assays are usually most sensitive at the extinction peak of the chromophore produced. If this is not known the spectrum should be determined before commencing the assay;
(iv) the reference cuvette should contain all the reagents other than the substance being assayed in the same concentrations as in the test cuvettes.
(v) the reference cuvette and its contents require, if anything, more care because any error in these will be reflected in all of the values obtained;
(vi) assays should normally be performed in duplicate and individual values, not means, should be plotted. This procedure allows the experimenter to omit erroneous extinction values justifiably from the calibration curve if the duplicate value lies on the curve produced from the other extinction values;
(vii) the best line through the points should be drawn, not necessarily the

Table 8.1 Common colorimetric assays.

Substance	Reagent	Wavelength (nm)
Inorganic phosphate	Ammonium molybdate; H_2SO_4; 1,2,4-aminonaphthol; $NaHSO_3$; Na_2SO_3	600
Amino acids	(a) Ninhydrin	570 (proline 420)
	(b) Cupric salts	620
Peptide bonds	Biuret (alkaline tartrate buffer, cupric salt)	540
Phenols, tyrosine	Folin (phosphomolybdate, phosphotungstate, cupric salt)	660 or 750 (750 more sensitive)
Protein	(a) Folin	660
	(b) Biuret	540
	(c) BCA reagent (Bicinchoninic Acid)	562
	(d) Coomassie Blue	595
Carbohydrate	(a) Phenol, H_2SO_4	Varies e.g. glucose 490 xylose 480
	(b) Anthrone (anthrone, H_2SO_4)	620 or 625
Reducing sugars	Dinitrosalicylate, alkaline tartrate buffer	540
Pentoses	(a) Bial (orcinol, ethanol, $FeCl_3$, HCl)	665
	(b) Cysteine, H_2SO_4	380–415
Hexoses	(a) Carbazole, ethanol, H_2SO_4	540 or 440
	(b) Cysteine, H_2SO_4	380–415
	(c) Arsenomolybdate	Usually 500–570
Glucose	Glucose oxidase, peroxidase, o-dianisidine, phosphate buffer	420
Ketohexose	(a) Resorcinol, thiourea, ethanoic acid, HCl	520
	(b) Carbazole, ethanol, cysteine, H_2SO_4	560
	(c) Diphenylamine, ethanol, ethanoic acid, HCl	635
Hexosamines	Ehrlich (dimethylaminobenzaldehyde, ethanol, HCl)	530
DNA	Diphenylamine	
RNA	Bial (orcinol, ethanol, $FeCl_3$, HCl)	665
α-oxo acids	Dinitrophenylhydrazine, Na_2CO_3, ethyl acetate	435
Sterols	Liebermann-Burchardt reagent (acetic anhydride, H_2SO_4, chloroform)	625
Steroid hormones	Liebermann-Burchardt reagent	425
Cholesterol	Cholesterol oxidase, peroxidase, 4-amino-antipyrine, phenol	500

best line through the origin and the other points. This is because it is just as feasible to obtain an erroneous value for reagent blanks as for any other sample;

(viii) the calibration curve may vary between batches of both reagent and standard and hence a new calibration curve should be constructed each time the assay is used, unless there is evidence to suggest that the previously obtained curve is still valid. Hence all calibration curves cannot be used as standard curves;

(ix) calibration curves should never be extrapolated beyond the highest extinction value measured. Thus it may sometimes be expedient to prepare a curve with extinction values up to 1.0. However, it is always more accurate to repeat the assay at a concentration which falls within the most accurate region of the calibration curve. This is usually from 0.0 to 0.6 because of the logarithmic nature of the extinction scale;

(x) if the extinction value of a sample lies outside the range of the calibration curve, the sample should be re-assayed at a concentration that falls within this range. Alternatively if the system obeys the Beer-Lambert law, it may be possible to use cuvettes of shorter optical light path;

(xi) in colorimetric assays, the reproducibility of the methods is far more important than the Beer-Lambert relationship. As with all analytical procedures, it is the reproducibility of the assay in the experimentalist's hands that determines the accuracy of the final data, not what the schedule or reference states.

Qualitative Analysis Visible and ultraviolet spectra may be used to identify classes of compounds in both the pure state and in biological preparations e.g. proteins, nucleic acids, cytochromes, chlorophylls. This technique may also be used to indicate chemical structures and intermediates occurring in a system. However, for really precise analysis the refinement of infra-red spectrophotometry is required (Section 8.3).

Quantitative Spectrophotometric Analysis A number of important classes of biological compounds may be measured semi-quantitatively using ultraviolet and visible spectrophotometers, e.g. proteins at 280 nm and nucleic acids at 260 nm (Section 3.2.2). Corrections may be made for the presence of impurities by also measuring the extinction at wavelengths where the impurities absorb more than the compound under investigation and applying the appropriate algebraic formula, e.g. the Morton and Stubbs correction for the amount of vitamin A in saponified extracts of natural oils, and the $E_{280/260}$ ratio for protein estimation in the presence of nucleic acids. The amount of two components with overlapping spectra, such as chlorophylls *a* and *b* in ether, may be estimated if their extinction coefficients at two wavelengths are known. For *n* components, extinction data at *n* wavelengths are required. When measuring the concentration of large DNA molecules at 260 nm the interference of *Rayleigh light scattering* may be corrected for by extrapolation from the extinction values obtained at non-absorbent regions of the DNA spectrum (e.g. 330 to 430 nm).

Enzyme Assays and Kinetic Analysis Ultraviolet and visible spectrophotometry provide the most commonly used assays of enzymes and their substrates (Sections 3.4 and 3.5).

Difference Spectra A *difference spectrum* is the difference between two absorption spectra. It may be obtained indirectly by subtracting one absolute absorption spectrum from another, for example by subtracting the absorption spectrum of ubiquinol from that of ubiquinone as shown in Fig. 8.5. In practice, a difference spectrum is usually obtained directly by using one compound, (say ubiquinol), in the reference cuvette, whilst measuring the absorption spectrum of the other compound, (ubiquinone), resulting in the

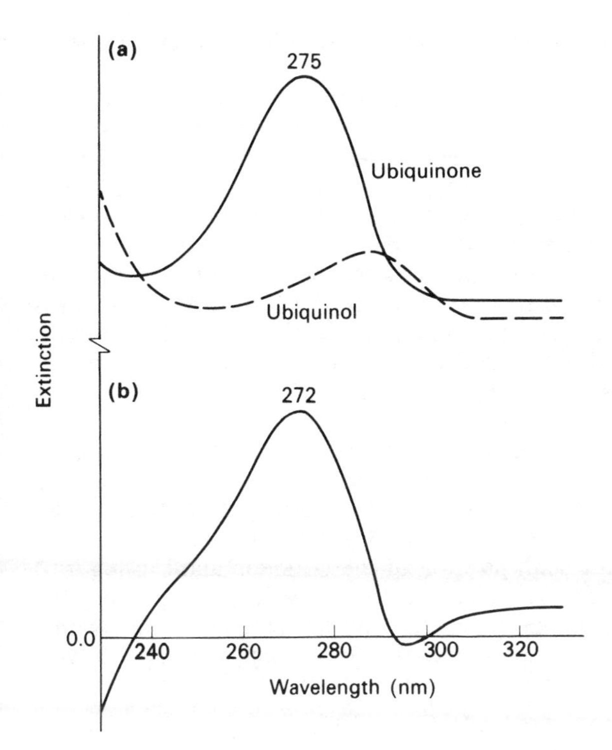

Fig. 8.5 **(a)** Absolute and **(b)** difference spectra of ubiquinone and ubiquinol.

difference spectrum of ubiquinone minus ubiquinol. The major advantage of *difference spectrophotometry* is that it enables the detection of relatively small extinction changes in a system which has a large extinction, for example changes in the oxidation state of components of the respiratory chain in intact mitochondria and chloroplasts.

Figure 8.5b presents the difference spectrum of ubiquinone minus ubiquinol, and illustrates a number of important points about difference spectra:

(i) difference spectra may contain negative extinction values;
(ii) both absorption peaks and minima are often displaced and extinction coefficients are different from those of absolute absorption peaks;
(iii) the points of zero extinction in the difference spectrum, equivalent to those wavelengths where both the reduced and oxidised forms of the compound exhibit identical extinctions, are known as *isobestic points*, and are used in checking for the presence of interfering material.

A more complex example of a difference spectrum is that of the cytochrome a_3CO complex minus cytochrome a_3 which may be obtained by using anaerobic mitochondria in the reference cuvette and the same system poisoned with carbon monoxide in the test cuvette, since cytochrome a_3 is the terminal electron carrier and is the only component that reacts with carbon monoxide.

A more specific use of the term difference spectrum, very commonly used in studies on respiratory components, means the reduced minus oxidised difference spectrum, for example the difference spectrum of cytochrome *c* corresponds to the cytochrome c_{red} minus cytochrome c_{ox} difference spectrum. Such a difference spectrum for a suspension of mitochondria can also be obtained by the *reversal technique*. This technique involves measuring the change in extinction at each wavelength when the preparation passes from the aerobic to the anaerobic state. The spectrum obtained is a combined difference spectrum for cytochromes, a, a_3, b, c, c_1, NAD^+ and flavoprotein. Shoulders on the peaks observed in difference spectra obtained at room temperature may be resolved into distinct peaks at − 196°C i.e. by measuring so called *low temperature difference spectra*.

Binding spectra, or more precisely *substrate binding difference spectra*, can sometimes be used to study the extent of interaction between an enzyme and its substrate. Generally the binding of a substrate to a haem group containing a ferric ion in the high spin form perturbs the spectrum by displacing water from the sixth position of the ferric ion and causing it to change to the low spin state, which is readily followed spectrophotometrically. For example the binding of a drug (substrate) to liver microsomal monooxygenase (mixed function oxidase) causes a blue shift of the cytochrome P450 component of the enzyme from 420 nm to 390 nm.

Protein Structural Studies The spectrum of a chromophore depends on the polarity of its environment. It therefore follows that if a change in the polarity of the solvent in which a protein is dissolved changes the spectrum of a constituent amino acid chromophore without changing the conformation of the protein, the amino acid residue must be in a position accessible to the solvent: i.e. at the surface of the protein. This is called *solvent perturbation*. Perturbing solvents commonly used include solutions of dimethyl sulphoxide, dioxane, glycerol, mannitol, sucrose, and polyethylene glycol. Also, if the unfolding of a polypeptide (denaturation) exposes a tyrosine in an internal (hydrophobic) environment to the external solvent (hydrophilic environment) the effects of pH, temperature and ionic strength on protein denaturation may be studied. Again if a protein–protein interaction or the binding of a ligand involves a tyrosine, these interactions may be studied, for example the binding of a substrate or inhibitor to the active site of an enzyme. Such studies may be extended by the use of *reporter groups* which are artifical chromophores attached to a relevant region of the protein, for example the interaction of a metal with the active site of an enzyme can be demonstrated by attaching a reporter group such as dimethylaminobenzene or arsanilic acid to the active site and studying the spectral changes on addition of the appropriate metal ion.

Nucleic Acid Structural Studies The extinction at 260 nm of double stranded DNA in solution increases (hyperchromicity) when it is heated through its transition temperature due to denaturation of its helical structure

(Section 5.2.2). The reverse occurs on renaturation of the DNA by cooling. Thus the effects of pH, temperature and ionic strength on the secondary structure of DNA may be studied. Solvent perturbation of nucleic acid spectra by replacement of normal water with 50% D_2O only changes the spectral components due to unpaired nucleotides. So the effect of 50% D_2O on the spectrum of an RNA allows the estimation of the fraction of unpaired bases, e.g. in tRNA.

Action Spectra An *action spectrum* may be shown as a plot of a physiological (non-extinction) parameter against wavelength. Even for complex biological systems such a spectrum often corresponds to the absorption spectrum of a single key compound. For example, plotting the rate of oxygen evolution by green plant tissue against the wavelength of light used to irradiate the system results in a graph similar to the spectrum of the chlorophylls.

Turbidometry and Nephelometry Very dilute suspensions may be assayed by *turbidometry*, that is by measuring their apparent extinction at a wavelength which is not absorbed. Fortunately, scattering of radiation varies much less with wavelength than does absorption. Relationships are, however, non-linear and standardisation of this technique is difficult since the particle size is critical, but it may be used to obtain an estimate of the concentration of bacterial cells at 600 nm and of proteins in solution (Section 3.2.2). In contrast, *nephelometry* measures the intensity of radiation *scattered* by a suspension and is commonly used for estimating the concentration of microorganisms (Note: strictly speaking, nephelometry is *not*, therefore, a spectrophotometric technique.)

8.3 Infra-red (IR) spectrophotometry

For an asymmetric molecule containing n atoms, the theory of molecular vibrations predicts $3n-6$ fundamental vibrations of which $2n-5$ cause bond deformations and $n-1$ cause bond stretching. Vibrations causing a change in *dipole moment*, i.e. a charge displacement, are observed in the infra-red region. The other vibrations may be detected in *Raman spectra*. Infra-red spectra of molecules are absolutely specific. This allows the molecules to be characterised and the term *finger-print* is used to describe the characteristic infra-red pattern observed. It is not difficult to assign the absorption bands of simple molecules such as carbon dioxide and water, but biochemists are usually interested in large complex molecules. A molecule with 50 atoms will have 144 fundamental modes of vibration. However, *infra-red spectra* do not always appear as complicated as might be expected from theory. Certain bands, which regularly appear near the same wavelength, may be assigned to specific molecular groupings, in the same way as for chromophores which absorb in the ultraviolet and visible region. These *group frequencies* are of great value in diagnostic work. (Since infra-red spectra are vibrational spectra, infra-red spectroscopists tend to work in frequency units Hz, rather than wavelength.) Fortunately a particular group does not always absorb at precisely the same frequency since its molecular environment does have some influence. If this did not occur the infra-red region would be far less useful to the structural biochemist. Thus it is possible to distinguish between the C – H vibrations of $-CH_2$ and those of $-CH_3$. When the bonding force between two atoms increases, for example in double bond formation, the stretching frequency increases, i.e. the wavelength absorbed decreases.

Despite attempts to use infra-red spectroscopy for the study of biological macromolecules and membranes, the major use of this technique in biochemistry is in research into the structure of purified molecules of intermediate size such as drugs. It is usually used in conjunction with such techniques as nuclear magnetic resonance spectroscopy and mass spectrometry. Interfacing infra-red spectrophotometry with gas–liquid chromatography results in a powerful technique for analysing drug metabolites (GC/IR). Another biological application of infra-red spectroscopy is *infra-red gas analysis*. This provides a convenient and sensitive means of detecting and measuring differences in the concentration of gases such as carbon dioxide, carbon monoxide and acetylene in biological samples. Its most common application is to study carbon dioxide metabolism during photosynthesis and respiration in plants.

8.4 Circular dichroism (CD) spectroscopy

8.4.1 Principles

Information on the three-dimensional structure (conformation) of macromolecules in solution can be obtained by studying their absorption of polarised light, using *circular dichroism (CD) spectroscopy*. CD spectroscopy measures the differential absorption of right (R) and left (L) circularly polarised light as a function of wavelength. The older technique of *optical rotary dispersion* (ORD) spectroscopy measures the ability of an optically active chromophor to rotate the plane of polarisation of light as a function of wavelength. CD and ORD are manifestations of the same phenomenon and they are merely two ways of investigating the interaction of polarised light with optically active molecules. Since ORD spectrometers were available before CD spectrometers, early work in this field yielded ORD spectra. However, the relative simplicity of CD spectra makes CD analysis superior to ORD analysis, so that nowadays ORD is rarely used. In addition CD analysis possesses superior ability for the resolution of bands due to electronic transitions of different optically active centres.

A beam of monochromatic light consists of electromagnetic waves oscillating in all planes perpendicular to the direction of that beam. *Plane polarised light*, that is light consisting of waves oscillating in a single plane, may be obtained by passing a beam of monochromatic light through a Nicol prism or a polarising screen (Polaroid). Circularly polarised light may be obtained by superimposing two plane polarised light waves of the same wavelength and amplitudes but differing in phase by one quarter of a wavelength and in their planes of polarisation by 90°. Such light can be either right (R) or left (L) circularly polarised depending on the relative positions of the peaks of the two component plane polarised waves. Superimposition of R and L waves of equal amplitude and wavelength results in plane polarised light. Similarly plane polarised light can be resolved into R and L waves.

Asymmetric (chiral) molecules i.e. those that cannot be superimposed on their mirror image, interact with plane polarised light in such a way that the R and L waves are differentially absorbed and refracted. As a consequence, the resultant R and L waves combine to give a beam of polarised light in a different plane to that of the incident beam. The observed angle of rotation is referred to as the specific rotation $[\alpha]_\lambda$. Moreover, the differential absorption of the R and L waves leads to them having different amplitudes so that the transmitted beam is elliptically rather than circularly polarised. Ellepticity (θ) is often measured rather than extinction:

$$\theta = 2.303\ \Delta E \frac{180}{4\pi} \qquad 8.4$$

$$= 33\ \Delta E \text{ degrees} \qquad 8.5$$

where ΔE = the differences in extinction for L and R waves

Usually a CD spectrum is a plot of the variation of ellipticity with wavelength.

8.4.2 Instrumentation

The main components of a CD spectrometer are illustrated in Fig. 8.6. Usually L and R circularly polarised radiation is produced from a single monochromator by passing plane-polarised light through an electro-optic modulator. This is a crystal subject to alternating currents that transmits either the R or L component of light dependent on the polarity of the electric field to which it is exposed. The photomultiplier detector produces a voltage in proportion to the ellipticity of polarisation of the combined beam falling on the photomultiplier.

8.4.3 Applications

Protein Conformation The CD spectrum of a protein can provide information about the relative amounts of the major types of secondary structure (α, β and random coil) within the protein in solution.

Currently CD spectra cannot provide detailed information about the specific tertiary structure of proteins because the relevant theory required for their precise interpretation has not been developed to include the effects of either conformations other than α, β and random coil, or disulphide bridges, aromatic side chains and prosthetic groups. However, before deciding that

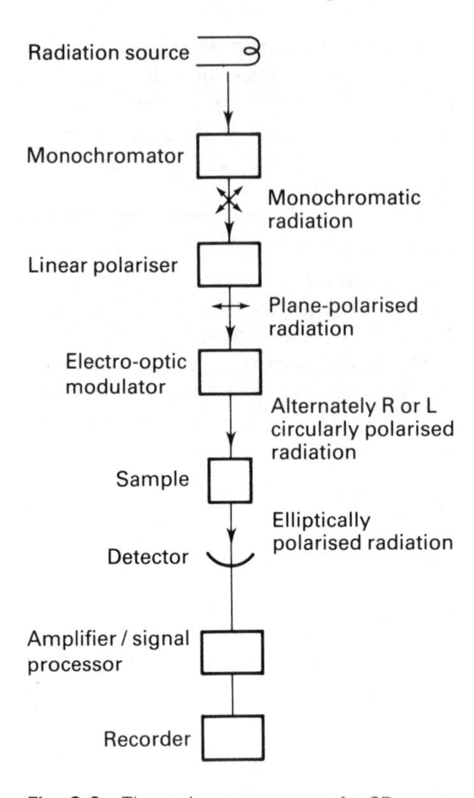

Fig. 8.6 The main components of a CD spectrometer.

CD is of no use in the investigation of protein structure, it should be noted that high resolution X-ray diffraction data, used to elucidate specific tertiary structures, are available for only relatively few crystallisable proteins and that such data are expensive, difficult and perhaps impossible to obtain for most proteins in the foreseeable future.

CD spectra of the α-helix, β conformation and the random coil of poly-L-amino acids are known and are distinctly different from each other. These have been used as standards in calculating the percentage of each form of secondary structure in proteins from their CD spectra, using a curve-fitting computer programme.

However, more importantly, the CD of a macromolecule is very sensitive to conformational changes. Even if the CD spectrum of a protein is far too complex for the determination of its structure, it is usually possible to study almost any interaction with the protein. For example; CD spectra may be used to determine the binding constants of substrates, cofactors, inhibitors or activators of any enzyme. Thus the binding of 3-cytidilic acid to the active site of pancreatic ribonuclease changes the CD of a tyrosine remote from that site. This shows that the binding of this inhibitor causes a conformational change in a distant part of the enzyme. Since denaturation involves the conversion of α and β forms to the random coil, it may be readily and sensitively followed by CD spectroscopy.

Conformation of Nucleic Acids The CD spectrum of a single stranded nucleic acid may be calculated fairly accurately from a knowledge of its nearest-neighbour frequency. Thus any differences between the calculated and measured CD spectrum must be due to variation in structure, such as double-strandedness. CD theory for double-stranded nucleic acids is not yet complete. However, the CD spectrum of double-stranded DNA seems to be independent of base composition in the wavelength range usually studied. The large increase in the CD spectrum of mononucleotides when they are linked to form even short-chain oligonucleotides originally provided evidence that hydrophobic interactions between the stacked bases are essential in stabilising the double-stranded structure of DNA.

Although all nucleotides are optically active, the CD of polynucleotides is greatly increased when they adopt a helical conformation. Thus CD spectra are frequently used to study changes in the structure of nucleic acids such as: the loss of helicity of single stranded nucleic acids as a function of temperature or pH; structural changes on binding cations and proteins; the effect of binding an amino acid to its appropriate tRNA; transitions between single and double stranded nucleic acids; DNA–histone interactions in chromatin and the structure of rRNA in the ribosome.

8.5 Spectrofluorimetry

8.5.1 Principles

Fluorescence is the phenomenon whereby a molecule, after absorbing radiation, emits radiation of a longer wavelength. Thus a compound may absorb radiation in the ultraviolet region and emit visible light. This increase in wavelength is known as the *Stokes' shift*.

At room temperature most organic molecules are in the ground state (S_0V_0 in Fig. 8.1b). Absorption of photons elevates an electron in these molecules to a higher energy state (S_1, S_2, etc.) in less than 10^{-15} seconds. After absorption, energy is lost very rapidly by collision degradation (as heat), resulting in the energy of the excited molecules falling rapidly to that of minimal vibrational energy in the lowest excited state (S_1V_0 in Fig. 8.1b). The energy emitted from these molecules in regaining the ground state within a period of less than 10^{-8} seconds gives rise to a fluorescent peak, showing the Stokes' shift. Although many organic molecules absorb in the ultraviolet and visible regions, only a few fluoresce. Whereas knowledge of the structure of an organic molecule permits predictions about its absorption spectrum, structural information cannot be used to predict that a compound will fluoresce. However, whereas flexible molecules such as aliphatic compounds tend to photodissociate rather than fluoresce, aromatic molecules containing delocalised π-electrons sometimes fluoresce. Fortunately for the biochemist a relatively high proportion of biological molecules do fluoresce.

Since the emitted radiation can have a number of closely related values depending on the final vibrational and rotational energy levels attained, fluorescence spectra are band spectra. They are usually independent of the wavelength of the radiation absorbed and have a mirror image relationship to the absorption peak with the greatest wavelength.

Obviously, fluorescence spectra can supply information only about events which take less than 10^{-8} seconds to occur. The *quantum efficiency, Q*, which is equal to the number of quanta fluoresced divided by the number of quanta absorbed, is usually independent of the wavelength of the activating light. At low concentrations, the intensity of fluorescence (I_f) is related to the incident radiation (I_0) by the following simple relationship:

$$I_f = I_0 2.3\, \epsilon_\lambda c d Q, \qquad \text{i.e. } I_f \propto c \qquad 8.6$$

where c = concentration of the fluorescing solution (molar)
d = light path in the fluorescing solution (cm)
ϵ_λ = molar extinction coefficient for the absorbing material at wavelength λ (in units of $dm^3\ mol^{-1}\ cm^{-1}$)

Simple electronics are adequate in fluorimeters due to the direct relationship between concentration and fluorescence intensity (equation 8.6). Spectrofluorimetry is most accurate at very low concentrations when absorption spectrophotometry is least accurate, for example 100 pg of catecholamines, or NADH, may be detected by spectrofluorimetry, as compared to the 100 μg of the catecholamines, serotonin and adrenaline detectable by absorption spectrophotometry. The sensitivity of fluorimeters is usually easily adjusted over a large range by amplification of the current produced in the photocell circuit. Spectrofluorimeters enable the utilisation of great spectral selectively since, due to the Stokes' shift, two monochromators may be used, one selecting the activating wavelength and the other the fluorescent wavelength. No reference cuvette is required, but a calibration curve must be constructed.

Disadvantages of spectrofluorimetry include its susceptibility to pH, temperature and solvent polarity, and the virtual impossibility of predicting whether a compound will fluoresce. The major problem with fluorimetry is *quenching*, whereby the energy, which could be emitted as fluorescence, is transferred to adjacent molecules. It should also be realised that detergents, stopcock grease, filter paper and some laboratory tissues may cause interference in fluorimetric assays by the release of strongly fluorescing materials.

8.5.2 Instrumentation

The main components of a complete spectrofluorimeter are indicated in Fig. 8.7, and include: a continuous spectrum source, for example a mercury lamp or Xenon arc; a monochromator (M_1) for irradiating the specimen with any chosen wavelength; a second monochromator (M_2) which, under conditions of constant irradiating wavelengths, enables the determination of the fluorescent spectrum of the specimen; and a detector which is usually a

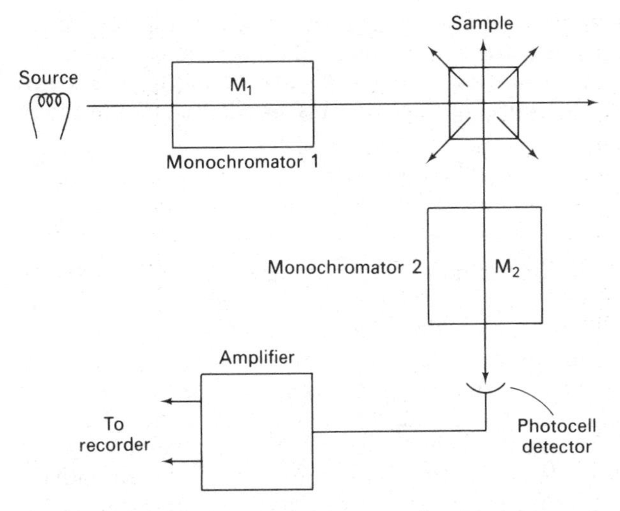

Fig. 8.7 The basic component of a spectrofluorimeter, set up for 90° illumination.

sensitive photocell, for example a red sensitive photomultiplier for wavelengths greater than 500 nm. Since the intensity of fluorescence may decrease by 10 to 50% if the sample temperature falls from 30°C to 20°C, adequate temperature control is essential for accurate fluorimetric estimations.

Pre-filter absorption reduces the amount of radiation reaching fluorescent molecules furthest from the light source, and post-filter effects reduce the amount of fluoresced radiation escaping from the cuvette (Fig. 8.8a). Use of microcells reduces both pre-filter and post-filter effects which occur with concentrated solutions.

Instead of 90° illumination as illustrated in Fig. 8.7, front face illumination (FFI), which is essential for suspensions, may be used to prevent filter

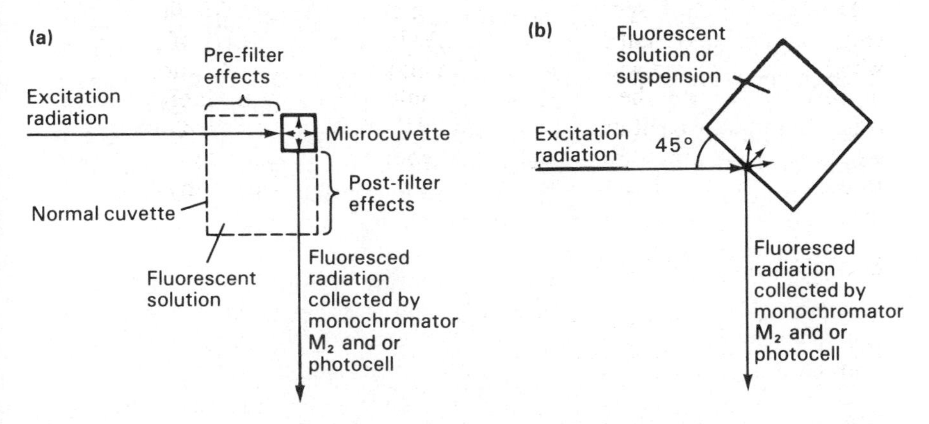

Fig. 8.8 Reduction of filter effects: **(a)** using microcuvettes and **(b)** using front face illumination.

effects. FFI only requires cuvettes with one optical surface so normal spectrophotometer cuvettes may be used. The geometry of the system is arranged so that fluorescence may be measured through the cuvette face that allowed entry of the excitation radiation (Fig. 8.8b). FFI is usually less sensitive than 90° illumination.

8.5.3 Applications

Qualitative Analysis The determination and comparison of both excitation and fluorescence spectra of a compound may help to identify it. The effect of pH and solvent composition on the fluorescence of a compound, and the polarisation of its fluorescence, may contribute information about its structure. The measurement of the *phosphorescence* and *phosphorescence lifetime* of a compound may contribute information to its identification. Phosphorescence is a light emission phenomenon which involves intersystem crossing to the lowest triplet state. Phosphorescence may take a relatively long time to decay and occurs at longer wavelengths than fluorescence.

In chromatography and electrophoresis amino acids and peptides may be detected by coupling their primary amino groups to dansyl chloride or o-phthalaldehyde to produce fluorescence. The latter conjugates fluoresce so intensely blue that the total oligopeptide fingerprint of only 10^{-5}g protein may be determined. The extrinsic fluor acridine orange can be used to determine the strandedness of polynucleotides since it shows a different Stokes' shift when bound to double and single stranded poly-nucleotides; fluorescing green and red respectively.

Quantitative Analysis This is really the major use of fluorimetry in biochemistry, because of the accuracy, reproducibility and sensitivity of the technique, at concentrations too low for absorption spectral analysis. Typical applications include the assay of vitamin B_1 in foodstuffs, NADH, hormones such as cortisol and oestradiol, drugs such as lysergic acid and barbiturates in blood, organophosphorous pesticides and carcinogens. Chlorophyll, cholesterol, porphyrins and some metal ions may also be assayed spectrofluorimetrically. The use of internal standards, i.e. measuring the fluorescence of an unknown quantity of pure compound before and after addition of a known quantity of standard, permits both self and contaminant quenching to be allowed for. Ethidium bromide binds tightly to DNA resulting in a large increase in fluorescence and allows the amount of DNA to be estimated. This is particularly useful in conjunction with gel electrophoresis (Section 7.5).

Instead of studying a molecule by measuring its intrinsic fluorescence it is often possible to couple or bind the molecule to a fluorescent probe or fluor, and to measure the so-called *extrinsic fluorescence* of the conjugate produced (similar to the use of reporter groups in absorption spectrophotometry, Section 8.2.3). Ideally the fluor should be tightly bound at a specific site in the molecule and its fluorescence should be sensitive to environmental conditions. Obviously the fluor should not affect the system being studied.

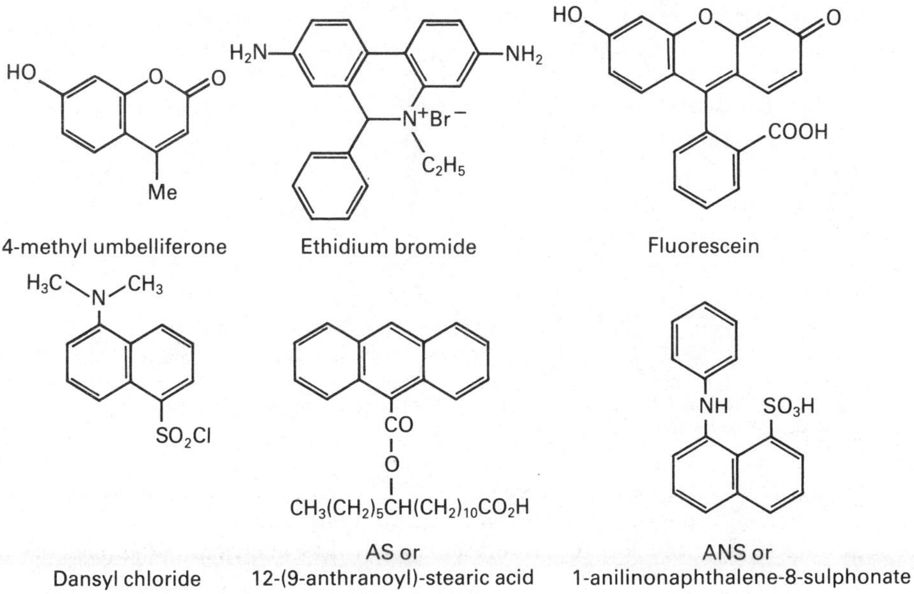

4-methyl umbelliferone | Ethidium bromide | Fluorescein

Dansyl chloride | AS or 12-(9-anthranoyl)-stearic acid | ANS or 1-anilinonaphthalene-8-sulphonate

Fig. 8.9 Structure of some fluorescent probes.

The structures of some fluorescent probes are illustrated in Fig. 8.9.

Quin-2, which preferentially binds Ca^{2+} may be used to measure unbound cytoplasmic calcium because the fluorescence of quin-2 increases about five-fold when it binds Ca^{2+}. Cells are loaded with the membrane-permeable quin-2 acetoxymethyl ester and cytoplasmic esterates hydrolyse the ester and leave the membrane-impermeant quin-2 anion trapped in the cytoplasm. Calibration curves for this fluorescence assay are linear over the range 10^{-4} to 10^{-2} M Ca^{2+}. Apparently the recently developed probes fura-2 and indo-1 are thirty times more sensitive than quin-2 in the fluorescent assay of Ca^{2+}.

The fluorescent probe quin-1 can be similarly used to monitor intracellular pH changes since it shows a thirty-fold increase in fluorescence as the pH changes from 5 to 9.

Enzyme Assays and Kinetic Analysis (Section 3.4.3) Group specific hydrolases may be readily assayed by measuring the rate of appearance of fluorescence at 450 nm of the anion of 4-methylumbelliferone when the enzyme acts upon an ether or ester derivative of 4-methylumbelliferone. The absorption and fluorescence spectra of these compounds are indicated in Fig. 8.10a and Fig. 8.10b respectively, which indicates that if the assay system is irradiated with 350 to 400 nm light virtually all of the fluorescence measured around 450 to 500 nm is due to the anion product of the enzyme. The fluorimetric assay of β-galactosidase using fluorescein di-(β-D-galactopyranoside) as substrate is so sensitive that the presence of a single enzyme molecule can be detected. This allows the number of molecules of this enzyme

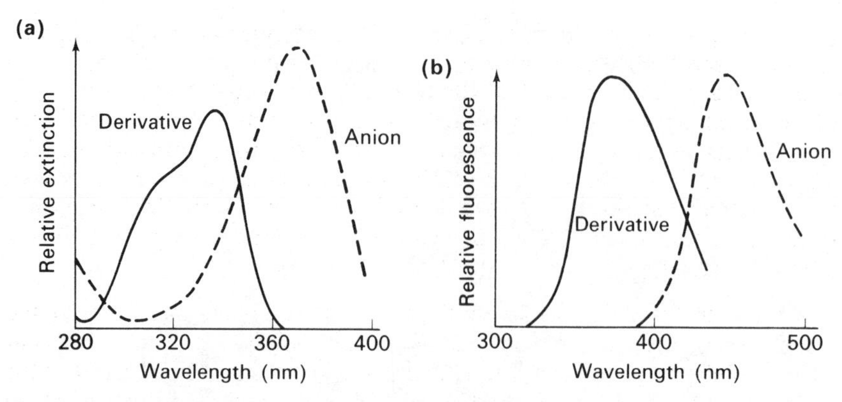

Fig. 8.10 Spectra of the methylumbelliferone anion and derivatives of 4-methylumbelliferone at pH 10. **(a)** Absorption spectra; **(b)** fluroescence spectra.

in a single bacterial cell to be determined and hence the induction of synthesis of the enzyme in individual cells of a bacterial population.

Since NADH and NADPH fluoresce, spectrofluorimetry can be applied to numerous metabolic reactions that are or can be coupled to the oxidation of NADH or NADPH, or the reduction of NAD^+ or $NADP^+$ (Section 3.4.3). Owing to the sensitivity of the technique, an enzyme can often be studied under conditions similar to those found *in vivo*, for example low concentrations of substrate, enzyme and cofactors. Sometimes exogenous cofactors need not be added. Instead, the kinetics of oxidation and reduction of the endogenous material in intact organelles or cells may be measured, for example mitochondrial NADH.

Studies of Protein Structure Proteins that contain fluorescent chromophores such as tryptophan and FAD fluoresce. The binding of substances such as inhibitors, coenzymes, and allosteric effectors close to such intrinsic fluors can be measured by the changes in their fluorescence spectra. This in turn may give information about the conformation, denaturation and aggregation of the proteins. These studies may be extended to proteins lacking suitable intrinsic fluors by coupling them to extrinsic fluors such as anilinonaphthalene-8-sulphonate (ANS), dansyl chloride and derivatives of fluorescein or rhodamine (Fig. 8.9). Examples include the study of conformational changes in enzymes on binding their substrates and determination of bound fatty acids in so-called purified bovine serum albumin.

Fluorescent Probes and Studies on Membrane Structure Since the fluorescent properties of a molecule vary with its mobility and the polarity of its environment, such properties may be monitored in the vicinity of a fluorescent probe by measuring changes in its fluorescence. Various fluorescent probes have been designed to sample specific locations in both artificial and natural membranes. ANS and *N*-methyl-2-anilino-6-naphthalene sulphonate (MNS) contain both charged and hydrophobic areas and locate at the water-lipid interface. Studies using such compounds with membranes yield information about that interface. By incorporating phospholipids containing 12-(9-anthroanoyl)-stearic acid and 2-(9-anthroanoyl)-palmitic acid into membranes, information about the region 0.5 nm and 1.5 nm respectively from the phosphate headgroups of the lipid bilayer may be obtained. Such experiments not only yield information about the basic structure of biological membranes, but allow the effects of temperature and biological phenomena on membrane structure to be studied. The use of the ANS probe has shown that structural changes occur in mitochondrial membranes during energy transduction.

Fluorescence Bleaching Recovery (FBR) This is a technique for studying the motion of specific components in a complex mixture such as fluorescent labelled phospolipids in biological membranes. A region containing the fluor is exposed to a pulse of high intensity radiation to irreversibly bleach the fluor. This region is then monitored with low intensity radiation to detect the re-emergence of fluorescence as the bleached and unbleached molecules interdiffuse.

The major application of FBR is in the study of the lateral diffusion of extrinsically labelled molecules in biological membranes, for example rhodopsin in the photoreceptor membrane. Other applications include the study of the polymerisation of proteins such as actin. Studies of the diffusion of fluorescent labelled proteins micro-injected into cells have indicated the enhanced viscosity of cytoplasm and that 85% of the actin in amoebae is bound to another protein which prevents its polymerisation into filaments.

Energy Transfer Studies Sometimes the energy emitted by one fluor (the donor) can be absorbed by another fluor (the acceptor) by resonance energy transfer. This process requires that the fluorescence spectrum of the donor overlaps the absorption spectrum of the acceptor and that the fluors be situated close enough together. The efficiency of energy transfer is a function of the distance between the donor and acceptor fluors. The efficiency of transfer may be measured either as a quenching of the fluorescence of the donor fluor by the acceptor, or as the intensity of fluorescence of the acceptor when irradiated in the presence and absence of the donor.

Energy transfer measurements of particular distances in proteins utilise intrinsic fluors such as tryptophan, or extrinsic fluors attached to amino acids, sulphydryl groups, sugars, or fluorescent analogues of substrates, inhibitors, cofactors, or phospholipids. The measurements obtained are accurate only to about $\pm$ 0.5 nm and include the localisation of metals in metalloproteins; measurement of the extent of conformational changes in enzymes on binding their substrates; the distances between various pairs of proteins in the ribosome; and the 3-dimensional structure of tRNAs.

Fluorescence Depolarisation Studies A fluorescent material may be irradiated with polarised light by placing a polariser between the excitation monochromator (M_1 in Fig. 8.7) and the sample. The fluoresced radiation is then either partially polarised or remains totally unpolarised. Hence the name, *fluorescence depolarisation*. This may be detected by placing a second polariser between the emission monochromator (M_2 in Fig. 8.7) and the detector.

The rotation of the absorber chromophore and energy transfer between chromophores are the two main factors increasing fluorescence depolarisation. Under conditions of high viscosity and high chromophore concentration energy transfer is mainly detected but at low viscosity and low chromophore concentration the effects of molecular motion predominate.

This technique is particularly useful in measuring changes in the mobility of either molecules or parts of molecules. Extrinsic fluors must usually be introduced into proteins and nucleic acids because macromolecules move relatively slowly resulting in the lifetimes of their intrinsic fluors being too short. Particular applications of this technique include the measurement of the binding of fluorescent substrates, inhibitors and cofactors to enzymes by means of their subsequent lack of mobility; the binding of fluorescent labelled antigen to its antibody due to the reduced mobility of the antigen-antibody complex; the association and dissociation of multi-subunit proteins such as lactate dehydrogenase and chymotrypsin; and the measurement of the viscosity of living cells.

Microspectrofluorimetry Combination of a microscope with a spectrofluorimeter and fibre optics allows the measurement of the fluorescence of single bacterial cells binding fluorescent antibodies and the fluorescent intensity of subcellular structures. This procedure is used to detect malignant cells which tend to contain more nucleic acid than normal cells and hence take up more of the fluorescent probe acridine orange.

The Fluorescence Activated Cell Sorter (FACS) uses the light emitted by cells labelled with a fluorescent antibody flowing through a fine capillary to trigger their physical separation from unlabelled cells (Section 1.6.4).

8.6 Luminometry

8.6.1 Principles

Luminometry is a photometric technique in which light, emitted as a result of a chemical reaction (*luminescence*) in contrast to the result of a physical reaction (fluorescence or atomic emission; Sections 8.5 and 8.7), is measured in a luminometer. It is not really a spectroscopic technique since it does not usually involve a monochromator, but it is discussed here as a technique involving light emission that is increasingly important to biologists.

When a chemical reaction yields a fluorescent product in an excited state light is emitted as the excited electrons regain the ground state (chemiluminescence), as illustrated in Fig. 8.1c. The chemiluminescent spectrum of a reaction such as that of luminol with oxygen to produce 3-aminophthalate is the same as the fluorescent spectrum of the product. Bioluminescence is characterised by light emission produced in enzyme catalysed reactions (Section 3.4.4). The colour of the light emitted is characteristic of the origin of the luciferase, and varies between greenish yellow (560 nm) and red (620 nm). Under favourable conditions the quantum yield is nearly 100%.

8.6.2 Instrumentation

Luminometers are relatively simple photometers, complicated only slightly by the need to amplify and record the signal from the photocell. Obviously temperature control is important because of the temperature sensitivity of enzyme catalysed reactions. The main components of a luminometer (Fig. 8.11) include a photomultiplier tube with a well stabilised high-voltage power supply to ensure sensitive, reproducible measurement of light emission; a direct current amplifier with a wide range of sensitivity and linear response; and a reaction chamber which allows temperature control, adequate mixing of reactants and protection from extraneous light.

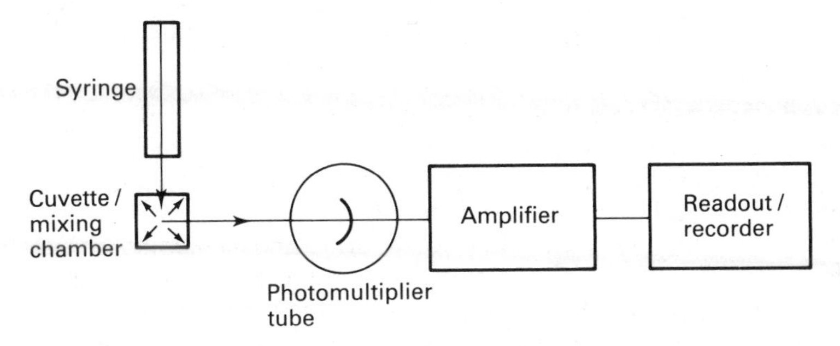

Fig. 8.11 A diagram of the main components of a simple luminometer.

8.6.3 Applications

The Firefly Luciferase System (for Determining ATP Concentration) This ATP assay is rapid to perform and can be as accurate as spectrophotometric and fluorimetric assays but it is much more sensitive (10^{-15} M). The linear range of the assay is 10^{-12} to 10^{-6} M ATP. This assay system may be extended to the determination of ADP, AMP and c-AMP in a single assay by successive additons of the appropriate enzymes. These include pyruvate kinase, to convert ADP to ATP, adenylate kinase to convert AMP to ADP and phosphodiesterase to convert c-AMP to AMP. In principle all enzymes and metabolites participating in ATP conversion reactions may be assayed by this method, including the enzymes creatine kinase, hexokinase and ATP-sulphurylase, and reactions involving creatine phosphate, glucose, GTP, PEP and 1,3 diphosphoglycerate.

The Bacterial Luciferase System The major uses of this system are in the determination of NADH, NADPH and $FMNH_2$ in the 10^{-9} to 10^{-12} M range, which is much more sensitive than the corresponding spectrophotometric and spectrofluorimetric assays. However, the NADPH assay is 20 times less sensitive than that for NADH.

The Aequorin System for Measuring Calcium Concentration The phosphoprotein aequorin isolated from luminescent medusae (jellyfish) is used as an intracellular calcium indicator despite the development of calcium specific electrodes, although its position is being challenged by metallochrome dyes such as Arsenazo III. In the presence of calcium ions the practically non-fluorescent yellow aequorin is converted into a blue fluorescent protein (BFP) with the emission of light. The bioluminescent spectrum of this reaction is identical to the fluorescent spectrum of BFP:$2Ca^{2+}$ and is different from that of BFP:Ca^{2+}.

The advantages of this system include its ease of use, high sensitivity to calcium, relative specificity for calcium and its lack of toxicity when injected into living cells. Its disadvantages include the scarcity and large molecular size of aequorin; its consumption during the reaction; the non-linear relationship of light emitted to calcium concentration; the sensitivity of the reaction to chemical environment and the limited speed with which it can follow very rapid changes in calcium concentration.

Luminol and Chemiluminescence Luminol and its derivatives can undergo chemiluminescent reactions with high efficiency, so enzymatically generated hydrogen peroxide may be detected by the production of light at 430 nm in the presence of luminol and microperoxidase (Section 3.4.4).

Low levels of compounds such as hormones, drugs and metabolites in biological fluids are often measured by competitive binding assays. These assays utilise the ability of certain proteins such as antibodies and cell receptors to bind specific ligands with high affinity (Section 4.7). Often the compound to be assayed is allowed to compete with a limited amount of radio-labelled or enzyme-labelled compound or analogue for a limited number of binding sites on the protein under standard conditions. Alternatively if the standard compound is labelled with a luminol derivative, the unbound labelled material can be separated from the binding protein and assayed by its chemiluminescence. Then the amount of (unlabelled) compound in the original sample is determined from a calibration curve relating concentration of unlabelled compound to concentration of unbound labelled compound under standard conditions. Under appropriate conditions 10^{-12} M of a compound can be detected.

Polymorphonuclear leucocytes exhibit chemiluminescence whilst phagocytosing, due to the generation of singlet molecular oxygen. This luminescence has been used to study the effects of pharmacological and toxicological agents on these and other phagocytic cells.

8.7 Atomic/flame spectrophotometry

8.7.1 Principles

Volatilisation of atoms either in a flame or electrothermally causes them to emit and absorb light of specific wavelengths. Atomic/flame spectrophotometry takes advantage of the specificity of these line spectra to determine the amounts of a specific element present. *Emission flame spectrophotometry* measures the emission of light of a specific wavelength by atoms in a flame. *Atomic absorption spectrophotometry* measures the absorption of a beam of monochromatic light by atoms in a flame or alternatively by atoms heated electrothermally in a graphite furnace. The energy absorbed is proportional to the number of atoms present in the optical path.

The most readily emitted and absorbed wavelengths in the spectrum of an element are those in which the energy change is minimal, such as the orange radiation of the D line transition of the sodium spectrum, involving the small 3s–3p transition (Fig. 8.1a). Since the transitions available to the electrons in any atom are specified by the available energy levels and the occupied energy levels, atomic spectra are absolutely characteristic of the element involved. Theoretically, for simple atoms one could deduce their electronic structure from their line spectra. The wavelengths of radiation emitted on volatilisation of elements in a flame may be readily resolved into simple line spectra in a spectroscope, spectrograph or direct reading spectrophotometer (which use as detectors the human eye, photographic emulsion and photoelectric cells, respectively).

The amount of radiation emitted is proportional to the number of excited atoms present, which depends on the temperature and composition of the flame. Standard solutions must always be used to calibrate the system. Flame composition is very important since sodium gives a very high background emission. Sodium is usually measured first, in order that a similar amount may be added to all the standard solutions. Chloride usually gives rise to the most volatile salts and hence the solutions studied should contain excess hydrochloric acid. Alkali metals cause an enhancement of the emission of other atoms whereas phosphate, silicate and aluminate give rise to non-dissociable salts and hence suppress the emission of calcium and magnesium. Fortunately this may be relieved by the addition of lanthanum or strontium salts. *Cyclic analysis* maybe used to allow for interference by other ions in a complex mixture. First each (interfering) component in the sample is estimated. Then the standards for all the components to be assayed are prepared such that they contain the estimated concentration of interfering components. By repeating all of the estimations a number of times and modifying the overall composition of the standards each time, such that composition of all other components in the standard solution approaches that in the unknown, constant and accurate values for each component may be obtained. Usually only one or two such cycles of determining and making up of standards are required. Because of flame instability, assays are usually made in triplicate. Calibration curves should be checked or reconstructed when the assays are carried out. A more accurate procedure is that of *bracketing* in which a standard solution, containing almost the same concentration of the element as the sample solution, is assayed before and after the sample solution. Whenever possible, internal standards should be used. Lithium is often used as a general internal standard. Since even well cleaned, best quality, glassware absorbs and releases metal ions, samples and standards are best stored in polythene bottles whenever possible.

When assaying metal in biological samples it is usual to remove organic molecules by ashing. Care must be taken to ensure that the more volatile elements are not inadvertently sublimed off. This may be achieved by ashing at relatively low temperatures in a stream of oxygen or by liquid ashing, i.e. by oxidative digestion of the sample in hydrogen peroxide/concentrated sulphuric acid solution. Selenium sulphate is added as a catalyst and lithium sulphate to raise the boiling point (cf. Kjeldahl analysis, Section 3.2.2).

8.7.2 Instrumentation for atomic emission spectrophotometry

The basic components of an atomic (flame) emission spectrophotometer are indicated in Fig. 8.12. *Nebulisers (atomisers)* are usually of the scent spray type, in which a forced stream of air passes over a capillary tube dipping into the test solution. Large and small drops of solution are produced which in direct injection systems pass with the air stream into the burner. Unfortunately, large drops may not remain in the hottest part of the flame long enough for their constituents to be volatilised and excited. This may be avoided by passing the drops through a cloud chamber which removes large drops from the air stream by gravitation. Sodium and potasium can be assayed at 1500°C which can be attained by combustion of a mixture of air and natural gas. Calcium can be assayed at this temperature but 2000–2500°C is preferable. Magnesium and iron require a mixture of air and acetylene to reach the necessary 2500°C. For simple routine analysis of sodium, potassium and calcium, the monochromator may be replaced by a simple filter, but a monochromator is required for more precise work. The most accurate instruments have a resolution of 0.1 to 0.2 nm over the range 200 to 1000 nm. The wavelengths used in the analysis of a number of metals are listed in Table 8.2 along with their detection limits. Photocells are usually used, but unfortunately the instability of the emission of most flames decreases the potential accuracy of the photocell. Multi-channel polychromators are used in some routine procedures to measure the emission from up to six elements at one time.

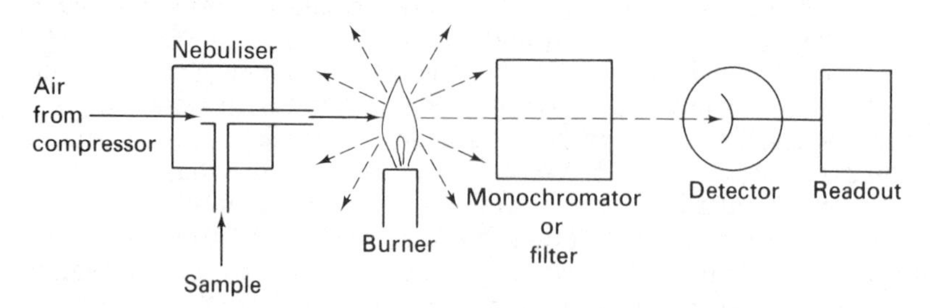

Fig. 8.12 The main components of an atomic emission (flame) spectrophotometer.

8.7.3 Instrumentation for atomic absorption spectrophotometry

In order to produce a beam of radiation with a very narrow band width, either a source of white light plus a double monochromator, or *a hollow cathode discharge lamp* is used. The discharge lamps are specific to the element being assayed. A sample of the element to be excited held in a metal cup, a tungsten anode carrier gas at a low pressure and a high voltage are used to produce an arc spectrum of the element. Recently electrodeless discharge lamps which are easier to use, brighter, longer lasting and more sensitive have become commercially available.

Nebulisers and flames are similar to those used in atomic emission spectrophotometry. In order to increase the optical path length of the sample, burners producing a 10 cm flame are used. Instruments with both single and double beam optics are available and the latter may include a *chopper* in the light path and a suitable circuit, which prevents stray light from being recorded in the detector system. The most useful wavelengths occur between 190 and 850 nm.

8.7.4 Flameless atomic absorption spectrophotometry

1 to 100 mm^3 of sample or standard analyte is deposited in a graphite tube in the presence of an inert gas and the temperature raised electrothermally to 3000°C either continuously or in discrete steps. Monochromatic light specific to the element being determined is produced by either a hollow cathode discharge lamp or an electrodeless discharge lamp and shone through the graphite tube. The absorption of this light is measured continuously as the temperature is raised. If the system is controlled by a programmable microcomputer the results may be obtained on a visual display unit as the superimposition of the absorption and temperature profiles of the graphite tube contents as a function of time. This allows the optimum conditions for the analysis to be found and used in future analytical programmes.

The flameless technique is often more than 100 times as sensitive as flame atomic absorption spectrophotometry (Table 8.2). Because the flameless system does not use combustible gases it may be safely left to run unattended which allows the relative slowness of such a system compared to flame spectrophotometry to be offset by automation.

8.7.5 Applications of flame spectrophotometry

Emission flame *filter photometers* are used routinely to assay sodium, potassium and calcium. Full scale deflection can be readily obtained for sodium and potassium, using standard solutions containing 1 and 3 parts per million respectively. Emission flame spectrophotometers are used to assay more than twenty elements in biological samples, particularly calcium, magnesium, and manganese. Absorption flame spectrophotometers are usually more sensitive than atomic emission spectrophotometers, although not for the alkali metals, and can detect less than 1 part per million of more than twenty elements. The reproducibility of the technique is about 1% and the best working range is 20 to 200 times the detection limit (Table 8.2). Flame photometry is used most extensively in clinical biochemistry laboratories to assay the composition of body fluids such as blood, urine, saliva, milk and cerebrospinal fluid. Departure of the composition of such fluids from the range of normal values can often be used in the diagnosis of particular clinical states, and the monitoring of such information can indicate the success or otherwise of a course of treatment. Similar monitoring of the body fluids of animals is used in physiological and pharmacological research. Sodium, potassium, calcium, magnesium, cadmium and zinc levels are measured directly, but copper, iron, lead and mercury must first be extracted from biological fluids. These techniques are also used extensively in soil and plant analyses after suitable extraction of metal salts from the soil or plant samples. After suitable ashing procedures have been carried out, these techniques may also be used to assay the metal composition of macromolecules, organelles, cells and tissues.

8.7.6 Atomic fluorescence spectrophotometry

This technique is based on an emission phenomenon which relies upon excitation of atoms by radiation instead of by thermal excitation. It is analogous to molecular fluorescence, but occurs with atoms in the vaporised state, not molecules in solution. An intense beam of light is required, but it needs to be less spectrally pure than that required for atomic absorption spectrophotometry, since only the resonant wavelengths will be absorbed and hence may lead to fluorescence. Modulation of the detector amplifier to the same frequency as the primary source prevents the direct emission from the flame being recorded.

Although currently limited to a few metals, this technique offers great refinement in the limits of their detection, for example zinc and cadmium may be detected at levels as low as 1 and 2 parts 10^{-10} respectively.

8.8 Electron spin resonance (ESR) spectrometry

8.8.1 Principles

This is a technique for detecting *paramagnetism*, i.e. the magnetic moment associated with an unpaired electron, and hence is sometimes called *electron paramagnetic resonance* (EPR). The technique may be used for detecting transition metal ions and their complexes, free radicals and excited states.

Since electrons possess both spin and charge they behave like magnets, i.e. they possess a *magnetic moment*. In the presence of an external magnetic

Table 8.2 The detection limits for various elements in emission and absorption flame spectrophotometry, flameless absorption spectrophotometry, and ion-selective electrodes.

Element	Emission		Absorption			Ion-selective electrode
	Detection limit (parts per million)	Wavelength (nm)	Detection limit (parts per million)		Wavelength (nm)	Detection limit (parts per million)
			(flame)	*(flameless)*		
Calcium	0.005	442.7	0.1	0.00007	442.7	0.02
Copper	0.1	324.8	0.1	0.0001	324.8	0.0006
Iron	0.5	372.0	0.2	0.0001	248.3	
Lead			0.5	0.0002	283.3	0.21
Lithium	0.001	670.7	0.03	0.0001	670.7	
Magnesium	0.1	285.2	0.01	0.00001	285.2	
Manganese	0.02	403.3	0.05	0.00004	279.5	
Mercury			10.0	0.018	253.8	
Potassium	0.001	766.5	0.03	0.00003	766.5	0.04
Sodium	0.0001	589.0	0.03	0.00001	589.0	0.02
Strontium	0.01	460.7	0.06	0.0001	460.9	

field they can exist in two states: either aligned parallel with the field in a low energy state or antiparallel to it in a high energy state. In order for an electron to change from the low energy state to the high energy state it must absorb the appropriate quantum of energy. If the electron is unpaired it is possible to apply electromagnetic radiation to make spin reversal (*resonance*) occur, if:

$$h\nu = g\beta H \qquad 8.7$$

where h = Planck constant
ν = frequency of applied radiation
g = a constant, called the *spectroscopic splitting factor*
β = magnetic moment of the electron, called the *Bohr magneton*
H = applied magnetic field

In a magnetic field of the order of one tesla (10^4 gauss) the appropriate quantum of energy is obtained from radiation in the microwave region of the electromagnetic spectrum, giving rise to the phenomenon known as *electron spin resonance* (ESR). The above formula (8.7) indicates that the frequency of the microwave radiation absorbed is a function of the paramagnetic species (β) and the applied magnetic field. However, it is usual practice to irradiate the sample with a constant microwave frequency appropriate to the species being studied and vary the applied magnetic field until resonance occurs, resulting in a peak absorption of the microwave frequency. Such a peak in an ESR spectrum corresponds to a paramagnetic species. The area under the peak is a measure of the concentration of that species, which may be quantitatively determined if a standard containing a known concentration of unpaired electrons is available. For a delocalised electron, g = 2.0023, but for localised electrons, especially those in transition metal atoms, g varies, and its precise value gives information about the nature of the bonding in the environment of the unpaired electron within the molecule. When resonance occurs, the absorption peak is broadened due to *spin-lattice interactions*, i.e. interaction of the unpaired electron with the rest of the molecule. This gives further information about the structure of the molecule.

Hyperfine splitting of an ESR absorption peak, caused by interaction of the unpaired electron with adjacent nuclei, yields information about the spatial location of atoms in the molecule. Proton (^{1}H) hyperfine splitting for free radicals lies in the region of 0 to 3×10^{-3} tesla and yields data analogous to those obtained from high resolution nuclear magnetic resonance (NMR) studies (Section 8.9). Thus ESR and NMR are complementary techniques.

8.8.2 Instrumentation

The basic components of an ESR spectrometer are illustrated in Fig. 8.13. Electromagnets generating fields of 50 to 500 millitesla with a uniformity of 1 in 10^6 are required for accurate work. Most experiments are conducted at around 330 millitesla, in conjunction with an *auxiliary sweep* of 10 to 100 millitesla. A Klystron oscillator produces the monochromatic microwave radiation, usually with a wavelength of about 3×10^{-2} m (9000 MHz).

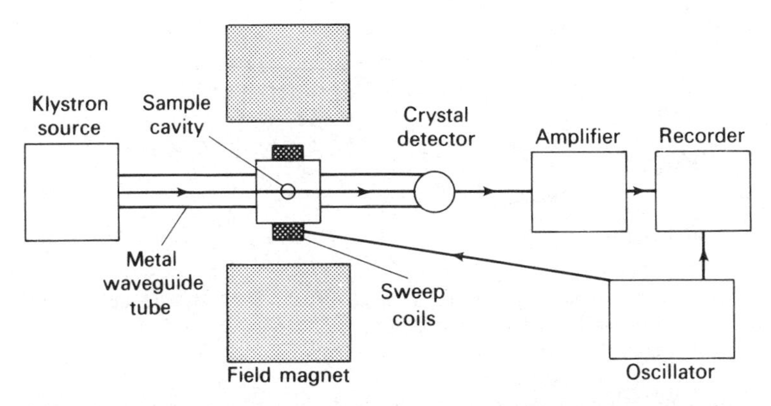

Fig. 8.13 The main components of an electron spin resonance spectrometer.

Samples must be in the solid state so biological samples are usually frozen in liquid nitrogen. Maximum precision is obtained by plotting the first differential of the absorption, that is the rate of change of absorption, as a function of the rate of change in field strength (dA/dH), not the absorption (A), against the applied magnetic field (H). Thus, instead of symmetrical absorption peaks as in Fig. 8.14a, ESR spectra contain non-symmetrical peaks adjacent to non-symmetrical troughs as in Fig. 8.14b. Such a peak-trough pair is called a *line* in ESR spectroscopy. However because there is rarely more than one unpaired electron in a molecule, a typical ESR spectrum contains fewer than ten lines which are not closely spaced.

8.8.3 Applications

Electron spin resonance spectrometry is one of the main methods used to study metalloproteins, particularly those containing molybdenum (e.g. xanthine oxidase), copper (e.g. cytochrome oxidase and copper blue enzymes) and iron (e.g. cytochromes, ferredoxin). Both copper and non-haem iron, which do not absorb radiation in the visible and ultraviolet regions, possess ESR absorption peaks in one of their oxidation states. Hence the appearance and disappearance of their ESR signals are used to monitor their activity in the multienzyme systems of intact mitochondria and chloroplasts as well as in isolated enzymes. In metalloproteins, the metal atom has a characteristic number of ligands coordinated to it in a definite geometrical arrangement. These ligands are frequently amino acid residues of the protein. Studies using ESR have provided evidence that this geometry is frequency distorted from that of model systems. It has been suggested that this distortion may be related to the particular biological function of the metalloprotein. Electron spin resonance is also extensively used to study the free radicals produced by irradiation of biological material.

The technique of ESR has been extended greatly by *spin labelling*. This involves the attachment of a stable and unreactive free radical such as

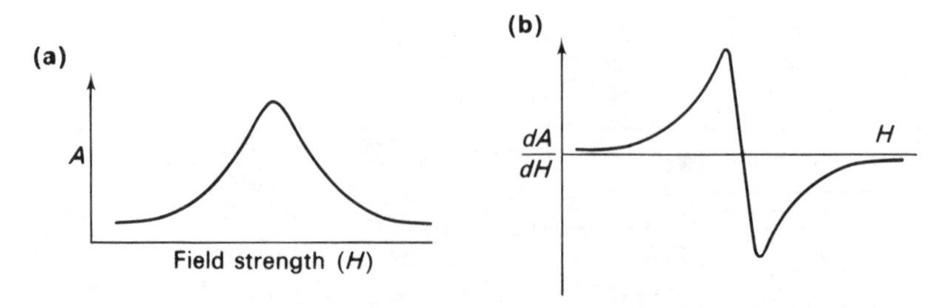

Fig. 8.14 Electron spin resonance spectra: **(a)** a plot of absorbance (A) against field strength (H); and **(b)** a plot of the first differential of A, (dA/dH) against H.

2,2,6,6-tetramethylpiperidine-1-oxyl (TEMPOL), to biological macromolecules which lack unpaired electrons. Thus, by spin labelling glycerophosphatides with this stable nitroxide free radical, the lateral diffusion of the labelled molecules in a membrane and also their *flip-rate* between the inner and outer surfaces of a lipid bilayer may be studied.

8.9 Nuclear magnetic resonance (NMR) spectrometry

8.9.1 Principles

This is a technique for detecting atoms which have nuclei that possess a magnetic moment. These are usually atoms containing an odd number of protons in their nuclei. In the same way that pairs of electrons in the same atomic orbital have opposite spin and no resultant magnetic moment, pairs of protons in a nucleus do not have a magnetic moment. However, an odd proton in a nucleus imparts a magnetic moment to the molecule which can interact with an applied magnetic field. This interaction is the basis of nuclear magnetic resonance spectrometry. An unpaired proton in such a magnetic field can exist in one of two states: either in a low energy state with the nuclear spin aligned parallel with the field, or in a higher energy state antiparallel to the field. In order to change from the low energy state to the high energy state, such nuclei must absorb the appropriate quantum of energy. In a magnetic field of several hundred millitesla (several thousand gauss) such nuclei absorb radiation in the radiowave region of the electromagnetic spectrum, giving rise to the phenomenon known as *nuclear magnetic resonance* (NMR). Most studies are conducted using the lightest isotope of hydrogen, ^{1}H (hence the term *proton magnetic resonance*, PMR), but ^{13}C, ^{15}N, ^{19}F and ^{31}P isotopes are also used in biochemical studies. These nuclei possess a nuclear spin value of 1/2. Other nuclei common in biological molecules, such as ^{12}C, ^{16}O and ^{32}S, possess zero spin and hence lack a magnetic moment and do not give NMR signals.

The frequency of the radiowaves absorbed during NMR depends on both the isotope being studied and the magnitude of the applied magnetic field. Usually, the absorption of a monochromatic radiowave frequency by the sample is measured as the applied magnetic field is varied, rather than using a constant magnetic field and variable radiation frequency. Thus NMR spectra are usually plotted as energy absorbed against the magnetic field strength applied (not wavelength or frequency). The range of magnetic field scanned is very small compared to the total magnetic field applied. Hence the total field applied and the frequency of the radiowaves absorbed are stated on such spectra. A radiowave frequency of 40 MHz is commonly used to achieve resonance of the ^{1}H nucleus.

The actual field experienced by a proton in a particular compound depends upon its molecular environment since the applied field induces secondary fields (15 to 20 $\times$ 10^{-4} tesla) by interacting with the bonding electrons adjacent to the proton. If this induced field opposes the applied field, a slightly higher applied field must be used to make the nucleus resonate. Such nuclei are said to be *shielded*. The magnitude of the shielding decreases with increasing electron-withdrawing power of nearby substituents. If the induced field augments the applied field, the nucleus is said to be *deshielded* and a smaller applied field is required to achieve resonance. Such displacements in the spectral lines are known as *chemical shifts*. The scale usually used for measuring such shifts records them in parts per million of the applied magnetic field with reference to the absorption peak of a reference compound. Usually tetramethylsilane, $Si(CH_3)_4$, is used as internal reference in organic solvents, and sodium trimethylsilylpropanesulphonate, $(CH_3)_3SiCH_2CH_2SO_3\text{-}Na^+$, in aqueous systems. Most modern NMR spectrometer scales are calibrated in tau (τ) units. On this scale the $Si(CH_3)_4$ peak occurs at τ 10, since most proton shifts occur between τ 0 and τ 10. As a result of chemical shifts the presence of protons in functional groups such as ^{1}H-O-, ^{1}H-CH, ^{1}H-CH_2 may be detected by the occurrence of particular peaks in the NMR spectrum. In addition, the relative number of ^{1}H atoms in each type of chemical group within the sample is indicated by the intensity of the appropriate absorption peaks, i.e. by the areas under those peaks.

Nuclear magnetic resonance spectra are further complicated by interaction between like or different spins through the bonding electrons. This *spin-spin coupling* may extend to nuclei which are 4 to 5 bonds apart. It is manifested as a splitting of the absorption peaks already separated by the chemical shift (*hyperfine splitting*).

Like IR and ESR spectra, NMR spectra are of great use in the study of chemical structures. The chemical shift yields qualitative information and the intensity of any peak yields quantitative information about groups present in the sample. In addition, the hyperfine structure yields information about adjacent nuclei and the magnitude of the hyperfine splitting yields information about geometrical relationships within the molecule. However, as with IR spectra, NMR spectra are complex and require expert interpretation.

8.9.2 Instrumentation

The basic components of an NMR spectrometer are similar to those illustrated in Fig. 8.13 for an ESR spectrometer. Electromagnets weighing 10^3 to 10^4 kg, and stable to better than $\pm 0.1\%$ are used to produce fields of 1 to 10 tesla. An auxiliary sweep generator is used to vary the magnetic field over about 10^{-2} tesla. A radio frequency transmitter in place of a Klystron source produces the monochromatic radiation used to irradiate the sample. For PMR the sample must be dissolved to a relatively high concentration in a solvent which lacks protons, such as D_2O or $CDCl_3$. To minimise variations in the magnetic field the sample is contained in a tube of high precision diameter and is usually rotated at high speed by an air turbine. The absorption signal, detected by a radio receiver, is amplified and recorded.

Computer enhancement (*computer averaging techniques*, CAT), in which the read-out from 10^1 to 10^3 spectra of the same sample are superimposed, minimises background noise and is very useful for studying weakly absorbing biological samples.

8.9.3 Applications

The main use of NMR is for the study of the molecular structure of relatively simple organic molecules. Thus structural information relevant to the biological action of the antibiotics, gramicidin and valinomycin has been obtained from NMR studies. In addition the effects of alamethicin and cholesterol on the mobility of lecithin, in both artificial and erythrocyte membranes, has been studied using NMR as has the relative motion of different parts of the fatty acid side chains of lecithin in lipid bilayers using lecithins substituted in different positions with ^{19}F and ^{13}C.

The use of NMR to study biological macromolecules is basically very difficult. Molecules such as proteins may contain several thousand protons, all of which resonate in a similar region. Only with the introduction of high resolution instruments and the development of computer techniques has the potential of NMR begun to be realised. Even so, there appears to be an upper molecular weight limit of 20 000 daltons for obtaining complete hyperfine structural information. Thus in a study of the catalytic mechanism of ribonuclease by NMR, it was possible to identify proton resonance from the four histidine residues present in the molecule since their NMR signals were shifted from those of the main bulk of protons. It was possible to identify which two of the four were involved in the catalytic site. In fact, this information was already known from chemical and X-ray diffraction studies, but it demonstrates the possibilities. The main advantage of NMR over X-ray diffraction is that it is capable of monitoring conformational changes of 100 pm. Thus it is invaluable in the study of small conformational changes accompanying membrane dynamics and the binding of enzymes to substrates, drugs to receptors, and antigens to antibodies.

NMR has also been particularly useful in the study of phosphate metabolism. The sensitivity of ^{31}P NMR is such that the concentration of inorganic phosphate, AMP, ADP, ATP, phosphocreatine and metabolic sugar phosphates can be measured and hence their metabolism studied in living cells and tissue. Intracellular and extracellular inorganic phosphate concentrations may be determined because the chemical shift of inorganic phosphate varies with pH. Development of NMR focusing techniques and larger, scanning NMR machines allows the study of whole animals including man. Studies on the effects of exercise on normal, athletic and unhealthy individuals have shown that some healthy people can function even when their ATP level has been depleted by 60% and that others can sustain a muscle pH decrease from the normal range of 7.0–7.4 to 6.0. In particular, detailed studies are revealing patients with previously unknown disorders such as mitochondrial myopathies.

8.10 Mass spectrometry

8.10.1 Principles

Mass spectrometry is widely used to elucidate the molecular structure of biological compounds. It is based upon the fact that a moving ion may be deflected by a magnetic field to an extent that is dependent upon its mass and velocity. Ions of a larger momentum are deflected less than ones of lower momentum, whilst a mixture of ions of different mass but constant velocity will be deflected in proportion to their mass. In a mass spectrometer, molecules of a compound are ionised, either by ejection of an electron or capture of a proton, to give the *parent molecular ion*, the energy of which is such that some fragmentation occurs to give a series of *fragment ions*. Knowledge of the mass of the molecular ion and its major fragment ions is frequently sufficient to enable the structure of the parent compound to be uniquely deduced. The method is very sensitive and uses as little as 10^{-6} to 10^{-9} g of material. A mass spectrum is a plot of the abundance of the fragment and molecular ions against mass and hence is not at all like the various types of electromagnetic spectra discussed elsewhere in this Chapter.

The majority of ions produced during the initial ionisation procedure have a single positive charge, i.e. one electron is removed from the molecule or fragment so that the mass to charge ratio (m/e) is numerically equal to the mass. Thus the ions produced differ only in their mass. Occasionally, however, molecules lose more than one electron and multicharged ions are produced. When producing ions by electron bombardment (Section 8.10.2) the degree of fragmentation of the molecule depends on the energy of the bombarding electrons. At low energies (1 to 2×10^{-18} J), only one electron is removed from the molecule. The resulting positively charged molecular ion has a m/e value corresponding to the molecular weight of the parent compound. At the usual operating electron beam energy of 10^{-17} J, the molecules are cleaved into positively charged fragments of different masses. The way in which the compound fragments to give a mass spectrum is characteristic of the compound being analysed and is called the *fragmentation pattern* (cf. IR and NMR fingerprints, Sections 8.3 and 8.9 respectively). From this pattern the structure of the molecules can be deduced. Tables have been compiled for the fragmentation patterns of a wide range of compounds and reference to these is particularly helpful in elucidating the structure of an unknown compound. Most modern instruments have built-in data systems which enable the experimental fragmentation data to be compared with stored reference data.

Mass spectra show a series of peaks or lines corresponding to the m/e values of the positive ions produced from the compound. The height of the peaks corresponds to the relative abundance of the ions. A reference ion of similar m/e value to that of the parent ion is used to calibrate the mass axis (abscissa) of the spectrum. The *parent ion* is the peak with the greatest mass, although it is not necessarily the most abundant (*base peak*). Ion intensities in a mass spectrum are usually recorded as percentages of the intensity of the base peak.

Carbon dioxide is ionised and fragmented to yield CO_2^+, CO^+, O_2^+, O^+ and C^+, resulting in a mass spectrum with major peaks at m/e values of 44, 28, 32, 16 and 12 respectively. Other minor peaks are also present corresponding to the other natural isotopes of the above fragments, for example $^{13}C^+$ (m/e, 13) and $^{13}CO_2^+$ (m/e, 45).

8.10.2 Instrumentation

The basic components of a double focusing mass spectrometer, essential for high resolution studies, are illustrated in Fig. 8.15. It is essential that the analysis is carried out under a very high vacuum in order to minimise the loss or transformation of ions by collision with any un-ionised molecules present. The major factor which governs the amenability of a sample to mass spectrometry is its vapour pressure at the temperature of the ion source. If the compound has a vapour pressure of approximately 1.3 Pa (1.9×10^{-4} p.s.i.) at 100 to 200°C, it can be injected into a reservoir attached to the ion source. As the pressure in the reservoir is higher than that in the ion source, which is under a vacuum of about 10^{-5} Pa (1.5×10^{-9} p.s.i.), a small stream of sample vapour is drawn through a minute aperture into the ionisation chamber. Samples of lower volatility or small amounts of other samples may be introduced directly into the ionisation chamber by means of a quartz probe, so that the small amount of vapour produced enters the ionisation chamber directly.

The most common means of ionising organic compounds is by *electron bombardment*. The electrons are obtained by thermionic emisson from a heated tungsten filament and are accelerated by application of a 50 to 100 V potential. If the mass spectrometer is coupled to a gas-liquid chromatograph, a lower accelerating voltage is used to avoid the ionisation of the helium carrier gas. In all cases, the energy associated with the electrons is sufficient to cause both ionisation and fragmentation of the organic compound so that

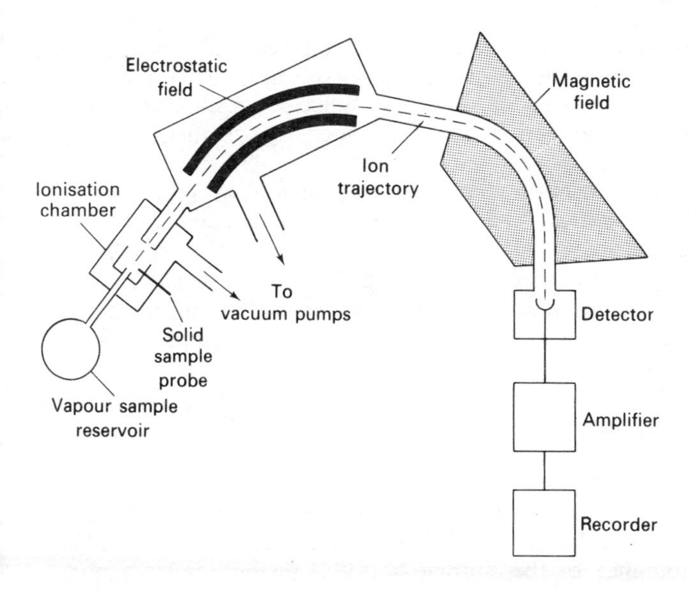

Fig. 8.15 The basic components of a mass spectrometer.

fragment ions predominate and the molecular ion is weak. An alternative form of ionisation is *chemical ionisation* in which ionisation of the organic compound is induced by ions of simple hydrocarbon molecules such as methane and butane. The hydrocarbon is introduced into an ionisation chamber at a pressure of up to 200 Pa (3×10^{-2} p.s.i.) and the ions so produced are immediately mixed with the molecules of the organic compound. Proton transfer from the hydrocarbon ions (e.g. CH_5^+ and $C_4H_{11}^+$) takes place with high efficiency to give the M + 1 ion of the organic compound. Ion fragmentation is less common than in electron bombardment, so chemical ionisation spectra are simpler than, and complementary to, electron ionisation spectra since they give more detail of the molecular ion. *Field ionisation*, which uses potentials of up to 10 kV, resembles chemical ionisation, in that it too gives rise predominantly to the molecular ion of the organic compound.

The ions are separated in an analyser according to their m/e values by being accelerated to constant velocity in a vacuum by a series of negatively charged electrostatic plates and are then deflected from their initial trajectory by a magnetic field. For ions carrying the same charge the amount of deflection is dependent on the mass of the ion, so that the ions of smallest mass undergo the greatest degree of deflection. By varying either the magnetic field or the accelerating voltage the ion trajectories may be varied so that fragment ions of a given m/e value impinge on the collector slit of the detector. The number of ions of a given mass impinging on the detector is a measure of the abundance of that particular ion.

The detector is usually either a *simple electrode* (Faraday cup) or an *electron multiplier*. The ion current is amplified and recorded. Due to the vast range of ion intensities occurring in a single mass spectrum, it is common practice to record them at a number of preselected sensitivities, usually varying 300 fold, in order to determine the intensities of all of the peaks accurately.

The ability of a mass spectrometer to separate two ions of similar mass, m_1 and m_2 is called the *resolving power*, $m_1/(m_2 - m_1)$. Resolving power is a function of the detector slit width, but the narrowness of the slit width that can be used is limited by the corresponding decrease in sensitivity of the instrument, due to the reduction in number of ions falling on the detector electrode.

8.10.3 Applications

The earliest biochemical uses of mass spectrometry were in the study of metabolic pathways. Substrates enriched with isotopes such as ^{15}N or ^{18}O, were fed to animals and the end products of their metabolism were isolated. The relative abundance of the isotope (ratio of uncommon to common isotope) in the different metabolic products was determined, with the mass spectrometer, by comparing the intensities of the parent ions (containing the common and uncommon isotopes) of either the metabolites directly, or their degradation products, for example $H_2{}^{18}O$ and $H_2{}^{16}O$. This technique is still used for ^{15}N and ^{18}O, but wherever possible it has been superseded by simpler and less costly assays using radioactive isotopes (Chapter 9), for example ^{14}C, ^{3}H and ^{32}P.

The major use of mass spectrometry in biochemistry remains the determination of chemical structure and hence the identification of a compound, i.e. qualitative analysis of small amounts of relatively complex organic molecules. Because all elements have a non-whole number atomic weight, the accurate determination of the molecular ion (to 4 places of decimals) gives a unique molecular formula. By studying the accompanying fragment ions and their relative abundance, functional groups, and frequently the whole molecular structure, can be deduced e.g. steroids, ubiquinones and triglycerides.

The interpretation of the structure of any compound by mass spectrometry necessitates having the compound in a relatively pure state. It is, therefore, particularly advantageous to couple a mass spectrometer to a gas-liquid chromatograph (Section 6.4.1). A mixture of compounds may be applied to the GLC and separated into the individual components. As each compound emerges after a different time interval from the GLC, it is fed into the mass spectrometer, analysed and its fragmentation pattern recorded in the usual manner. This requires rapid scanning mass spectrometers, for example the m/e range 10^1 to 10^3 should be scanned in 2 to 5 seconds. Such rapidly scanning instruments allow the identification of samples inadequately resolved by GLC, by scanning both the front and rear of overlapping peaks. Excellent mass spectra can often be obtained from less than 10^{-7} g of compound present in the sample originally introduced into the GLC. The maximum gas flow into the ion source of the mass spectrometer is 1 $mm^3 s^{-1}$ if a high vacuum is to be maintained. Thus it is usual to remove the bulk of the carrier gas from the outlet of the GLC, before passing the sample into the spectrometer, by taking advantage of the rapid diffusion of carrier gas through very fine sinters.

Mass spectrometry has been used to determine the amino acid sequence of oligopeptides derived from protein hydrolysates and other sources. Originally the peptide was made more volatile by acetylation and permethylation. The peptide bonds were then most susceptible to cleavage on electron bombardment. Moreover, fragmentation occurred from the *C*-terminal end of the peptide. Since no two commonly occurring amino acids have the same molecular weight (other than leucine and isoleucine), the difference in mass between adjacent major peaks in the spectrum unambiguously identifies the *C*-terminal amino acid residue of the heavier fragment ion. More recently *fast atom bombardment* (FAB) has been used to sequence oligopeptides without derivatisation. A paste of the peptide and sodium chloride is introduced into the ionisation chamber and bombarded with excited argon atoms, resulting in ionisation and fragmentation of the peptide. Computers have been programmed to determine the amino acid sequence of the parent ion by determining the mass difference between each of the major peaks of such a mass spectrum. The limits of this approach in determining the primary structure of proteins are governed only by the rate at which suitable oligopeptides may be isolated from the proteins and by the total mass of the oligopeptide.

8.11 Tabular summary of spectroscopic techniques

Type	Principles involved	Main uses
Visible and ultraviolet spectrophotometry	Energy transitions of bonding and non-bonding outer electrons of molecules, usually delocalised electrons	Routine qualitative and quantitative biochemical analysis including a vast number of colorimetric assays, enzyme assays and kinetic studies. Difference spectra, action spectra, turbidometry and nephelometry
Infra-red spectrophotometry	Atomic vibrations involving a change in dipole moment	Qualitative analysis and fingerprinting of purified molecules of intermediate size. Infra-red gas analysis is important in plant physiology
Circular dichromism	Differential absorption of right and left circularly polarised light by optically-active chromophores.	Research into the conformation of proteins and nucleic acids and changes in such conformation under various environmental conditions

continued

Type	Principles involved	Main uses
Spectrofluorimetry	Absorbed radiation emitted at longer wavelengths	Routine quantitative analysis, enzyme analysis and kinetics and detection of changes in protein conformation. More sensitive at lower concentrations than visible and UV absorption spectrophotometry. Qualitative analysis
Luminometry	The emission of radiation by excited fluors formed as a result of a chemical reaction	Quantitative assay of ATP, NAD^+, Ca^{2+} and any compound or enzyme that can be linked to their production. More sensitive than fluorimetry
Flame spectrophotometry (emission and absorption)	Energy transitions of outer electrons of atoms after volatilisation in a flame	Qualitative and quantitative analysis of metals, particularly in clinical biochemistry. Emission techniques; routine determination of alkali metals. The absorption technique extends the range of metals that may be determined and the sensitivity
Electron spin resonance spectrometry	Detection of magnetic moment associated with unpaired electrons	Research on metalloproteins, particularly enzymes and changes in the environment of free radicals introduced into biological assemblies, e.g. membranes
Nuclear magnetic resonance spectrometry	Detection of magnetic moment associated with an odd number of protons in an atomic nucleus	Research into the structure of organic molecules of molecular weight less than 20 000 daltons. Whole animal phosphate metabolism
Mass spectrometry	Determination of the abundance of positively ionised molecules and fragments	Qualitative analysis of small quantities of material (10^{-6} to 10^{-9} g) particularly in conjunction with gas-liquid chromatography. Mainly used in research, but has high potential for the rapid determination of the primary structure of peptides

8.12 Suggestions for further reading

Brown, S.B. (1980). *An Introduction to Spectroscopy for Biochemists*. Academic Press, London. (Contains comprehensive sections on the principles, instrumentation and chemical applications of many of the techniques considered in this chapter.)

DeLuca, M.A. (Ed.) (1978). Bioluminescence and Chemiluminescence. In *Methods in Enzymology*, Vol. 57. Academic Press, London. (Contains a comprehensive coverage of luminescence methodologies.)

Florkin, M. and Stotz, E.H. (Eds) (1967). Methods for the Study of Molecules. In *Comprehensive Biochemistry*, Vol. 3. Elsevier, Amsterdam. (Contains a comprehensive coverage of many of the techniques discussed in this chapter.)

Freifelder, D.M. (1982). *Physical Biochemistry-Applications to Biochemistry and Molecular Biology*, 2nd edition. Freeman & Co. Ltd., San Francisco. (Contains excellent readable accounts of the techniques discussed in this chapter.)

Gilbert, B. (1984). *Investigation of Molecular Structure: Spectroscopy and Diffraction Methods*, 2nd edition. Bell and Hyman, London. (A basic introduction to classical spectroscopy techniques.)

Knowles, P.F., Marsh, D. and Rattle, H.W.E. (1976). *Magnetic Resonance of Biomolecules*. John Wiley and Sons, London. (Contains a comprehensive coverage of ESR and NMR.)

Lakowicz, J.R. (1983). *Principles of Fluorescence Spectroscopy*. Plenum Press. (Starts with basic principles but contains topics of current interest.)

Moore, G.R., Radcliffe, R.G. and Williams, R.J.P. (1983). NMR and the biochemist. In *Essays in Biochemistry*, Vol. 19. Academic Press, London. (An excellent article, written for advanced students, giving a comprehensive account of NMR in biochemical studies.)

Williams, D.H. and Flemming, I. (1973). *Spectroscopic Methods in Organic Chemistry*, 2nd edition. McGraw Hill. (Contains further information on the more classical spectroscopic methods.)

9
Radioisotope techniques

9.1 The nature of radioactivity

9.1.1 Atomic structure

An atom is composed of a positively charged nucleus which is surrounded by a cloud of negatively charged electrons. The mass of an atom is concentrated in the nucleus even though it accounts for only a small fraction of the total size of the atom. Atomic nuclei are composed of two major particles, *protons* and *neutrons*. Protons are positively charged particles with a mass approximately 1850 times greater than that of an orbital electron. The number of orbital electrons in an atom must be equal to the number of protons present in the nucleus, since the atom as a whole is electrically neutral. This number is known as the *atomic number (Z)*. Neutrons are uncharged particles with a mass approximately equal to that of a proton. The sum of protons and neutrons in a given nucleus is the *mass number (A)*. Thus:

$$A = Z + N$$

where N = the number of neutrons present.

Since the number of neutrons in a nucleus is not related to the atomic number, it does not affect the chemical properties of the atom. Atoms of a given element may not necessarily contain the same number of neutrons. Atoms of a given element with different mass numbers (i.e. different numbers of neutrons) are called *isotopes*. Symbolically, a specific nuclear species is represented by a subscript number for the atomic number and a superscript number for the mass number, followed by the symbol of the element. For example:

$$^{12}_{6}C \quad ^{14}_{6}C \quad ^{16}_{8}O \quad ^{18}_{8}O$$

However, in practice it is more conventional just to cite the mass number (e.g. ^{14}C).

The number of isotopes of a given element varies; there are three isotopes of hydrogen, ^{1}H, ^{2}H and ^{3}H, seven of carbon ^{10}C to ^{16}C inclusive, and 20 or more of some of the elements of high atomic number.

9.1.2 Atomic stability and radiation

In general the ratio of neutrons to protons in the nucleus will determine if an isotope of a given element is stable enough to exist in nature. *Stable isotopes* for elements with low atomic numbers tend to have an equal number of neutrons and protons whereas stability for elements of higher atomic numbers is associated with a neutron to proton ratio in excess of one. Unstable isotopes or as they are more commonly known, *radioisotopes*, are in the main produced artificially, but some, including ^{40}K, occur in nature. Radioisotopes emit particles and/or electromagnetic radiation as a result of changes in the composition of the atomic nucleus. These processes, which are known as *radioactive decay*, result, either directly, or as a result of a decay series, in the production of a stable isotope.

9.1.3 Types of radioactive decay

There are several types of radioactive decay, only the most important of which are considered below.

Decay by Negatron Emission In this case a neutron is converted to a proton by the ejection of a negatively charged beta (β) particle called a *negatron* (β – ve):

$$\text{NEUTRON} \rightarrow \text{PROTON} + \text{NEGATRON}$$

To all intents and purposes a negatron is an electron, but the term negatron is preferred, although not always used, since it serves to emphasise the nuclear origin of the particle. As a result of negatron emission the nucleus loses a neutron but gains a proton. The N/Z ratio therefore, decreases while the atomic number (Z) increases by one and the mass number (A) remains constant. An isotope frequently used in biological work which decays by negatron emission is $^{14}_{6}C$:

$$^{14}_{6}C \rightarrow {}^{14}_{7}N + \beta\ -\text{ve}$$

Decay by Positron Emission Some isotopes decay by emitting positively charged β particles referred to as *positrons* (β + ve). This occurs when a proton is converted to a neutron:

$$\text{PROTON} \rightarrow \text{NEUTRON} + \text{POSITRON}$$

Positrons are extremely unstable and have only a transient existence. Once they have dissipated their energy they interact with electrons and are annihilated. The mass and energy of the two particles are converted to two gamma rays emitted at 180° to each other. This phenomenon is frequently described as *back to back emission*.

As a result of positron emission the nucleus loses a proton and gains a neutron, the N/Z ratio increases, the atomic number decreases by one and the mass number remains constant. An example of an isotope decaying by positron emission is $^{22}_{11}Na$:

$$^{22}_{11}\mathrm{Na} \rightarrow {}^{22}_{10}\mathrm{Ne} + \beta + \mathrm{ve}$$

Decay by Alpha Particle Emission Isotopes of elements with high atomic numbers frequently decay by emitting *alpha* particles (α). An α-particle is a helium nucleus in that it consists of two protons and two neutrons ($^{4}_{2}He^{2+}$). Emission of α-particles results in a considerable lightening of the nucleus, a decrease in atomic number of two and a decrease in the mass number of four. Isotopes which decay by α-emission are not frequently encountered in biological work. $^{226}_{88}$Radium ($^{226}_{88}Ra$) decays by α-emission to $^{222}_{86}$Radon ($^{222}_{86}Rn$) which is itself radioactive. Thus begins a complex *decay series* which culminates in the formation of $^{206}_{82}Pb$:

$$^{226}_{88}\mathrm{Ra} \rightarrow {}^{4}_{2}\mathrm{He}^{2+} + {}^{222}_{86}\mathrm{Rn} \rightarrow\rightarrow\rightarrow {}^{206}_{82}\mathrm{Pb}$$

Decay by Emission of Gamma Rays In contrast to emission of α- and β-particles, *gamma* (γ) emission involves electromagnetic radiation similar to, but with a shorter wavelength than, X-rays. These γ-rays result from a transformation in the nucleus of an atom (in contrast to X-rays which are emitted as a consequence of excitation involving the orbital electrons of an atom) and frequently accompany α- and β-particle emission. Emission of γ-radiation in itself leads to no change in atomic number or mass.

9.1.4 Radioactive decay energy

The usual unit used in expressing energy levels associated with radioactive decay is the *electron volt*. (One electron volt (eV) is the energy acquired by one electron in accelerating through a potential difference of 1 volt and is equivalent to 1.6×10^{-19}J). For the majority of isotopes the term million or mega electron volts (MeV) is more applicable. Isotopes emitting α-particles are normally the most energetic, falling in the range of 4.0 to 8.0 MeV, whilst β- and γ-emitters generally have decay energies less than 3.0 MeV.

9.1.5 Rate of radioactive decay

Radioactive decay is a spontaneous process and it occurs at a definite rate characteristic of the source. This rate always follows an exponential law. Thus the number of atoms disintegrating at any time is proportional to the number of atoms of the isotope present at that time. Expressed mathematically the exponential curve (Fig. 9.1) gives the equation:

$$-\frac{\delta N}{\delta t} = \lambda N \qquad 9.1$$

Thus the rate of change in the number of radioactive atoms is proportional

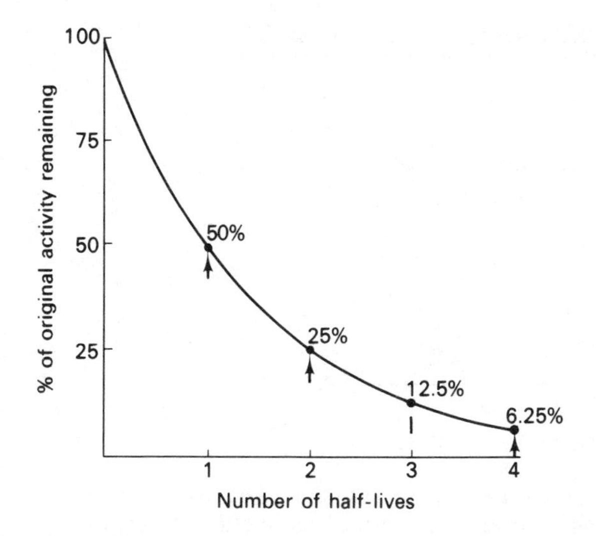

Fig. 9.1 Demonstration of the exponential nature of radioactive decay.

to the number of atoms present (N) multiplied by the decay constant (λ). This constant is a characteristic of a given isotope and is defined as the fraction of an isotope decaying in unit time (t^{-1}). By integrating equation 9.1 it can be converted to a logarithmic form:

$$\ln \frac{N_t}{N_o} = -\lambda t \qquad 9.2$$

where N_t = number of radioactive atoms present at time t
N_o = number of radioactive atoms originally present

In practice it is more convenient to express the decay constant in terms of *half life* ($T_{1/2}$). This is defined as the time taken for the activity to fall from any value to half that value. If N_t in equation 9.2 is equal to one half N_0 then t will equal the half life of the isotope. Thus:

$$\ln_{1/2} = -\lambda T_{1/2} \qquad 9.3$$

or

$$2.303 \log_{10}(1/2) = -\lambda T_{1/2} \qquad 9.4$$

or

$$T_{1/2} = \frac{0.693}{\lambda} \qquad 9.5$$

The values of $T_{1/2}$ vary widely from over 10^{19} years for ^{204}Lead (^{204}Pb) to 3×10^{-7} seconds for ^{212}Polonium (^{212}Po). The half lives of some isotopes frequently used in biological work are given in Table 9.1. Note that two important elements, oxygen and nitrogen, are missing from the table. This is because the half lives of radioactive isotopes of these elements are too short for most biological studies (^{15}O has a $T_{1/2}$ of 2.03 minutes while ^{13}N has a $T_{1/2}$ of 10.00 minutes).

Table 9.1 Half lives of some isotopes used in biological studies.

Isotope	Half life
^{3}H	12.26 years
^{14}C	5760 years
^{22}Na	2.58 years
^{32}P	14.20 days
^{35}S	87.20 days
^{42}K	12.40 hours
^{45}Ca	165 days
^{59}Fe	45 days
^{125}I	60 days
^{131}I	8.05 days
^{135}I	9.7 hours

9.1.6 Units of radioactivity

The International System of Units (SI system) uses, as the unit of radioactivity, the *Bequerel* (Bq). This is defined as one disintegration per second (1 d s^{-1}). However, to date this unit has not been widely adopted and the more widely used unit is still the *Curie* (Ci). This is defined as the quantity of radioactive material in which the number of nuclear disintegrations per second is the same as that in 1 g of radium, namely 3.7×10^{10} (3.7×10^{10} s^{-1} or 37 gBq). For biological purposes this unit is too large and the microcurie (μCi) and millicurie (mCi) are used. It is important to realise that the Curie refers to the number of disintegrations actually occurring in a sample (i.e. d s^{-1}) not to the disintegrations detected by the radiation counter which will generally be only a proportion of the disintegrations occurring and are referred to as *counts* (i.e. ct s^{-1}).

Normally in carrying out experiments with radioisotopes a *carrier* of the stable isotope of the element is added. It therefore becomes necessary to express the amount of radioisotope present per unit mass. This is the *specific activity* which may be expressed in a number of ways including disintegration rate (d s^{-1} or d min^{-1}), count rate (ct s^{-1} or ct min^{-1}) or Curies (mCi or μCi) per unit of mass of mixture (units of mass are normally either moles or grammes). An alternative method of expressing specific activity, which is not very frequently used, is *atom per cent excess*. This is defined as the number of radioactive atoms per total of 100 atoms of the compound.

9.1.7 Interaction of radioactivity with matter

Alpha Particles have a very considerable energy (3 to 8 MeV) and all the particles from a given isotope have the same amount of energy. They react with matter in two ways. Firstly, they may cause *excitation*. In this process energy is transferred from the α-particle to orbital electrons of neighbouring atoms, these electrons being elevated to higher orbitals. The α-particle continues on its path with its energy reduced by a little more than the amount transferred to the orbital electron. The *excited* electron eventually falls back to its original orbital, emitting energy as *photons* of light in the visible or near visible range. Secondly, α-particles may cause *ionisation* of atoms in their path. When this occurs the target orbital electron is removed completely. Thus, the atom becomes ionised and an *ion pair*, consisting of a positively charged ion and an electron, results. Because of their size, slow movement and double positive charge, α-particles frequently collide with atoms in their path. Therefore they cause intense ionisation and excitation and their energy is rapidly dissipated. Thus, despite their initial high energy, α-particles are not very penetrating.

Negatrons Compared with α-particles, negatrons are very small and rapidly moving particles which carry a single negative charge. They interact with matter to cause ionisation and excitation exactly as with α-particles. However, due to their speed and size, they are less likely than α-particles to interact with matter and therefore are less ionising and more penetrating than α-particles.

Another difference between α-particles and negatrons is that, whereas for a given α-emitter all the particles have the same energy, negatrons are emitted over a *range of energy*, i.e. negatron emitters have a characteristic *energy spectrum*. The maximum energy level (E_{max}) varies from one isotope to another, ranging from 0.018 MeV for ^{3}H up to 4.81 MeV for ^{38}Cl. The reason for negatrons of a given isotope being emitted within an energy range was explained by Pauli in 1931 when he postulated that each radioactive event occurs with an energy equivalent to E_{max} but that the energy is shared between a negatron and a *neutrino*. The proportion of total energy taken by the negatron and the neutrino varies for each disintegration. Neutrinos have no charge and negligible mass and do not interact with matter.

Gamma Rays are electromagnetic radiation and therefore have no charge or mass. Thus they rarely collide with neighbouring atoms and travel great distances before dissipating all their energy (i.e. they are highly penetrating). They interact with matter in many ways. The three most important ways lead to the production of *secondary electrons* which in turn cause excitation and ionisation. In *photoelectric absorption*, low energy γ-rays interact with orbital electrons, transferring all their energy to the electron which is then ejected as a *photoelectron*. The photoelectron subsequently behaves as a negatron. In contrast, *Compton scattering*, which is caused by medium energy γ-rays, results in only part of the energy being transferred to the target electron which is ejected. The γ-ray is deflected and moves on with reduced energy. Again the ejected electron behaves as a negatron. *Pair production* results when very high energy γ-rays react with the nucleus of an atom and all the energy of the γ-ray is converted to a positron and a negatron.

9.2 Detection and measurement of radioactivity

9.2.1 Absolute and relative counting

There are two bases upon which radioactive measurement may be quantified. Either every radioactive disintegration occurring in a sample is counted giving rise to *absolute counting*, or a fixed proportion of the disintegrations are counted giving rise to *relative counting*. In absolute counting either a technique which is 100% efficient must be used or, more likely, the efficiency of the counting method must be determined by using appropriate standards, and the absolute count rate computed. In relative counting it is assumed that all samples are counted with the same *counting efficiency*. This should be checked and if it is found to be an incorrect assumption, absolute counting must be used.

There are two main methods of detecting and quantifying radioactivity. These are based on the ionisation of gases, or on the excitation of solids or solutions. A third method of detecting radioactivity, which depends upon the ability of radioactivity to expose photographic emulsions, is becoming increasingly important in both qualitative and quantitative work and is the basis of autoradiography.

9.2.2 Methods based upon gas ionisation

The Effect of Voltage upon Ionisation As an energetic charged particle passes through a gas, its electrostatic field dislodges orbital electrons from atoms sufficiently close to its path and causes ionisation. The ability to induce ionisation decreases in the order:

$$\alpha > \beta > \gamma \qquad (10\,000{:}100{:}1)$$

Accordingly, α- and β-particles may be detected by gas ionisation methods, but these methods are poor for detecting γ-radiation. If ionisation occurs between a pair of electrodes enclosed in a suitable chamber, a pulse (current) flows, the magnitude of which is related to the applied potential and the number of radiation particles entering the chamber (Fig. 9.2). The various regions of Fig. 9.2 will now be considered.

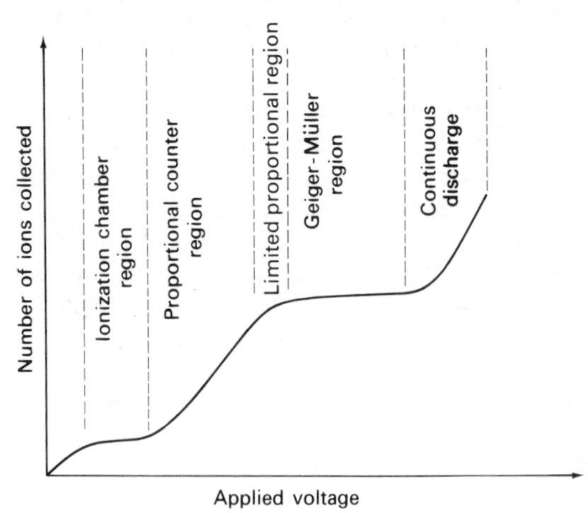

Fig. 9.2 Effect of voltage on pulse flow.

In the *ionisation chamber region* of the curve, each radioactive particle produces only one ion pair per collision. Hence the currents are low, and very sensitive measuring devices are necessary. This method is little used in quantitative work, but various types of *electroscopes*, which operate on this principle, are useful in demonstrating the properties of radioactivity. At a higher voltage level than that of the simple ionisation chambers, electrons resulting from ionisation move towards the anode much more rapidly; consequently they cause secondary ionisation of gas in the chamber resulting in the production of *secondary ionisation electrons* which cause further ionisation and so on. Hence from the original event a whole torrent of electrons reach the anode. This is the principle of *gas amplification* and is known as the *Townsend avalanche effect* after its discoverer. As a consequence of this gas amplification, current flow is much greater. As can be seen in Fig. 9.2, in the *proportional counter region* the number of ion pairs collected is directly proportional to the applied voltage until a certain voltage is reached when a plateau occurs. Before the plateau is reached there is a region known as the *limited proportional region* which is not often used in detection and quantification of radioactivity and hence will not be discussed.

The main drawback of counters which are manufactured to operate in the *proportional region* is that they require a very stable voltage supply since small fluctuations in voltage result in significant changes in amplification. Proportional counters are particularly useful for detection and quantification of α-emitting isotopes, but it should be noted that relatively few such isotopes are used in biological work.

In the *Geiger-Müller region* all radiation particles, including weak β-particles, induce complete ionisation of the gas in the chamber. Thus the size of the current is no longer dependent on the number of primary ions produced. Since maximal gas amplification is realised in this region, the size of the output pulse from the detector will remain the same over a considerable voltage range (the so-called *Geiger-Müller plateau*). The number of times this pulse is produced is measured rather than its size. Therefore it is not possible to discriminate between different isotopes using this type of counter.

Since it takes a finite time for the ion pairs to travel to their respective electrodes, other ionising particles entering the tube during this time fail to produce ionisation and hence are not detected, thereby reducing the counting efficiency. This is referred to as the *dead time* of the tube and is normally 100 to 200 μs. When the ions reach the electrode they are neutralised. Inevitably some escape and produce their own ionisation avalanche. Thus, if unchecked, a Geiger-Müller tube would tend to give a *continuous discharge*. To overcome this, the tube is *quenched* by the addition of a suitable gas, which reduces the energy of the ions. Common quenching agents are ethanol, ethyl formate and the halogens.

Geiger-Müller counting used to be the main method employed in the quantification of radioisotopes in biological samples. Now, scintillation counting (Section 9.2.3) is more popular. Nonetheless Geiger-Müller counting is still used and a brief resumé of some aspects of its use is appropriate.

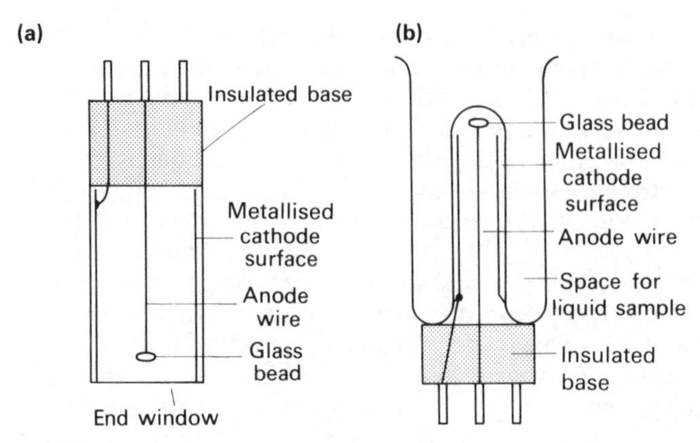

Fig. 9.3 The more important types of Geiger-Müller tubes. **(a)** End window tube; **(b)** annular-well liquid tube.

Counter Tubes A wide range of Geiger-Müller tubes is available, the most important of which are illustrated in Fig. 9.3. The *end window tube* (Fig. 9.3a) is the most widely used type. Thick end window (glass) tubes are quite robust, but can only be used for high energy β-emitters (e.g. ^{32}P) which emit particles with sufficient energy to penetrate the glass window and cause ionisation of the gas in the tube. More delicate thin end window tubes (mica or mylar) are necessary for detection of weak β-emitters (e.g. ^{14}C). The *annular well* tube (Fig. 9.3b) is used for liquid samples but, because these tubes are constructed of glass, they are suitable only for strong β-emitters. With very soft β-emitters (e.g. ^{3}H) and α-emitters, even very thin end window tubes will absorb most, if not all, of the particles before they enter the sensitive volume of the tube. This can be overcome by using a *windowless* tube through which a gas mixture such as 2% butane in helium is continually passed. This is known as *continuous gas-flow counting* or *windowless counting*.

Tube Characteristics A plot of count rate from a given source against increasing potential applied to the Geiger-Müller tube is shown in Fig. 9.4. The characteristic curve at first shows a sharp rise in counting rate, the starting potential, where only the most energetic β-particles cause the tube to go into discharge. At the threshold voltage this rise begins to level off into a *plateau* of perhaps 300 volts length. The precise length of the plateau is a characteristic of a particular tube. At higher voltages a second sharp rise in the count rate occurs. Here the potential across the detector electrodes is so high that spontaneous discharges occur in the tube, and these are not caused by radioactivity. The tube should never be operated in this *continuous discharge region*, otherwise it will soon be irreparably damaged.

Instrumentation The range of commercially available Geiger-Müller counters is very large and extends from simple, cheap counters suitable for the demonstration of radioactivity to highly sophisticated systems for

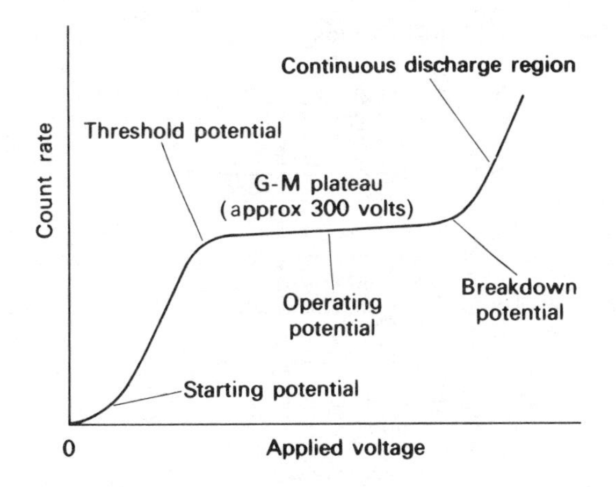

Fig. 9.4 The effect of applied voltage on count rate using a Geiger-Müller tube.

quantification. Despite the trend towards scintillation counting as the major method of quantifying isotopes used in biological work, Geiger-Müller counters still have some specific applications. Many experiments using radioisotopes involve the separation of labelled metabolites by chromatographic or electrophoretic methods. These metabolites may be located by exposing the paper or plates to photographic film (Section 9.2.4) and, once located, may be eluted for counting. This is both laborious and time consuming but, the whole process can be greatly speeded up and made more accurate by using chromatograph scanners based on thin-end window or windowless Geiger-Müller tubes.

9.2.3 Methods based upon excitation

As outlined in Section 9.1.7, radioactive isotopes interact with matter in two ways causing ionisation, which forms the basis of Geiger-Müller counting, and excitation. The latter effect leads to the emission by the excited compound, known as the *fluor*, of photons of light. This *fluorescence* can be detected and quantified. The process is known as *scintillation* and when the light is detected by a photomultiplier, forms the basis of *scintillation counting*. The electric pulse which results from the conversion of light energy to electrical energy in the photomultiplier is directly proportional to the energy of the original radioactive event. This is a considerable asset of scintillation counting since it means that two, or even more, isotopes can be separately detected and measured in the same sample provided they have sufficiently different emission energy spectra (see below).

Types of Scintillation Counting There are two types of scintillation counting which are illustrated diagrammatically in Fig. 9.5. In *solid* or *external scintillation counting* the sample is placed adjacent to a crystal of fluorescent material. The crystal which is normally used for γ-isotopes is

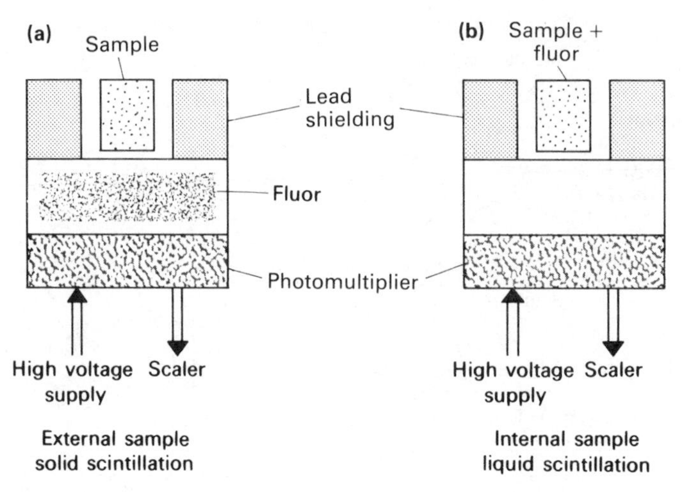

Fig. 9.5 Diagrammatic illustration of **(a)** solid (external) and **(b)** liquid (internal) scintillation counting methods.

sodium iodide, whilst for α-emitters zinc sulphide crystals are preferred and for β-emitters organic scintillators such as anthracene are used. The crystals themselves are placed near to a photomultiplier which in turn is connected to a high voltage supply and a scaler (Fig. 9.5a). Solid scintillation counting is particularly useful for γ-emitting isotopes. This is because, as explained in Section 9.1.7, γ-rays are electromagnetic radiation and only rarely collide with neighbouring atoms to cause ionisation or excitation. Clearly, in a crystal the atoms are densely packed making collisions more likely. Conversely, solid scintillation counting is generally unsuitable for weak β-emitting isotopes such as ^{3}H and ^{14}C since even the highest energy negatrons emitted by these isotopes would have hardly sufficient energy to penetrate the walls of the counting vials in which the samples are placed for counting. Since many of the isotopes used in radioimmunoassay (Section 4.5) are γ-emitting isotopes, solid scintillation counting is frequently used in biological work.

In *liquid* or *internal scintillation counting* (Fig. 9.5b), the sample is mixed with a *scintillation cocktail* containing a solvent and one or more fluors. This method is particularly useful in quantifying soft β-emitters such as ^{3}H, ^{14}C and ^{35}S which are frequently used in biological work. For these isotopes liquid scintillation counting is the usual counting method. Thus the remainder of this section will place particular emphasis on this technique, though it should be pointed out that most of what follows applies equally to solid scintillation counting used in the quantification of γ-emitters.

Energy Transfer in Liquid Scintillation Counting A small number of organic solvents fluoresce when bombarded with radioactivity. The light emitted is of very short wavelength and is not efficiently detected by most available phototubes (Fig. 9.6). However, if a compound is dissolved in the solvent which can accept the energy from the solvent and itself fluoresce at a longer wavelength, then the light can be more efficiently detected. Such a

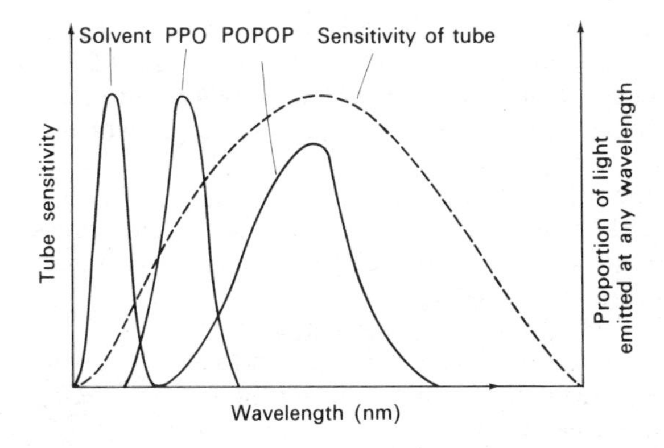

Fig. 9.6 Emission spectra of various fluors in relation to sensitivity of phototubes.

compound is known as a *primary fluor* and the most frequently used example is 2,5-diphenyloxazole (PPO). Unfortunately the light emitted by PPO is not detected with very high efficiency but this can be overcome by including a *secondary fluor* or *wavelength shifter* such as 1,4-di-[2-(5-phenyl-oxazoyl)]-benzene (POPOP). Thus, the energy transfer process becomes:

RADIATION → (SOLVENT → EXCITED) (EXCITED → PRIMARY FLUOR) (SECONDARY FLUOR → EXCITED) → LIGHT

The question obviously arises as to why a primary fluor *and* a secondary fluor are necessary when it is the latter which emits light at the best wavelength for detection. The answer is simply that the solvent cannot transfer its energy directly to the secondary fluor.

PPO and POPOP were among the original fluors used in liquid scintillation counting and remain a favourite choice. However compounds such as 2-(4′-*t*-butyl-phenyl)-5-(4″-biphenylyl)-1,3,4-oxadiazole (BUTYL-PBD) are popular since they act as primary fluors but emit light at a suitable wavelength for detection thereby making a secondary fluor unnecessary. BUTYL-PBD does however, have some practical disadvantages for certain sample types and care needs to be taken in its use.

Advantages of Scintillation Counting The very fact that scintillation counting is widely used in biological work indicates that it has several advantages over Geiger-Müller counting. These advantages include:

(i) the rapidity of fluorescence decay (10^{-9} s) which, when compared to dead time in a Geiger-Müller tube (10^{-4} s) means much higher count rates are possible;

(ii) much higher counting efficiencies particularly for low energy β-emitters. Whilst 5% efficiency for ^{3}H would be about the best that could be expected with a Geiger-Müller tube, over 50% efficiency is routine in scintillation counting. This is partly due to the fact that the negatrons do not have to travel through air or pass through an end-window of a Geiger-Müller tube thereby dissipating much of the energy before causing ionisation but interact directly with the fluor. Thus energy loss before the event which is counted is minimal;

(iii) the ability to accommodate samples of any type, including liquids, solids, suspensions, gels, and chromatograms whereas Geiger-Müller tubes and counters are specific to one type of sample or require different accessories to cope with different sample types;

(iv) the general ease of sample preparation (see below);

(v) the ability to count separately different isotopes in the same sample which is not possible in Geiger-Müller tubes and which means *dual labelling experiments* can be carried out (see below).

Disadvantages of Scintillation Counting It would not be reasonable, having outlined some of the advantages of scintillation counting, to disregard the disadvantages of the method. Fortunately, however, most of the inherent disadvantages have been overcome by improvement in instrument design. These disadvantages include the following.

(i) The cost per sample of scintillation counting can be significantly higher than for Geiger-Müller counting. However, other factors including versatility, sensitivity, ease and accuracy outweigh this factor for most applications.

(ii) At the high voltages applied to the photomultiplier, electronic events occur in the system which are independent of radioactivity but which contribute to a high background count. This is referred to as *photomultiplier noise* and can be partially reduced by cooling the photomultipliers. Since temperature affects counting efficiency, cooling also presents a controlled temperature for counting which may be useful. Low noise photomultipliers, however, have been designed to provide greater temperature stability in *ambient temperature* systems. However, cooling alone does not reduce photomultiplier noise to acceptable levels and other methods have been developed. These include firstly the use of a *pulse height analyser* which can be set so as to reject, electronically, most of the noise pulses which are of low energy (*the threshold* or *gate setting*). The disadvantage of the technique is that this also rejects the low energy pulses resulting from low energy radioactivity (e.g. ^{3}H). The second method of reducing noise, which is incorporated into most scintillation counters, is to use *coincidence counting*. In this system two photomultipliers are used. These are set in coincidence such that only when a pulse is generated in both tubes at the same time is it allowed to pass to the *scaler*. The chances of this happening for a pulse generated by a radioactive event is very high compared to the chances of a noise event occurring in both photomultipliers during the, so-called, *resolution time* of the system which is generally of the order of 20 nanoseconds. In general, this system reduces photomultiplier noise to a very low level.

(iii) The greatest disadvantage of scintillation counting is *quenching*. This occurs when the energy transfer process described earlier suffers interference. Quenching can be any one of three kinds.

(*a*) *Optical quenching* occurs if dirty scintillation vials are used. These will absorb some of the light being emitted before it reaches the photomultiplier. Care is therefore necessary to use clean vials which are handled carefully to ensure no grease or dirt is transferred to the vial surface.
(*b*) *Colour quenching* occurs if the sample is coloured and results in light emitted being absorbed within the scintillation cocktail before it leaves the sample vial. When colour quenching is known to be a major problem, it can be reduced as outlined later.
(*c*) *Chemical quenching*, which occurs when anything in the sample interferes with the transfer of energy from the solvent to the primary fluor or from the primary fluor to the secondary fluor, is the most difficult form of quenching to accommodate. In a series of homogenous samples (e.g. $^{14}CO_2$ released during metabolism of ^{14}C-glucose and trapped in alkali which is then added to the scintillation cocktail for counting) chemical quenching may not vary greatly from sample to sample. In these cases relative counting using sample counts (ct min^{-1}) can be compared directly. However, in the majority of biological experiments using radioisotopes, such homogeneity of samples is unlikely and it is not sufficiently accurate to use relative counting (i.e. ct min^{-1}). Instead, an appropriate method of standardisation must be used. This requires the determination of the counting efficiency of each sample and the conversion of ct min^{-1} to absolute counts (i.e. d min^{-1}) as described later. It should be noted that quenching is not such a great problem in solid (external) scintillation counting.

(iv) *Chemiluminescence* can also cause problems during liquid scintillation counting. It results from chemical reactions between components of the samples to be counted and the scintillation cocktail and produces light emission unrelated to excitation of the solvent and fluor system by radioactivity. These light emissions are generally low energy events and are rejected by the threshold setting of the photomultiplier in the same way as photomultiplier noise. Chemiluminescence, when it is a problem, can usually be overcome by storing samples for some time before counting, to permit the chemiluminescence to decay.

(v) *Phospholuminescence* results from components of the sample, including the vial itself, absorbing light and re-emitting it. Unlike chemiluminescence, which is a once-only effect, phospholuminescence will occur on each exposure of a sample to light. Samples which are pigmented are most likely to phosphoresce. If this is a problem, samples should be dark-adapted prior to counting and the sample holder should be kept closed throughout the counting process.

Using Scintillation Counting for Dual Labelled Samples A feature of the scintillation process is that the size of electric pulse produced by the conversion of light energy in the photomultiplier is directly related to the energy of the original radioactive event. Since different β-emitting isotopes have different energy spectra it is possible to quantify separately two isotopes in a single sample provided their energy spectra are sufficiently different. Examples of pairs of isotopes which have sufficiently different energy spectra are ^{3}H and ^{14}C, ^{3}H and ^{35}S, ^{3}H and ^{32}P, ^{14}C and ^{32}P, ^{35}S and ^{32}P. The principle

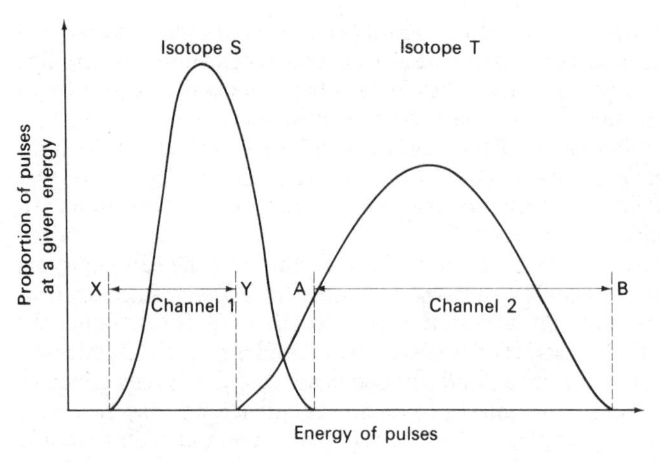

Fig. 9.7 Diagram to illustrate the principle of counting dual labelled samples.

of the method is illustrated in Fig. 9.7, where it can be seen that the spectra of two isotopes (S and T) only slightly overlap. By using one pulse height analyser set to reject all pulses of an energy below X (threshold X) and to reject all pulses of an energy above Y (*window Y*) and a second analyser with a threshold of A and a window of B, it is possible to completely separate the two isotopes. A pulse height analyser consisting of a threshold and window is known as a *channel*. Thus, to be able to count two isotopes in a single sample a two channel apparatus is necessary or, alternatively, each sample must be counted twice in a single channel apparatus with the channel set differently each time. With three and four channel instruments it is possible to count separately three isotopes in a given sample provided their energy spectra do not significantly overlap. The use of dual and triple labelling techniques has proved especially useful in studies on DNA/RNA hybridisation, RNA synthesis using a DNA template (transcription) and in studies on mechanisms of ribosomal protein synthesis as well as in studies on complex metabolic pathways (e.g. steroid synthesis).

Determination of Counting Efficiency As outlined above, a major problem encountered in scintillation counting is that of quenching which makes it necessary to determine the counting efficiency of some, if not all, of the samples in a particular experiment. This can be done by one of several methods of *standardisation*, all of which apply to both solid and liquid scintillation counting, though again in this section emphasis is placed on the latter.

(*i*) *Internal standardisation* The sample is counted (A ct min^{-1}), removed from the counter and a small amount of standard material of known disintegrations per minute (B d min^{-1}) is added. The sample is then recounted (C ct min^{-1}) and the counting efficiency of the sample calculated:

$$\text{Counting efficiency} = 100\frac{(C-A)}{B}$$

It is obviously necessary in this method to use an internal standard which contains the same isotope as the one being counted and also to ensure that the standard itself does not act as a quenching agent. Suitable ^{14}C-standards include ^{14}C-toluene, ^{14}C-hexadecane, ^{3}H-benzoic acid and ^{3}H-water (benzoic acid and water are themselves quenching agents and must only be used in very small amounts).

Internal standardisation is simple and reliable and corrects adequately for all types of quenching. Carefully carried out it is the most accurate way of correcting for quenching. On the other hand it demands very accurate pipetting in adding the standard and it is time-consuming, since each sample must be counted twice. It also means that the sample cannot be recounted in the event of error, since it will be contaminated with the standard. Moreover, time elapses between the first and second count and change in sample quenching characeristics can also occur which can lead to considerable inaccuracies.

(*ii*) *External standardisation* The determination of counting efficiency involves the use of a γ-emitting *external standard* which is built into the scintillation counter and necessitates the preparation of a calibration curve. There are many ways in which the calibration can be carried out and the best way varies according to the particular brand of counter being used. One particular method is described to illustrate the principle of external standardisation. First, a blank vial containing only scintillator solution is counted (S ct min^{-1}) in the presence of a γ-emitting external standard which is positioned outside, but adjacent to the vial. The blank is removed from the counter and a small amount of quenching agent added. Suitable quenchers include aniline, acetone and Sudan III. The blank is then recounted (T ct min^{-1}) and the count recorded in the quenched blank is expressed as a percentage of the count of the unquenched blank ($100[T/S]$). The blank is then removed from the counter, more quenching agent added, and counted again. This is repeated several times so that a series of results ($100[T/S]$) is obtained, ranging from low quench to high quench. A second vial is then prepared which contains a known amount of radioactive standard (X d min^{-1}). This is counted (Y ct min^{-1}) and the counting efficiency determined ($100[Y/X]$). This sample is removed, an amount of the quenching agent added and the counting efficiency again determined. This process is repeated using exactly the same amounts of quenching agent as in the blank. In this way the counting efficiency ($100[Y/X]$) over a range of quenching is determined. From these two sets of data (i.e. $100[T/S]$ and $100[Y/X]$ a calibration curve, Fig. 9.8, is prepared.

When counting experimental samples, it is necessary to count first in the absence of the external standard (D ct min^{-1}) then in its presence (E ct min^{-1}). A blank containing no radioactivity is also counted in the presence of the external standard (F ct min^{-1}). The percentage of the external standard which has been detected in each experimental sample is then calculated:

$$100\frac{(E-D)}{F}$$

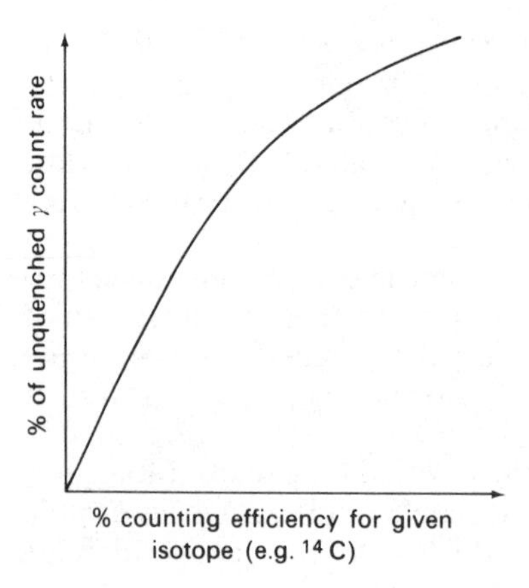

Fig. 9.8 Calibration curve for determination of counting efficiency using external standardisation.

The counting efficiency is then read from the calibration curve.

Many scintillation counters have fully automated external standard facilities, therefore this technique has the advantage over internal standardisation that the counting efficiency for each sample can be determined automatically. Furthermore there are no pipetting errors, the composition of each sample remains unaltered and a single calibration curve suffices for all types of quenching. It is important to note that a calibration curve only applies for a given isotope being counted in a given scintillator solution at given counter settings (i.e. it is necessary to prepare different calibration curves for different isotopes or for different scintillator solutions). Despite the above advantages external standardisation is probably the least accurate method of determining counting efficiency, particularly with highly coloured quenched samples. Factors which contribute to this inaccuracy include variation in vial wall thickness leading to variation in γ-ray penetration and slight variations in the placing of the γ-source in relation to the sample vial.

(*iii*) *Channels ratio* The effect of quenching, as shown in Fig. 9.9, is to decrease the average energy level of the γ-spectrum. The higher the degree of quenching, the higher is the resulting decrease in the β-spectrum. This fact is made use of in the channels ratio method of determining counter efficiency. Like external standardisation, this method involves the preparation of a calibration curve. Again there are many different approaches to the preparation of the calibration data and one example using a two channel counter is described. (It is preferable to have a two channel counter, but this is not essential since each sample can be counted twice at different channel settings in a single channel apparatus.) One channel is set to cover the whole of the unquenched β-energy spectrum (channels 1 and 2 respectively in Fig. 9.9). A sample is prepared containing a known amount of standard (P d min^{-1}) and counted in each channel (Q ct min^{-1} in channel 1 and R ct min^{-1} in channel 2). The counting efficiency in channel 1 is calculated ($100[Q/P]$), as is the channels ratio. This is the ratio of counts recorded in channel 2 and channel 1

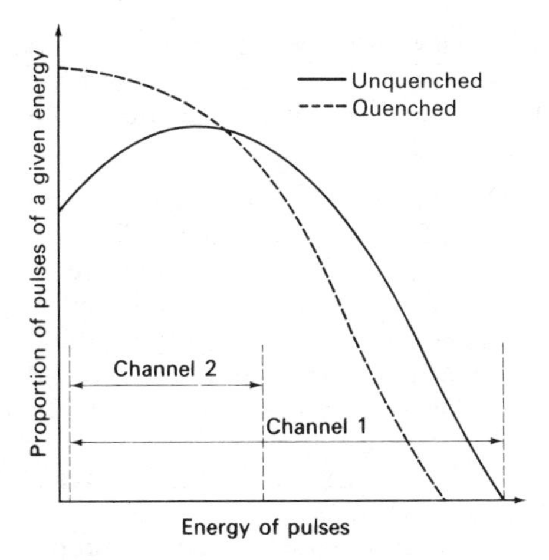

Fig. 9.9 The effect of quenching of a β-energy spectrum.

(*R/Q*). The sample is removed from the counter, a small amount of quenching agent added and the sample recounted. This is repeated many times and the counting efficiency in channel 1 and the channels ratio calculated after each increment of quenching agent. A calibration graph (Fig. 9.10) is then prepared which is used when determining the counting efficiency of experimental samples. The channels ratio for each sample is determined and the counting efficiency read from the graph. As with external standardisation, the calibration curve applies only to samples containing the same isotope, the same scintillator, etc., as used in preparation of the curve.

The channels ratio method is suitable for all types and even high degrees of quenching. Furthermore, in a two channel counter only one count is required and, therefore, this method is less time-consuming than either internal or external standardisation. It is also, in practice, an acceptably accurate

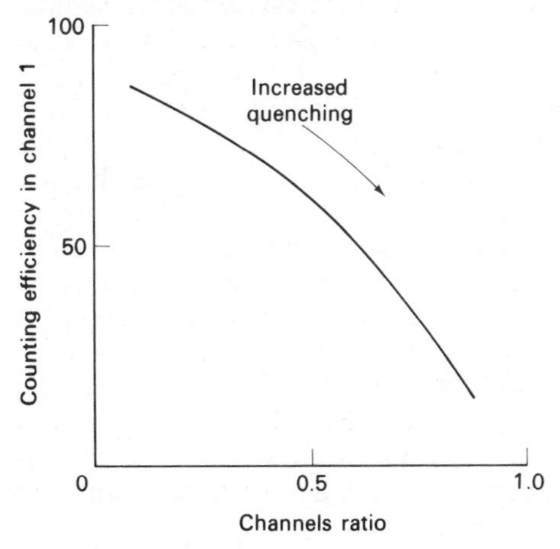

Fig. 9.10 Channels ratio quench correction curve.

method of determining counting efficiency, provided care is taken in preparation of the calibration curve. However, it is notoriously inaccurate at low sample count rates and in very highly quenched samples when, in both cases, internal standardisation should be used.

Instrumentation The range of scintillation counters now available is very large and cannot be considered fully here. For instance, models available range from single channel, single sample, manual counters to multichannel, fully automated counters capable of sequentially counting up to 400 samples. Most manual instruments can be adapted for use as solid or liquid scintillation counters and are therefore useful in small laboratories and for demonstration purposes. However, most workers use automated systems which may include such facilities as the following.

(*i*) *Temperature control* As mentioned previously cooling can be advantageous but ambient temperature systems are cheaper.

(*ii*) *Automatic programming* Many laboratories use a variety of isotopes and many instruments are capable of automatic programming so that manual alteration of channel settings from one isotope to another is unnecessary.

(*iii*) *Data analysis*. Here, too, the choice is very varied. Data can be presented by a lister which simply records the counts detected in a given time for each sample. More sophisticated systems have an inbuilt computing system which will, if suitably programmed, automatically correct for quenching and record actual d min^{-1}. Most modern systems fall into this category.

Sample Preparation It is impossible here to give details of all aspects of sample preparation for scintillation counting. However, major considerations are outlined below and the reader is referred to books cited at the end of the chapter for further details.

(*i*) *Sample vials* In solid scintillation counting, sample preparation is easy and only involves transferring the sample to a glass or plastic vial (or tube) compatible with the counter. In liquid scintillation counting sample preparation is more complex and starts with a decision on the type of sample vial to be used. These may be for glass, low potassium glass (with low levels of ^{40}K which reduces background count) or polyethylene. The latter are cheaper but are not suitable for cleaning and re-use whilst glass vials can be re-used many times provided they are thoroughly cleaned. Polyethylene vials give better light transfer and result in slightly higher counting efficiencies, but are inclined to exhibit more phosphorescence than glass vials. The recent trend towards mini vials which use far smaller volumes of expensive scintillation cocktails is worthy of comment. Such mini vials are usually inserted inside normal vials so that they are compatible with the counter sample holder. This can lead to variation in the juxtaposition of sample and photomultiplier unless each mini vial is held in exactly the same position within its carrier vial. Great care must be taken to prevent huge inaccuracies which can result from such variation. Some scintillation counters now have the facility to accommodate both normal sized and mini-vials directly and in such cases mini vial placing is not a problem.

(*ii*) *Scintillation cocktails* Toluene based cocktails are the most efficient, but will not accept aqueous samples since toluene and water are immiscible and massive quenching results. A second solvent mixed with toluene can overcome this problem. Alternatively cocktails based on 1,4-dioxane which can accommodate up to 20% water can be used. If more than 20% aqueous content is required then compounds such as Triton X-100 can be added which, when mixed with water, form a gel suitable for counting. Such gels are also suitable for counting suspensions of particulate samples e.g. scrapings from TLC plates. It must be emphasised that due to problems of chemiluminscence, it is essential that the highest purity solvents and solutes should be used in all liquid scintillation counting and that samples should be dark-adapted to overcome both phospho- and chemiluminescence where these are problematical.

(*iii*) *Volume of cocktail* It should be noted that the efficiency of scintillation counting varies with sample volume though this is less of a problem in modern counters. Nevertheless, care should be taken that sample vials in a given series of counts contain the same volume of sample and that all instrument calibration is done using the same sample volume as for experimental samples.

(*iv*) *Overcoming major colour quenching* If colour quenching is a problem it is possible to bleach samples before counting. Care should be taken, however, since bleaching agents such as hydrogen peroxide can give rise to chemiluminescence in some scintillation cocktails.

(*v*) *Tissue solubilisers* Solid samples, such as plant and animal tissues, may be best counted after solubilisation by strong bases. These tissue solubilisers include Hyamine 10-X hydroxide, NCS solubiliser and Soluene. The sample is added to the counting vial containing a small amount of solubiliser and digestion is allowed to proceed. When digestion is complete, scintillation cocktail is added and the sample counted. Again chemiluminescence is a problem often associated with the use of tissue solubilisers.

(*vi*) *Combustion methods* A suitable alternative to bleaching of coloured samples or digestion of tissues is the use of combustion techniques. Here samples are combusted in an atmosphere of oxygen, usually in a commercially available combustion apparatus. Thus samples containing ^{14}C would be combusted to $^{14}CO_2$ which is collected in a trapping agent such as sodium hydroxide and then counted whilst ^{3}H-containing samples are converted to $^{3}H_2O$ for counting.

As indicated earlier, only important considerations in sample preparation are discussed above and details are not given. However, it is worthy of comment that almost any type of radioactive sample containing β-emitting isotopes can be prepared for counting in a liquid scintillation counter by one method or another, including cuttings from paper chromatograms or membrane filters, again illustrating the versatility and importance of this technique of quantifying radioactivity.

Cerenkov Counting If a β-emitter has a decay energy in excess of 0.5 MeV, then this causes water to emit a bluish white light usually referred to as *Cerenkov light*. It is possible to detect this light using a typical liquid

Table 9.2 Some isotopes suitable for Cerenkov counting.

Radioisotope	E_{max} (MeV)	% of spectrum above 0.5 MeV	Counting efficiency (%)
^{22}Na	1.39	60	30
^{32}P	1.71	80	40
^{36}Cl	0.71	30	10
^{42}K	3.5	90	80

scintillation counter. Since there is no requirement for organic solvents and fluors, this technique is relatively cheap, sample preparation is very easy, and there is no problem of chemical quenching. Cerenkov counting, even though it was originally discovered in 1910, is only now gaining in popularity. Table 9.2 shows some isotopes which are suitable for this detection method. Most work has been done on ^{32}P which has 80% of its energy spectrum above the Cerenkov threshold and which can be detected at around 40% efficiency. It will be noted from Table 9.2 that, as the proportion of the energy spectrum above 0.5 MeV increases, then so too does the detection efficiency. So far, few workers have exploited the possibility of combining Cerenkov counting and scintillation counting for pairs of isotopes in a single sample, but this is obviously a possibility. It would simply be necessary to choose two isotopes, one of which has an E_{max} below 0.5 MeV, while the other has much of its energy above this threshold. In particular, it would seem useful to combine a soft β-emitter with a higher energy γ-emitter, since γ-emitters are readily detected by this method, whereas they are not usually efficiently counted by liquid scintillation counting. (Normally γ-emitters are counted by solid scintillation counting.)

9.2.4 Methods based upon exposure of photographic emulsions

Ionising radiation acts upon a photographic emulsion to produce a latent image much as visible light does. For a photograph a radiation source, an object to be imaged and photographic emulsion are required. For an *autoradiograph* a radiation source (i.e. radioactivity) emanating from within the material to be imaged (the object) is required along with a sensitive emulsion. The emulsion consists of a large number of silver halide crystals embedded in a solid phase such as gelatin. As energy from the radioactive material is dissipated in the emulsion, the silver halide becomes negatively charged and is reduced to metallic silver thus forming a particulate latent image. Photographic *developers* are designed to show these silver grains as a blackening of the film and *fixers* to remove any remaining silver halide. Thus a permanent image of the location of the original radioactive event remains. This process, which is known as *autoradiography*, is very sensitive and has been used in a wide variety of biological experiments. These usually involve a requirement to locate the distribution of radioactivity in biological specimens of different types. For instance the sites of localisation of a radiolabelled drug throughout the body of an experimental animal can be determined by placing whole body sections of the animal in close contact with a sensitive emulsion such as an X-ray plate. After a period of exposure the plate, upon development, will show an image of the section in tissues and organs in which radioactivity was present (Fig. 9.11). Similarly, radioactive metabolites isolated and separated by chromatographic or electrophoretic techniques during metabolic studies can be located on the chromatogram or electrophoretogram and the radioactive *spots* can subsequently be recovered for counting and identification (counting may be carried out on the original chromatogram by using a chromatogram scanner, the design of which is generally based on Geiger-Müller tubes, or by elution from the paper, plate or gel for counting by liquid scintillation counting).

The techniques of autoradiography have become more important with recent developments in molecular biology (Chapter 5). Consequently more detail is given below on some important aspects of the technique.

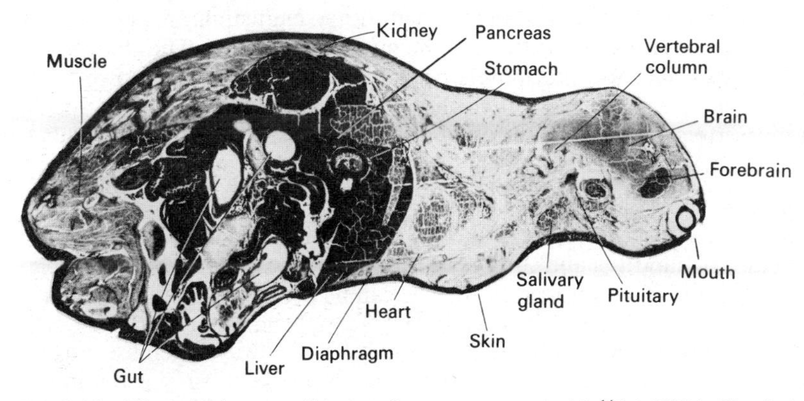

Fig. 9.11 Whole-body autoradiogram of a mouse treated with ^{14}C-L-DOPA. The dark areas indicate the presence of radioactive isotope and show high concentrations in the liver, pancreas, kidney, skin and forebrain. (Reproduced with the permission of Roche Products Ltd, Welwyn Garden City.)

Suitable Isotopes In general soft β-emitting isotopes (e.g. ^{3}H and ^{14}C) are most suitable for autoradiography particularly for cell and tissue localisation experiments. This is because, as a result of the low energy of the negatrons, the *ionising track* of the isotope will be short and a discrete image will result. This is particularly important if radioactivity associated with subcellular organelles is being located. For this, ^{3}H is the best radioisotope since its energy will all be quickly dissipated within the emulsion. Electron microscopy can then be used to locate the image in the developed film. For location within whole organisms or tissues, either ^{14}C or ^{3}H is suitable whilst more energetic isotopes (e.g. ^{32}P) are less suitable since their higher energy negatrons produce much longer track lengths and result in less discrete images which are not sufficiently discriminatory for microscopic location. Conversely for location of, for example, DNA bands in gel electrophoretograms ^{32}P is useful. In this case less energetic ^{14}C or ^{3}H negatrons would largely dissipate their energy within the gel (and in the wrapping around the gel which is usually necessary to prevent the gel sticking to the emulsion) thereby reducing sensitivity to a low level. However, the more energetic ^{32}P negatrons will leave the gel and produce a strong image.

Choice of Emulsion and Film A variety of suitable emulsions is available with different *packing densities* of the silver halide crystals. Care must be taken to choose an emulsion suitable for the purposes of the experiment, since the sensitivity of the emulsion will affect the resolution obtained. Manufacturers literature should be consulted and their advice sought if in any doubt. X-ray film is generally suitable for macroscopic samples such as whole body sections of small mammals, chromatograms or electrophoretograms. When light or electron microscopic detection of the location of the image in the emulsion is required (cellular and subcellular localisation of radioactivity) very sensitive films are required, as is a very close apposition of sample and film. In these cases a *stripping film technique* can be used in which the film is supplied attached to a support. It is stripped from this and applied directly to the sample. Alternatively liquid emulsions are prepared by melting strips of emulsion by heating them to around 60°C. The emulsion is then either poured onto the sample or the sample attached to a support is dipped into it. The emulsion is then allowed to set before being dried. Such a method is often referred to as a *dipping-film method* and is preferred when very thin films are required.

Background Accidental exposure to light, chemicals in the sample, natural background radioactivity (particularly ^{40}K in glass vials) and even pressure applied during handling and storage of film will cause a background *fog* (i.e. latent image) on the developed film. This can be problematical, particularly in high resolution work (e.g. involving microscopy) and care must be taken at all times to minimise its effect. Background will always increase during exposure time which should therefore always be kept to the minimum possible.

Time and Conditions of Exposure and Film Processing The time and conditions of exposure (e.g. room temperature or refrigerated) will depend upon the isotope, sample type, level of activity, film type and purpose of the experiment. The same applies to the processing of the film to display the image. Generally the process must be adapted to a given purpose and a great deal of trial and error is often involved in arriving at the most suitable procedures.

Preflashing The response of a photographic emulsion to radiation is not linear and usually involves a slow initial phase (lag) followed by a linear phase. Sensitivity of films may be increased by *pre-flashing*. This involves a millisecond light flash prior to the sample being brought into juxtaposition with the film and is often used in high resolution work.

Fluorography Many of the currently popular methods in molecular biology (Chapter 5) involve separation of macromolecules or fractions of macromolecules by gel electrophoresis. The separated macromolecules or fractions form bands in the electrophoretogram which must be located. This is often achieved by radiolabelling the macromolecules with ^{3}H or ^{14}C and subjecting the gel to autoradiography. Because these are soft β-emitters much of their energy is lost in the gel and long exposure times are necessary even when using very high specific activity sources. However, by infiltrating into the gel a fluor (e.g. PPO), drying the gel and then placing it in contact with a preflashed film, sensitivity can be increased by several orders of magnitude. This is because the negatrons emitted from the isotope will cause the fluor to become excited and emit light which will expose the film. Thus use is made of *both* the ionising and exciting effect of radioactivity in fluorography.

Intensifying Screens When locating ^{32}P labelled or γ-isotope labelled samples (e.g. ^{32}P-DNA or ^{125}I-protein fractions in gels) the opposite problem to that presented by low energy isotopes prevails. These much more penetrating particles and rays cause little exposure of the film since they penetrate right through it producing a poor image. The image can be greatly improved by placing, on the other side of the film from the sample, a thick *intensifying screen* consisting of a solid phosphor. Negatrons penetrating the film cause the phosphor to fluoresce and emit light which superimposes its image on the film. Again the increase in resolution can be of several orders of magnitude.

Quantification As indicated earlier autoradiography is usually used to *locate* rather than quantify radioactivity. However it is possible to obtain quantitative data directly from autoradiographs by using a densitometer which records the intensity of the image. This in turn is related to the amount of radioactivity in the original sample. There are many varieties of densitometers available and the choice made will depend on the purpose of the experiment.

9.3 Other practical aspects of counting radioactivity and analysis of data

9.3.1 Counter characteristics

Background Count Radiation counters of all types always register a count even in the absence of radioactive material in the apparatus. This may be due to such sources as cosmic radiation, natural radioactivity in the vicinity, nearby X-ray generators, and circuit noises. By use of the various methods already outlined and the use of lead shielding, this *background radiation* may be considerably reduced, but its value must always be recorded and accounted for in all experiments. Some commercial instruments have automatic background subtraction facilities.

Dead Time At very high count rates in Geiger-Müller counting, counts are lost due to the *dead time* of the Geiger-Müller tube. Correction tables are available and these should be used when necessary to correct for lost counts. Dead time is not a problem in scintillation counting.

Geometry Solid samples for Geiger-Müller counting are normally dried on to either stainless steel planchets or on to ground glass discs and counted with a thin end window tube. When counting such samples, it is important to standardise the position of the sample in relation to the tube, otherwise the fraction of the emitted radiation entering the tube may vary and hence so will the observed count. A second important *geometric effect* is that care should be taken to ensure that all the samples have the same surface area. Geometry problems are relatively insignificant in scintillation counting.

9.3.2 Sample and isotope characteristics

Self Absorption The problem of *self absorption* also applies mainly to Geiger-Müller counting. Radioactivity from a solid sample surface must traverse the air between the sample and the window and then pass through the window. Radioactivity from inside the sample must traverse the solid above it, then the air and then the window. It is usually found that particles emitted below a certain minimum distance are completely absorbed and never reach the counter. The sample is then said to be *infinitely thick*. Self absorption is primarily a problem with low energy β-emitters. The problem may be overcome either by ensuring that the samples are so thin that no self absorption occurs, i.e. the samples are *infinitely thin* samples or the samples are bulked with carrier material and thoroughly mixed so that the sample is infinitely thick.

Self absorption is a serious problem in the counting of radiactivity by scintillation counting only if the sample is particulate or is, for instance, stuck to a membrane filter. Care should be taken to ensure comparability of samples since the methods of standardisation outlined earlier will not correct for self absorption effects. Where homogeneity is not possible particulate samples should be digested or otherwise solubilised prior to counting.

Half Life The half life of an isotope (Section 9.1.5) may be short and, if so, this must be allowed for in the analysis of data.

Statistical Aspects The emission of radioactivity is a random process. This can readily be demonstrated by making repeated measurements of the activity of a long lived isotope, each for an identical period of time. The resulting counts will not be the same but will vary over a range of values with a clustering near the centre of the range. If a sufficiently large number of such measurements is made and the data plotted, a normal distribution curve will be obtained. For a single count therefore one cannot obtain a *true count*. However, from the *Poisson distribution equation* it can be shown that the mean value ($\bar{n}$) is the best estimate of the true value when large numbers of repeated counts are made. For practical purposes, the *standard deviation* (σ) can be defined as $\sqrt{\bar{n}}$. Only 31.7% of the values of a normal distribution curve are outside the limits $\bar{n} \pm \sigma$ and only 5.5% are outside $\bar{n} \pm 2\sigma$.

The foregoing comments apply to repeated measurements on a single sample. However, in radiotracer work, it is not usually convenient to do this and it is normal for only one or two counts on a given sample to be taken. In such cases there is no ($\bar{n}$) from which σ may be calculated. Fortunately, because of the nature of the Poisson distribution curve, the standard deviation of a single determination (n) may be taken as the square root of the total counts collected, i.e. $\sigma = \sqrt{\bar{n}}$. A related unit, the *relative standard deviation* (relative σ), is defined as $100/\sqrt{\bar{n}}$. Thus the larger the number of counts collected, the smaller will be the relative σ and to realise a specified relative σ it is necessary only to count until sufficient counts are collected.

Radiotracer experiments are usually designed to have a relative standard deviation of 1%. In order to achieve this it is necessary to collect 10 000 counts since $100/\sqrt{\bar{n}} = 1\%$ when $n = 10\,000$. From the single assay one could then state that there is 68.3% probability that the true count is 10 000 ± 100. For degrees of accuracy better than 1% it is necessary to collect more counts, while collecting less counts will give a lower degree of accuracy. It follows that in counting a series of samples, it is always better to count for a preset number of counts rather than a preset time since all samples will at least be counted to the same degree of accuracy.

9.3.3 Supply, storage and purity of radiolabelled compounds

There are several suppliers of radiolabelled compounds, the main one in Britain being Amersham International plc. The suppliers usually include details of the best storage conditions and quality control data for their products. In general, biologists will be using radiolabelled organic compounds which are subject to biodegradation. Care should be taken, therefore, to store such compounds under conditions where they are unlikely to be degraded. Usually this will mean the preparation of a stock solution of material of the required specific activity from the supplied material and keeping this deep frozen. However, some compounds do not keep for long periods of time even under these conditions and it is advisable to use appropriate chromatographic methods to check the purity frequently. The suppliers normally give appropriate chromatographic details such as solvents and locating reagents, along with the original quality control data.

9.4 Inherent advantages and restrictions of radiotracer experiments

Perhaps the greatest advantage of radiotracer methods over most other chemical and physical methods is their *sensitivity*. For example, the specific activity of carrier-free tritium is approximately 50 Curies per millimole. This means that a dilution factor of 10^{12} can be tolerated without jeopardising the detection of tritium-labelled compounds. It is thus possible to detect the occurrence of metabolic substances which are normally present in tissues at such low concentrations as to defy the most sensitive chemical methods of identification. A second major advantage of using radiotracers is that studies *in vivo* can be carried out to a far greater degree than by any other technique.

In spite of these significant advantages, certain restrictions have to be appreciated. Firstly, although they undergo the same reactions, isotopes may do so at different rates. This effect is known as the *isotope effect*. The different rates are approximately proportional to the differences in mass between the isotopes. The extreme case is the isotopes of hydrogen ^{1}H and ^{3}H, the effect being smaller for ^{12}C and ^{14}C and almost insignificant for ^{31}P and ^{32}P. Secondly, the amount of activity employed must be kept to a minimum necessary to permit reasonable counting rates in the samples to be analysed, otherwise the radiation from the tracer may elicit a response from the experimental organism and hence distort the results. A third consideration is that, in order to administer the tracer, the normal chemical level of the compound in the organism is automatically exceeded. The results are therefore always open to question.

9.5 Applications of radioisotopes in the biological sciences

9.5.1 Investigating aspects of metabolism

Metabolic Pathways Radioisotopes are frequently used for tracing metabolic pathways. This usually involves adding a radioactive substrate, taking samples of the experimental material at various times, extracting and chromatographically, or otherwise, separating the products. Radioactivity is then located either by using a Geiger-Müller chromatogram scanner or by obtaining an autoradiogram of the chromatogram. By identifying the labelled compounds, counting the radioactivity in each of them and plotting suitable graphs, it is possible to obtain considerable information on the metabolic pathways involved. Radioactivity detectors can also be attached to GLC or HPLC chromatographic columns to monitor radioactivity coming off the column during separation.

If it is suspected that a particular compound is metabolised by a particular pathway, then radioisotopes can also be used to confirm this. For instance, it is possible to predict the fate of individual carbon atoms of ^{14}C-acetate through the tricarboxylic acid cycle (TCA cycle). Methods have been developed whereby intermediates of the cycle can be isolated and the distribution of carbon within each intermediate can be ascertained. This is the so-called *specific labelling pattern*. Should the actual pattern coincide with the theoretical pattern then this is very good evidence for the operation of the TCA cycle.

Another example of the use of radioisotopes to confirm the operation, or otherwise, of a metabolic pathway is in studies carried out on glucose catabolism. There are numerous ways whereby glucose can be oxidised, the two most important ones in aerobic organisms being glycolysis followed by the TCA pathway and the pentose phosphate pathway. Frequently organisms or tissues possess the necessary enzymes for both pathways to occur and it is of interest to establish the relative contribution of each to glucose oxidation. Both pathways involve the complete oxidation of glucose to carbon dioxide, but the origin of the carbon dioxide in terms of the six carbon atoms of glucose is different (at least in the initial stages of respiration of exogenously added substrate). Thus it is possible to trap the carbon dioxide evolved during the respiration of specially labelled glucose (e.g. ^{14}C-6-glucose in which only the C-6 atom is radioactive and ^{14}C-1-glucose) and obtain an evaluation of the contribution of each pathway to glucose oxidation.

The use of radioisotopes in studying the operation of the TCA cycle or in evaluating the pathyway of glucose catabolism are just two examples of how they can be used in confirming metabolic pathways. Further details of these and other examples, including use of dual labelling methods, can be found in the various texts recommended in Section 9.8.

Metabolic Turnover Times Radioisotopes provide a convenient method of ascertaining turnover times for particular compounds. As an example, the turnover of proteins in rats will be considered. A group of rats is injected with a radioactive amino acid and left for 24 hours, during which time most of the amino acid will be assimilated into proteins. The rats are then sacrificed at suitable time intervals and radioactivity in organs or tissues of interest is determined. In this way it is possible to ascertain the rate of metabolic turnover of protein. Using this sort of method it has been shown that liver protein is turned over in 7 to 14 days, while skin and muscle protein is turned over every 8 to 12 weeks and collagen is turned over at a rate of less than 10% per annum.

Studies of Absorption, Accumulation and Translocation Radioisotopes have been very widely used in studying the mechanisms and rates of absorption, accumulation and translocation of inorganic and organic compounds by both plants and animals. Such experiments are generally simple to perform and can also yield evidence on the route of translocation and sites of accumulation of molecules of biological interest.

Pharmacological Studies Another field where radioisotopes are widely used is in the development of new drugs. This is a particularly complicated process since, besides showing if a drug has a desirable effect, much more must be ascertained before it can be used in the treatment of clinical conditions. For instance, the site of drug accumulation, the rate of accumulation, the rate of metabolism and the metabolic products must all be determined. In each of these areas of study, radiotracers are extremely useful, if not indispensable. For instance, autoradiography on whole sections of experimental animals (Section 9.2.4 and Fig. 9.11) yields information on the site and rate of accumulation while typical techniques used in metabolic studies can be used to follow the rate and products of metabolism.

9.5.2 Analytical applications

Enzyme and Ligand Binding Studies Virtually any enzyme reaction can be assayed using radiotracer methods as outlined in Section 3.4.5 provided that a radioactive form of the substrate is available. Radiotracer-based enzyme assays are more expensive than other methods, but frequently have the advantages of a higher degree of sensitivity. Radioisotopes have also been used in studying the mechanism of enzyme action and in ligand binding studies (Section 3.7).

Isotope Dilution Analysis There are many compounds present in living organisms which cannot be accurately assayed by conventional means, since they are present in such low amounts and in mixtures of similar compounds. Isotope dilution analysis offers a convenient and accurate way of overcoming this problem and avoids the necessity of quantitative isolation. For instance, if the amount of iron in a protein preparation is to be determined, this may be difficult using normal methods, but it can be done if a source of ^{59}Fe is available. This is mixed with the protein and a sample of iron is subsequently isolated, assayed for total iron and the radioactivity determined.

If the original specific activity was 10 000 d min^{-1} per 10 mg and the specific activity of the isolated iron was 9000 d min^{-1} per 10 mg then the difference is due to the iron in the protein (x) i.e.

$$\frac{9000}{10} \times \frac{10000}{10 + x}$$

and

$$x = 1.1 \text{ mg}$$

This technique is widely used in, for instance, studies on trace elements.

Radioimmunoassay One of the most significant advances in biochemical techniques in recent years has been the development of radioimmunoassay (RIA). This technique is discussed in Section 4.5 and is not elaborated upon here.

Radiodating A quite different analytical use for radioisotopes is in the *dating* (i.e. determining the age) of rocks, fossil and sediments. In this technique it is assumed that the proportion of an element which is naturally radioactive has been the same throughout time. At the time of fossilisation or deposition the radioactive isotope begins to decay. By determining the amount of radioisotope remaining (or by examining the amount of a decay product) and from a knowledge of the half life, it is possible to date the sample. For instance, if the radioisotope normally composes 1% of the element and it is found that the sample actually contains 0.25% then two half lives can be assumed to have elapsed since deposition. If the half life is 1 million years then the sample can be dated as being 2 million years old.

For long term dating, isotopes with long half lives are necessary, such as ^{235}U, ^{238}U and ^{40}K, whereas for shorter term dating ^{14}C is widely used. It cannot be over-emphasised that the assumptions made in radiodating are sweeping and hence palaeontologists and anthropologists who use this technique can only give approximate dates to their samples.

9.5.3 Other applications

Molecular Biology Techniques Recent advances in molecular biology which have led to the advances in genetic manipulation have depended heavily upon use of radioisotopes in DNA and RNA sequencing, DNA replication, transcription, synthesis of cDNA, recombinant DNA technology and many similar studies. Many of these techniques are more fully discussed in Chapter 5.

Clinical Diagnosis Radioisotopes are very widely used in medicine, in particular for diagnostic tests. Lung function tests routinely made using ^{133}Xenon (^{133}Xe) are particularly useful in diagnosis of malfunctions of lung ventilation. Kidney function tests using ^{131}I-iodohippuric acid are used in diagnoses of kidney infections, kidney blockages or imbalance of function between the two kidneys. Thyroid function tests using ^{131}I are employed in the diagnosis of hypo- and hyper-thyroidism.

Various aspects of haematology are also studied by using radioisotopes. These include such aspects as blood cell lifetimes, blood volumes and blood circulation times, all of which may vary in particular clinical conditions.

Ecological Studies The bulk of radiotracer work is carried out in biochemical, clinical or pharmacological laboratories, nevertheless, radiotracers are also of use to ecologists. In particular, *migratory patterns* and *behaviour patterns* of many animals can be monitored using radiotracers. Another ecological application is in the examination of *food webs* where the primary producers can be made radioactive and the path of radioactivity followed throughout the resulting food web.

Sterilisation of Food and Equipment Very strong γ-emitters are now widely used in the food industry for sterilisation of prepacked foods such as milk and meats. Normally either ^{60}Co or ^{137}Ce is used, but care has to be taken in some cases to ensure that the food product itself is not affected in any way. Thus doses often have to be reduced to an extent where sterilisation is not complete but nevertheless food spoilage can be greatly reduced. ^{60}Co and ^{137}Ce are also used in sterilisation of plastic disposable equipment such as petri dishes and syringes, and in sterilisation of drugs which are administered by injection.

Mutagens Radioisotopes may cause *mutations*, particularly in microorganisms (Section 1.7). In various microbiological studies mutants are desirable, especially in industrial microbiology. For instance, development of new strains of a microorganism which produce higher yields of a desired microbial product frequently involve mutagenesis by radioisotopes.

9.6 Safety aspects of the use of radioisotopes

No chapter on radioisotopes would be complete without some consideration of the safety aspects of their use. Radiation is hazardous but, if handled with proper care and responsibility, it should be no more hazardous than work involving non-radioactive compounds. Regulations for the safe handling and use of radioisotopes are documented in such publications as *Code of Practice: Ionising Radiations in Research and Training*, HMSO.

In particular it should be noted that, except for establishments using only very small quantities of radioactivity, it is necessary to obtain a licence from the Department of the Environment. This will stipulate maximum holdings of radioactivity, the amounts of solid waste which can be disposed of along with normal refuse and the amount of liquid waste which can be disposed of down the drains. It will also stipulate that adequate records of use and disposal of each isotope are kept and periodically made available for inspection.

Precautions necessary in handling radioisotopes vary with the isotope, but certain safeguards should be taken at all times. For instance, all work should be carried out over trays lined with special absorbent paper, disposable gloves should be worn, a laboratory coat is essential and no mouth pipetting should be carried out. Consuming food and drink in laboratories where isotopes are used is forbidden as is the application of cosmetics and smoking. All cuts and grazes of the skin must be covered with a waterproof dressing. Another important point is that all spillages must be immediately attended to. This applies also to contamination of skin or clothes.

The topic of radiation protection and acceptable dosages for personnel using radioactive substances is a complex one and cannot be covered here. Readers who intend to use radiotracers are strongly advised to become fully conversant with radiation protection by reading the *Code of Practice* and other suitable publications as cited at the end of this Chapter.

9.7 Suggestions for further reading

Aronoff, S. (1958). *Techniques in Radiobiochemistry*. Iowa State University Press, Iowa. (Very comprehensive, if dated, account of biochemical uses of radioactivity.)

Dyer, A. (1974). *An Introduction to Liquid Scintillation Counting*. Heyden, London. (An excellent account of liquid scintillation counting including a good bibliography and an appendix listing liquid scintillation cocktails and sample solubilisation techniques.)

Gahan, P.B. (1972). *Autoradiography for Biologists*. Academic Press, London. (Comprehensive account of autoradiography techniques especially for subcellular localisation.)

Hendee, W.R. (1973). *Radioactive Isotopes in Biological Research*. John Wiley and Sons, New York. (A useful discussion of all aspects of radiotracer methods. Particularly good on radiation protection and dose rates.)

Noujaim, A.A., Ediss, C. and Weibe, L.I. (1976). *Liquid Scintillation*. Science and Technology. Academic Press, New York. (A comprehensive account of liquid scintillation counting and sample preparation).

Rogers, A. (1979). *Techniques of Autoradiography*. Elsevier, New York. (Covers, in depth, all aspects of autoradiography.)

Wang, C.H. and Willis, D.L. (1965). *Radiotracer Methodology in Biological Sciences*. Prentice-Hall, Englewood Cliffs, New Jersey. (Probably the most comprehensive and readable book on the theoretical and practical aspects of radioactivity.)

N.B. The above references are the best sources of general information currently available. Many are, however, somewhat dated. For specific aspects, particularly of recent developments, many instrument manufacturers produce comprehensive booklets. The *Review Series*, published by Amersham International plc., Amersham, Bucks, England is especially recommended.

10
Electrochemical techniques

10.1 Introduction

10.1.1 The range of electrochemical techniques

Many biochemical studies require the measurement of ions, such as H^+, Na^+, Ca^{2+}, NH_4^+, Cl^- in solution, or the ease with which certain substrates are oxidised or reduced in solution. These types of measurement rely upon the fact that if an inert metal electrode such as platinum is placed in such a solution, a potential will be set up at the surface of the electrode. A metal electrode and the solution in which it is immersed is referred to as a *half cell*. It is impossible to measure the potential of a half cell in isolation, but by connecting it to a *reference half cell* the potential of the half cell can be determined relative to the reference half cell. This is the basis of electrochemical techniques and of the construction of a number of electrodes designed to make specific measurements. These electrodes are generally inexpensive, portable and readily used without the need for extensive operator training. Unlike spectrophotometric methods, electrochemical measurements are not affected by cloudy or coloured solutions.

Since so many biological processes are sensitive to the hydrogen ion concentration (pH) of the surrounding medium (Section 1.2.1), pH measurements and adjustments are vital in virtually all biochemical experiments. pH values are most accurately measured by use of a glass electrode (Section 10.2) but rough estimates of pH can be obtained using pH-sensitive dyes which change their colour at a specific pH.

Ions other than the hydrogen ion can also be measured by special electrodes which are generally referred to as *ion-selective electrodes* (Section 10.3). These electrodes are available for many biologically important ions and can be used, for example, to measure the sodium ion concentration in samples as diverse as blood, urine and sea water and for measuring the concentration of ammonium ions in Kjeldahl digests (Section 3.2.2).

Many biologically important compounds are capable of existing in an oxidised and a reduced state. The cytochromes, for example, can exist as the ferrous or ferric form. The ease with which a compound can accept or donate an electron is expressed by its *oxidation-reduction (redox) potential* (Section 10.4). A compound which has a more negative oxidation-reduction potential than a second compound is capable of acting as an electron donor to the second compound and in the process will itself be oxidised. A series of consecutive oxidation-reduction reactions is the basis of a so-called *electron transport chain*. Such chains are found in mitochondria, chloroplasts, bacteria and microsomes, and are fundamental to the linking of catabolism and anabolism to energy utilisation, as well as to the oxidation of many endogenous and exogenous compounds. The value of a redox potential of a particular compound is most accurately measured by electrochemical means using a platinum electrode. A rough estimate of a redox potential can be obtained using redox dyes which are capable of gaining or donating an electron and, in the process, changing their colour. Redox dyes, however, are more commonly used to measure the *rate* of electron transport in mitochondria or chloroplasts.

Since oxygen utilisation or evolution is fundamental to so many biological processes, electrodes capable of measuring oxygen concentration and concentration changes are very valuable for biochemical studies. The *oxygen electrode* (Section 10.5) is commonly used in studies of mitochondria and chloroplasts but can be used for the study of any reaction in which oxygen is evolved or absorbed (Section 3.4.7).

Biosensors (Section 10.6) which use biological material to detect the presence of chemicals, are becoming increasingly important. Instead of merely responding to ions (as do ion-selective electrodes) they enable a wide range of molecules to be detected including, for example, drugs, glucose and urea. Biosensors contain biological material in the form of enzymes, cells or antibodies, which respond 'specifically' to a compound even at very low concentrations. For example, an enzyme in a biosensor may react with its substrate (which is the substance under test) to produce a pH change. Included as part of such a biosensor would be a pH glass electrode which would measure the resultant pH change.

HPLC is a most useful chromatographic technique which can separate many types of compounds (Section 6.8). As the compounds are eluted from a HPLC column, they are commonly detected by their fluorescence or ultraviolet absorption. However, if the compound can be readily oxidised or reduced, it can be detected by an *electrochemical detector* such as a wall jet electrode (Section 10.7). Such detectors are very sensitive and are the best type for some drugs (e.g. morphine) and the catecholamines (e.g. noradrenaline).

10.1.2 Electron transport processes

In a number of biological processes, a series of coupled oxidation-reduction reactions occurs in which one or more electrons are passed from one electron carrier to another to form an electron transport chain. In some instances a hydrogen atom is also transferred. Examples of the components of electron transport chains include the cytochromes, flavoproteins and certain quinones. The oxidised and reduced forms of each component may be represented by the general equation:

$$\text{reduced form} \rightleftharpoons \text{oxidised form} + ne^-$$

when n = number of electrons transferred.

When two such systems are coupled together, resulting in the oxidation of one and the reduction of the other, there is a change in the standard free energy of the coupled system which is related to the difference in the oxidation-reduction potentials (Section 10.4) of the two component reactions in accordance with the following equation:

$$\Delta G^{0\prime} = -nF\Delta E_0^{\prime} \qquad 10.1$$

where $\Delta G^{0\prime}$ = standard free energy change
$\Delta E_0^{\prime}$ = potential difference between the two participating redox systems providing n is the same for each system.

If $\Delta E_0^{\prime}$ is positive then $\Delta G^{0\prime}$ will be negative and the coupled reaction is said to be *exergonic*, i.e. free energy is released and the coupled reaction is thermodynamically favoured. In mitochondrial and bacterial respiration, this is true of all parts of the electron transport chain, and the release of energy in some of the coupled reactions is sufficient to drive ATP production, by a process known as *oxidative phosphorylation*. The individual components of the mitochondrial electron transport chain, and the sites at which ATP molecules are produced, are shown in Fig. 10.1.

Electron transport inhibitors are known which act specifically at one site in the chain to stop electron transport, but this can sometimes be restarted by the addition of an artifical or natural electron donor (Section 10.4.3). For example, succinate restarts electron transport which has been stopped by rotenone (Fig. 10.1).

Electron transport from NADH to molecular oxygen produces three

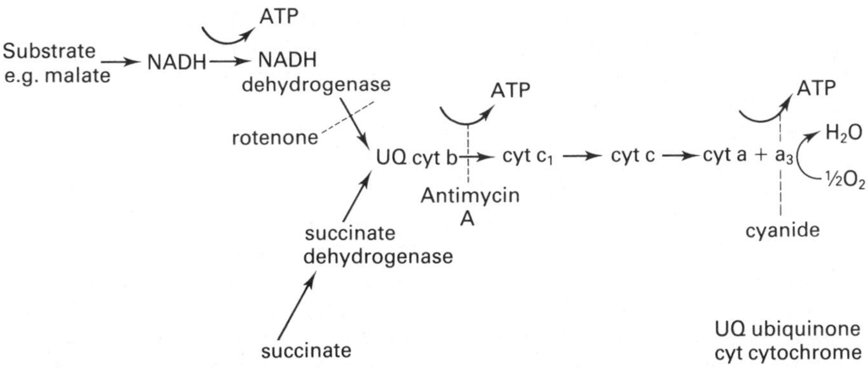

Fig. 10.1 The mitochondrial respiratory chain, showing the site of action of inhibitors, and sites of ATP production by oxidative phosphorylation.

molecules of ATP whilst electron transport from succinate bypasses the first site and produces only two. The number of molecules of ATP produced by electron transport from a particular substrate can be found by comparing the rate of phosphorylation with the rate of electron transport i.e. with the rate of oxygen uptake. Phosphorylation can be measured by the disappearance of either ADP or Pi (inorganic phosphate) and this can be used to calculate an ADP/O or P/O ratio. The uptake of oxygen can be most conveniently measured using an oxygen electrode.

As electrons and hydrogen atoms travel along the series of carriers in a transport chain, a pH (or proton) gradient builds up between the two faces of the mitochondrial inner membrane. This is called ΔpH. A charge (potential) also builds up on the membrane as electrons accumulate on the inner face of the membrane. This is known as $\Delta\psi$. Details of ΔpH and $\Delta\psi$ measurements are given in Section 10.2.3. Build up of ΔpH and $\Delta\psi$ is a consequence of the respiratory chain looping across the two faces of the membrane. According to the *Mitchell chemiosmotic theory*, ΔpH and $\Delta\psi$ are sources of power which can drive an enzyme, located in the membrane, known as *proton translocating ATPase* to form ATP from ADP and Pi. This enzyme is shown in Fig. 10.2 which also shows how the build up of ΔpH brought about by respiration can be used to drive the ATPase. ($\Delta\psi$ has a similar effect but it is not so easy to show this diagrammatically.)

Most mitochondria in tissues and carefully prepared samples of isolated mitochondria show so-called *respiratory control*. This means that if ATP cannot be synthesised because ADP levels are low, respiration slows down, so that NADH is not wasted. If ADP is added, then ATP synthesis can occur and electron transport speeds up. There is a close link between electron transport and oxidative phosphorylation and the mitochondria are said to be *tightly coupled*. The *respiratory control ratio* is defined as

$$\frac{\text{oxygen consumption with ADP}}{\text{oxygen consumption when ADP is used up.}}$$

It can be measured by using oxygen electrodes. A high respiratory control ratio shows that tightly coupled mitochondria are present, but a ratio of 1 shows that the mitochondria are *uncoupled*. Only tightly coupled mitochondria can be used for studies involving phosphorylation, e.g. for measuring ADP/O and P/O ratios.

Uncoupled mitochondria can be obtained by adding an artificial uncoupling agent such as 2,4-dinitrophenol. This renders the membrane permeable, and both $\Delta\psi$ and ΔpH are removed leaving no available power source to drive the ATPase enzyme, so oxidative phosphorylation stops. However, electron transport speeds up because, as respiration occurs, there is no build up of ΔpH and $\Delta\psi$, so the respiratory chain is not moving protons to a region already high in protons or moving electrons to an area already rich in electrons. The effect of uncouplers can be investigated with an oxygen electrode. This is also true of inhibitors of oxidative phosphorylation (e.g. oligomycin) which combine with the ATPase, either in the stalk region (F_0) (Fig. 10.2) or with the active site (F_1). Since the ATPase is not active, ΔpH

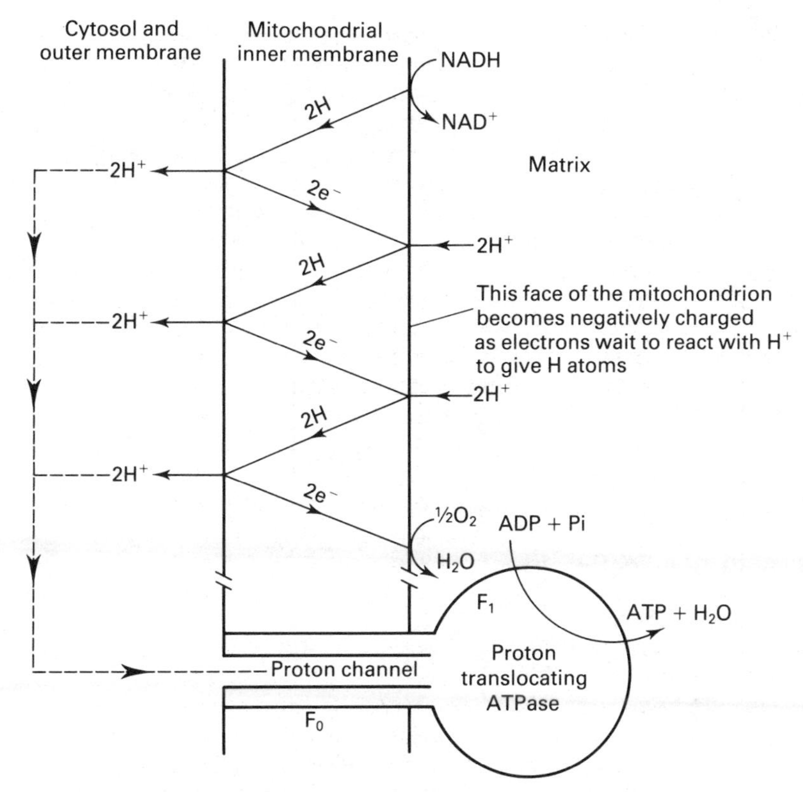

Fig. 10.2 Arrangement of the electron transport chain as a series of loops across the membrane, resulting in ATP formation.

and $\Delta\psi$ build up, slowing down electron transport.

Experiments designed to study the location of carriers within the membrane and to study phosphorylation can be carried out with *vesicles*. These are small fragments of the mitochondrial membrane which contain all the carriers and are in the form of closed sac-like compartments. It is possible to get *inverted vesicles* in which the outer face (of the inner mitochondrial membrane) is now the inner face of the vesicle. The way in which these vesicles interact with redox dyes, inhibitors of electron transport and antibodies to the electron carriers can then be studied. It is also possible to obtain small portions of the mitochondrial membrane that only contain one part of the electron transport chain (e.g. cytochrome oxidase), by treating mitochondria with either a detergent (including bile salts) and/or organic solvent followed by ammonium sulphate fractionation. The particles obtained can then be tested to see which inhibitors and redox dyes interact with them. Only cytochrome oxidase can be monitored with an oxygen electrode since only it uses molecular oxygen directly.

Chloroplasts are capable of *photosynthetic electron transport* (Fig. 10.3).

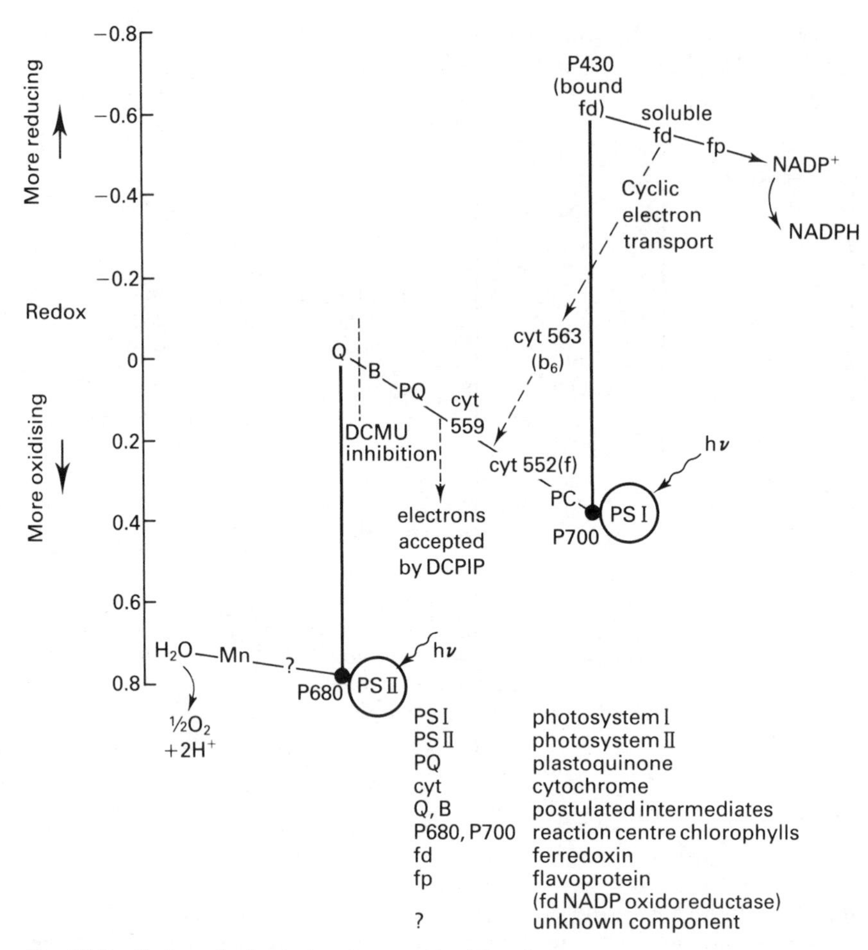

Fig. 10.3 Photosynthetic electron transport in chloroplasts.

At two places in this transport system, light energy is absorbed so that an electron can be moved from one carrier to another with a more negative redox. This is an *endergonic* process and requires the expenditure of energy. Electrons can travel from water to $NADP^+$ via PSII and PSI producing oxygen and NADPH: this is known as *non-cyclic electron transport* and produces ATP by *non-cyclic photophosphorylation*. Electrons can also cycle around PSI. There is then no release of oxygen from water or production of NADPH from $NADP^+$ but ATP is formed. This is *cyclic electron transport* which results in *cyclic photophosphorylation*.

Many aspects of photosynthetic electron transport and photosynthetic phosphorylation are similar to mitochondrial electron transport and oxidative phosphorylation. The same techniques can therefore be used to study chloroplasts. Thus there are inhibitors which act at specific sites in photosynthetic electron transport, e.g. diuron or DCMU (3(3,4-dichlorophenyl)-1,1-dimethylurea) which inhibits on the reducing side of PSII (Fig. 10.3). Redox dyes such as DCPIP (2,6-dichlorophenol indophenol) can interact with the photosynthetic carriers (Section 10.4.3). Artificial donors rather than water can also pass electrons to the photosynthetic electron transport chain (Section 10.4.3).

As in mitochondria, when electrons or hydrogen atoms travel along the series of carriers which span the chloroplast membrane, ΔpH and $\Delta\psi$ form. However, there is a fundamental difference from the situation in mitochondria. Protons move inwards in response to photosynthetic electron transport, not outwards as for respiratory electron transport. This means that the sign of ΔpH and $\Delta\psi$ are reversed when compared with mitochondria. The interior of the chloroplast is considerably more acid than the outside. $\Delta\psi$ is low but tends to be negative on the outside. As a result, reagents used to measure photosynthetic ΔpH and $\Delta\psi$ are not the same as those used for mitochondrial measurement (Section 10.2.3). The chloroplast ATPase is structurally very similar in its general shape and subunit structure to the mitochondria ATPase, but is orientated differently because of the reversal in the direction of ΔpH and $\Delta\psi$; the ATPase is on the outer face of the membrane, not the inner face.

Photosynthetic uncouplers such as NH_4^+ are known which have similar effects to respiratory uncouplers. They speed up the rate of photosynthetic electron transport but abolish phosphorylation because they carry protons across the membrane. Inhibitors of photophosphorylation are also known which, like the inhibitors of oxidative phosphorylation, combine with the ATPase at a specific site. One example is the antibiotic Dio-9.

The oxygen electrode is particularly valuable for studying photosynthetic electron transport systems. Noncyclic photosynthetic electron transport involves evolution of oxygen which can be monitored in this way. Photosynthetic electron transport in cyanobacteria can also be followed by using an oxygen electrode but other bacteria capable of photosynthesis cannot evolve oxygen. However, the techniques used for measuring ΔpH and $\Delta\psi$ in chloroplasts can be applied to photosynthetic bacteria.

Electron transport processes also occur in *microsomes* which come from the smooth endoplasmic reticulum. They contain a flavoprotein called NADPH-cytochrome P450 reductase and a special cytochrome called P450. Microsomes catalyse the hydroxylation of many kinds of substrates including fatty acids, steroids, squalene, some amino acids and many drugs such as phenobarbitone, amphetamines, morphine and codeine.

10.1.3 Principles of electrochemical techniques

Reference Electrodes Electrochemical methods which involve an electrode which responds to a particular experimental situation by giving a potential, require a second so-called *reference electrode* of constant potential to be present so that the difference between the two can be measured. A reference electrode is required when pH glass electrodes, ion-selective electrodes and redox electrodes are used.

Historically, one of the most important reference electrodes is the *standard hydrogen electrode* which contains an inert metal electrode (e.g. platinum coated with platinum black) in a solution of a fixed concentration of hydrochloric acid (which supplies H^+ ions) and hydrogen is bubbled through at 1 atmosphere pressure allowing the following equilibrium to be established:

$$\tfrac{1}{2}H_2 \rightleftharpoons H^+ + e^-$$

However, this electrode is highly inconvenient to use because the hydrogen has to be generated at a constant pressure and must be oxygen free and the platinum black is also readily contaminated. In practice, therefore, although oxidation reduction potentials are expressed relative to the standard hydrogen electrode, other reference electrodes are used.

Calomel electrodes (Fig. 10.4a) are a commonly used type of reference electrode. They consist of a solution of mercurous chloride (calomel) and

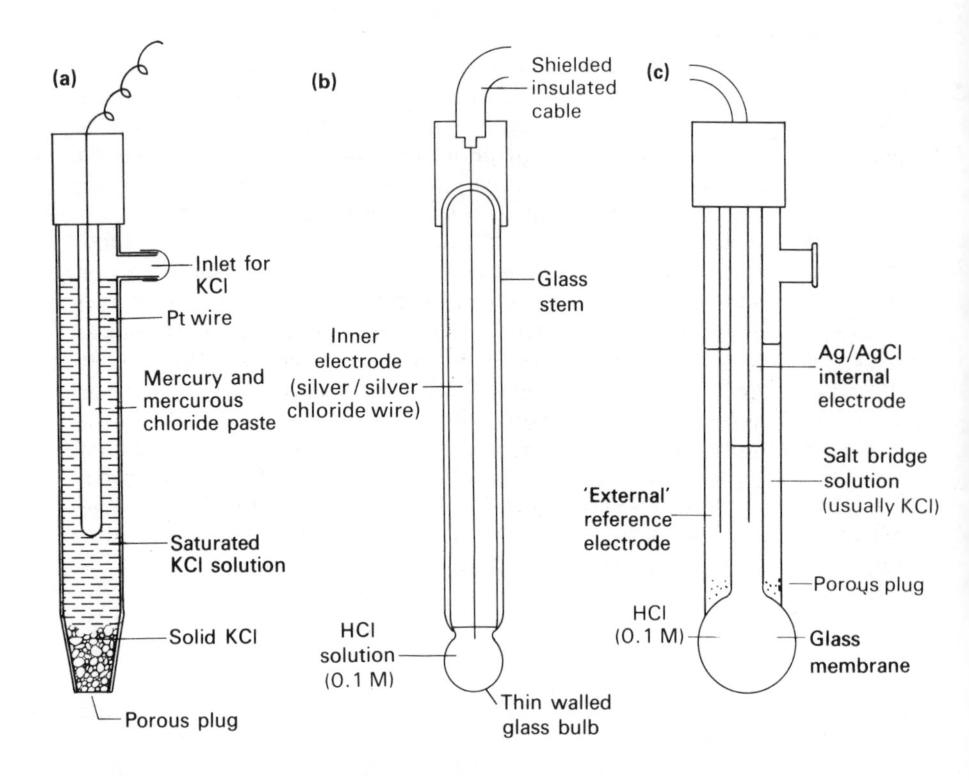

Fig. 10.4 Electrodes: **(a)** a calomel reference electrode; **(b)** glass electrode; **(c)** a combination electrode.

potassium chloride in contact with solid mercurous chloride and mercury. This part of the circuit can be written as:

$$\text{Hg} \;\Big|\; \text{Hg}_2\text{Cl}_2 \;\Big|\; \begin{matrix}\text{saturated}\\ \text{KCl}\end{matrix} \;\Big\|\; \text{test solution}$$

The double lines indicate the presence of a salt bridge.

An alternative to the calomel electrode is the *silver/silver chloride electrode*. A deposit of silver chloride is present on metallic silver in a chloride solution, e.g. KCl.

$$\text{Ag} \mid \text{AgCl} \mid \text{KCl} \parallel \text{test solution}$$

Another type of reference electrode based on redox is found in the *Ross pH electrode* and it is designed to cope with measurements in extremes of temperature.

Any reference electrode must be in contact with the test solution via a *liquid junction*. These generally involve potassium chloride which slowly diffuses out of the electrode and this gives the continuity. Unfortunately, the liquid junction is likely to give an unknown *junction potential* which cannot be completely eliminated. In using liquid junctions, care has to be taken to ensure that the potassium chloride solution diffuses out slowly rather than the test solution diffusing in. Although outward diffusion does involve some contamination of the sample, this is normally not important, but if either potassium or chloride is being measured, then a specially designed reference electrode called a *double junction reference electrode* must be used to prevent contamination.

There are several types of junction through which the KCl diffuses: *ceramic or fritted material, fibrous junction* and *sleeve junction*. Fritted material is a collection of small particles pressed closely together allowing some of the filling solution to leak through the gaps between the particles. Fibrous junctions can consist of woven fibres or of straight fibres, the latter giving an increased flow. The sleeve type reference electrode has a narrow ring-shaped junction formed by a gap between an outer sleeve and the inner body of the electrode. The space between the sleeve and the electrode widens above the tip and forms a reservoir for the fluid. Flow occurs in some areas of the narrow ring junction but not others. A sleeve junction is easier to clean than the other types (because the sleeve can be removed): it is also faster flowing which means it is less likely to get clogged.

The Nernst Equation This equation is relevant to equipment which produces a potential, e.g. pH glass electrodes and ion-selective electrodes. It describes electrode behaviour and can be expressed in the form:

$$\text{E} = \text{E}_x + 2.303 \frac{RT}{nF} \log_{10} \text{A} \qquad 10.2$$

or in a simplified form as:

$$\text{E} = \text{E}_x + \text{S} \log_{10} \text{C} \qquad 10.3$$

where E = the total potential (mV) developed between the sensing and the reference electrodes

E_x = a constant that depends mainly on the reference electrode

$2.303\frac{RT}{nF}$ (or S) = the *Nernst factor* or *slope*

n = the number of charges on the ion

A = the activity of the ion

C = the concentration of the ion

Activity is an important physico-chemical concept. It is the true measure of an ion's ability to affect chemical equilibria and reaction rates and is its *effective* concentration in solution. The relationship between activity (A) and concentration (C) is $A = \gamma C$ where γ is the *activity coefficient*. At low concentrations, the activity and concentration are equal. Equation 10.3 therefore applies only in dilute solution. At higher concentrations γ becomes less than 1 and hence activity less than concentration.

The Nernst equation (equation 10.2) shows clearly that the electrode response depends on both temperature and the number of charges on the ion. It can be shown that at 25°C, there is a 59.16 mV change for a ten-fold change in the activity of a monovalent ion, and a 29.58 mV change if the ion is divalent. An electrode is said to have Nernstian characteristics if it obeys Nernst's law. If the changes in potential relative to activity (slopes) are less than the theoretical values (in the operating range), then it indicates either interference from other ions or an electrode malfunction.

10.2 Measurement of pH by glass electrodes

10.2.1 Principles of operation

The most convenient and accurate way of measuring pH is by using a *glass electrode*, the action of which depends on ion-exchange in the hydrated layers formed on the glass surface. Glass consists of a silicate network amongst which are metal ions coordinated to oxygen atoms, and it is the metal ions which exchange with H^+. The glass electrode acts like a battery whose voltage depends on the H^+ ion activity of the solution in which it is immersed. The size of the potential (E) due to H^+ ions is given by the equation:

$$E = 2.303\frac{RT}{F}\log\frac{[H^+]_i}{[H^+]_o} \qquad 10.4$$

where $[H^+]_i$ and $[H^+]_o$ = the molar concentrations of H^+ ions inside and outside the glass electrode.

In practice $[H^+]_i$ is fixed and is generally equal to 10^{-1} since the electrode contains 0.1M HCl. Since $pH = -\log[H^+]$, it follows that the developed potential is directly proportional to the pH of the solution outside the electrode.

Glass electrodes are particularly useful because of the lack of interference from the components of the solution. On the whole, these electrodes are not readily contaminated by molecules in solution, and if other ions are present they do not cause significant interference. However, in very alkaline solutions they do respond to sodium. Inaccuracies also occur in very acid solutions.

A glass electrode (Fig. 10.4b) contains a thin glass membrane which is at the end of a glass tube or sometimes an epoxy body. Also present in the glass electrode is an internal reference electrode of the silver/silver chloride type (Section 10.1.3). A thin film of silver chloride is present on silver and immersed in 0.1M HCl. This internal electrode gives rise to a steady potential. The varying potential of the glass electrode has to be compared with a steady potential produced by an external reference electrode. This reference electrode can either be a separate probe (Fig. 10.4a) or built around the glass electrode giving a *combination electrode* (Fig. 10.4c). If a combination electrode is used, the level of the test solution must be high enough to cover the porous plug (liquid junction) but not so high as the level of the salt bridge solution (KCl) in the external electrode since it is essential for the KCl to slowly diffuse into the test solution.

Whatever reference electrode system is used, the measured voltage is the result of the difference between that of the reference and glass electrodes. In practice, however, there are other potentials present in the system. These include the so-called *asymmetry potential* which is poorly understood but which is present across the glass membrane even when the H^+ ion concentration is the same on both sides. Also included are the potentials due to the Ag-AgCl and to the liquid junction to the reference electrode which gives a potential because K^+ and Cl^- ions do not diffuse at exactly the same rate and therefore generate a small potential at the boundary between the sample and the KCl in the reference electrode.

A pH electrode is used in conjunction with a *pH meter*. This records the potential due to the H^+ ion concentration but is designed to take little current from the circuit. A large current flow would cause changes in ion concentration and hence pH changes; this is prevented by having a high resistance present.

The measured potential, E_G', of a glass electrode at 25°C is given by:

$$E_G' = E_{constant} + 0.059 \text{ pH} \qquad 10.5$$

where $E_{constant}$ is related to the various constant potentials present in the system.

As explained in Section 10.1.3, at 25°C there is a 59.16 mV change for a ten-fold change in a monovalent ion; this means for a change of 1 pH unit, there will be a 59.16 mV change. The pH meter, glass electrode and reference calomel electrodes are designed so that pH 7 gives a zero potential.

It can be seen from equation 10.2 that the potential produced is dependent on temperature (each pH unit change represents 54.20 mV at 0°C, and 61.54 mV at 37°C). This effect is entirely predictable and can be compensated for. The further away from pH 7 (the *isopotential point* where temperature has no effect on potential) the more important it is that this temperature compensation is applied accurately because of the accumulation of errors. Before pH measurements are taken, the equipment is standardised using two buffer solutions of widely differing pH. If the buffer solutions and the solution of unknown pH are at a different temperature then an error will occur.

Recently the Ross pH electrode has been developed. This gives a fast response with no drift and is very accurate, even in hot or cold solutions. Instead of using silver chloride or calomel electrodes which have a solid in equilibrium with a liquid, redox reference cells are used where both oxidised and reduced components are already in solution and the temperature coefficient can be reduced to almost zero. It can stay calibrated even after measuring boiling or iced samples, and is therefore useful for quality control in processes involving extreme temperatures.

It is important that the outer layer of glass on the glass electrode remains hydrated and so it is normally kept immersed in solution. The thin glass membrane is fragile and care must be taken not to break or scratch it or to cause a build up of static electric charge by rubbing it. Gelatinous and protein-containing solutions should not be allowed to dry out on the surface of the membrane as they would inhibit the response.

Electrodes in a variety of shapes and sizes are available for different applications. These include electrodes for the measurement of the pH of blood, the mouth, flat moist surfaces such as isoelectric focusing gels, the interior pH of solid samples and equipment for field work. Intracellular pH can also be measured using miniature probes.

10.2.2 The pH-stat

This is a form of automatic titrator which can be used to maintain a constant pH during a reaction which involves the production or removal of protons. The rate of the reaction can therefore be determined since a recorder draws a curve of the amount of reagent added, against time. A glass electrode, pH meter, recorder, controller and burette are necessary; the burette can either be an ordinary burette with a magnetic valve or, better still, a motor driven burette syringe. A controller is necessary in these circuits as it breaks the current to the burette motor when the end point is reached. The best sort of automatic titrators are arranged so that less and less reagent is added as the end point is approached. This avoids the danger of overshoot. The kind of pH glass electrode used should be very accurate and stable for kinetic work.

The pH-stat has some limitations in that, for example, the solution in the reaction vessel has to be constantly stirred and this may cause denaturation of protein and introduce carbon dioxide into solution. Other problems are the existence of an unknown junction potential and the tendency of the liquid from the burette tip (under the surface of the solution) to leak. The latter effect can be partly counteracted by making the density of the burette liquid lower than that of the reaction mixture.

The main use of the pH-stat is in studying enzyme kinetics, especially in the case of proteolytic enzymes (Section 3.4.7). It can also be used to follow degradation of peptides and nucleic acids, for bacteriological studies, and industrial bulk processes, e.g. penicillin production.

10.2.3 Measurement of ΔpH and $\Delta\psi$

These are often estimated either by using a *molecular probe* which responds to $\Delta\psi$ (or ΔpH) by showing an absorption change or fluorescence or by studying the *accumulation* of a compound which depends on $\Delta\psi$ (or ΔpH).

Measurement of $\Delta\psi$ Membrane potentials are often estimated using a fluorescent ion such as anilinonaphthalene sulphonate (ANS^-) (Fig. 8.9) which acts as a probe, but there are uncertainties about the mechanism of the probe's response to membrane potential. The probe must have the correct charge to respond (a cation if the interior is negative, or an anion if the interior is positive as in chloroplasts (Section 10.1.2)).

The movement of an ion in response to $\Delta\psi$ can be measured using an ion-selective electrode (Section 10.3). For example, inward movement of K^+ in mitochondria can be followed by a drop in external K^+ concentration. Valinomycin is an ionophore, i.e. a compound which carries an ion across a membrane and is used to transport the ion in response to $\Delta\psi$. K^+ ions are then distributed on either side of the membrane according to the appropriate form of the Nernst equation:

$$\Delta\psi = -2.303\frac{RT}{nF}\log\frac{[K^+]_i}{[K^+]_o} \quad 10.6$$

where i refers to inside and o to outside
$n = 1$ because K^+ is a monovalent ion

The K^+ ion gradient can be calculated from the fall in external K^+ and the volume of the matrix into which K^+ is being accumulated. Synthetic ions which can permeate bilayers can be used to replace K^+ and valinomycin; some of these can be made radioactive and used in the isotopic determination of $\Delta\psi$.

A method which does not involve separation of the organelle (or vesicle) from the medium is *flow dialysis*. This method uses a small flow cell consisting of two compartments separated by a membrane. In the upper compartment, the following are added: a radioactively-labelled permeant ion, a substrate and a buffer. In the experiment, organelles or vesicles are also added to the upper compartment, but a control experiment must also be carried out omitting them. Buffer is slowly passed through the lower compartment, collected in fractions and the amount of radioactive label measured in each fraction. If there is no biological material in the upper compartment, then the label will quickly diffuse into the lower compartment and appear in the early fractions. Organelles or vesicles will slow down the rate of diffusion because the label enters them. Detergent or uncouplers are added to check how much label has actually entered the organelles or vesicle due to $\Delta\psi$, and how much has been bound to proteins and is not therefore a response to $\Delta\psi$. From knowledge of these various measurements, $\Delta\psi$ can then be calculated. Carotenoids respond to $\Delta\psi$ changes by producing changes in light absorption. Photosynthetic organisms have carotenoids in the photosynthetic membranes where they act as natural $\Delta\psi$ probes showing changes in the 515 to 520 nm region of the spectrum. The response to $\Delta\psi$ is very rapid (nano seconds or less) and this enables the primary events of the light reactions of photosynthesis to be studied. Other artificial optical indicators of $\Delta\psi$ can be used but they are not embedded in the membrane.

Measurement of ΔpH Historically ΔpH was obtained using glass pH electrodes or pH internal indicators. Nowadays the more usual method of measuring ΔpH is by following the uptake of a radioactively-labelled weak acid (HA) for respiratory systems or weak base (B) for photosynthetic systems. The ionisation of these electrolytes is pH dependent in accordance with the Henderson-Hasselbalch equation (Section 1.2.2). Only the un-ionised form of the acid or base (HA or B respectively) can move through the membrane by passive diffusion. The accumulation of radioactivity on the opposite side of the membrane is therefore dependent upon the difference in pH between the external and internal environments, i.e. on ΔpH. In practice, the organelle is exposed to a solution of the labelled compound and then quickly separated either by high speed centrifugation or by centrifuging the organelle through silicone oil or by filtration through a cellulose filter. The amount of radioactivity is then measured. Some of the label will result from external contamination rather than that entering the vesicle or organelle. This can be determined and subtracted using a second non-permeant radioactive label. The amount of radioactive label which has entered is now known and the concentration of that label can be found by determining the internal water volume. For measurements of the internal water volume, [^{14}C] sucrose is used as an impermeable solute and tritiated water as the label which enters the volume. The external pH must also be measured using either an electrode or an externally situated pH indicator.

Sometimes instead of the weak bases being radioactively labelled, fluorescent molecules are used. Fluorescent amines such as 9-aminoacridine are quenched (Section 8.5) when absorbed and the extent of quenching can be used to determine the uptake of the base. Flow dialysis can also be used to detect ΔpH using radioactively-labelled acids and amines by a similar technique to that described for $\Delta\psi$ measurements. ^{31}P NMR may also be used to measure ΔpH by exploiting the fact that the phosphorus signal from phosphate alters as the neighbouring oxygen atoms gain or lose protons in response to pH (Section 8.9.3). Thus there are various ways of measuring ΔpH. In practice several methods are generally used because accurate and unambiguous results by any one method are difficult to obtain.

It is worthy of comment that for both ΔpH and $\Delta\psi$ measurements, several assumptions have to be made. The indicator compound is either assumed not to bind to the membranes at all (or to bind to a known extent), not to precipitate out, not to disturb the gradient it is used to measure, to enter by diffusion rather than by active transport and not to be metabolised. Clearly, it is expedient to check the validity of these assumptions whenever possible.

10.3 Ion-selective electrodes (ISE) and gas sensors

10.3.1 Introduction

The glass pH electrode is really a kind of ion-selective electrode in that it is sensitive to hydrogen ions. Electrodes have also been developed which are sensitive to other ions, for example Cl^-, NO_3^-, NH_4^+ and Na^+. The response is logarithmic; ten-fold changes in ion activity giving equal increments on the meter scale.

The various types of ion-selective electrodes are shown in Table 10.1. All (except gas-sensing) operate on ion movement causing a difference in potential. Modifications of these electrodes have been made, for example in the development of mini- and micro-electrodes. A combination of micro-electronics and ion-selective field effect transistors are called ISFETs and these miniaturised detectors probably will be much used in the future.

An ion-selective electrode responds to the *activity* of an ion. However, if the instrument is calibrated with a standard of known concentration then, provided the ionic strengths of the solutions are similar, the *concentration* of the test solution will be recorded. To ensure that the ionic strengths are similar, an *ionic strength adjustor* (ISA) may be added. ISAs contain a high concentration of ions and sometimes pH adjusters and decomplexing agents or agents to remove species which interfere with the measurement. If, however, some of the ion is not free but exists as a complex or an insoluble precipitate, then ion-selective electrodes will give a much lower reading than a method which detects all of the ions present. Thus atomic absorption spectroscopy (Section 8.7.3) measures concentration, and for an ion such as calcium which readily forms calcium phosphate, it will give a much higher reading than that obtained using ISE. The ISE results are significant, however, because often it is the free ions which are responsible for clinical/biological effects. Unlike measurements by atomic spectrophotometry, ISEs respond over a wide range of concentrations, do not destroy the test compound and are rapid in use. However for clinical use, where high precision is required and where the normal range of blood cations is so small, ISEs are less commonly used than atomic absorption or emission spectrophotometry (see Table 8.2).

An electrode may be ion-selective but not necessarily ion-specific. Manufacturers' instructions will give information about this and will also mention chemicals which can poison the electrode. As with pH glass electrodes, ion-selective electrodes can be fouled by protein forming a surface film.

Many ions can be measured directly by the use of an ISE or indirectly by titration. One form of indirect measurement is the use of the electrode as the

Table 10.1 Types of ion-selective and gas-sensing electrodes.

Operating principle	Composition	Ion or gas detected	Limitations or advantages
Glass ISE (also called rigid matrix)	Special glass e.g. more alumino silicate groups	Na^+	Not for anions or divalent cations
Liquid ion-exchange electrodes (liquid membrane)	An organic liquid ion-exchanger specific for that ion, a thin porous hydrophobic membrane	Ca^{2+}	Limited lifetime because ion-exchanger is lost
Neutral carrier liquid membrane electrodes	An organic solution of an ion specific complexing agent (ionophore) in an inert polymer matrix	K^+ (valinomycin as the ionophore)	Limited lifetime
Solid state electrodes	A crystal or solid as part of a membrane	Ag^+	Long lifetime because no liquid involved
Gas-sensing	Varies but includes a gas-permeable but ion-impermeable membrane	NH_3	Slower response

endpoint indicator of a titration. The electrode can be sensitive either to the species being determined or the titrant ion. Titrations are ten times more accurate than direct measurement since the procedure requires accurate measurement of a *change* in potential rather than the absolute *value* of a potential. This is an important point because ISEs are not intrinsically very accurate. For example, the determination of the concentration of calcium ions in solution is best carried out by titrating the solution with ethylenediaminetetraacetic acid (EDTA), which is a strong complexing agent for calcium, using a calcium electrode. Since the electrode responds to the logarithm of the concentration of calcium ions in solution, as the EDTA is added, a sharp end point is observed giving a precision of 0.1% or better.

At 25°C the potential of an ISE changes 59.16 mV for a ten-fold change in the activity of a monovalent ion and 29.58 mV for a divalent ion (Section 10.1.3). As with pH electrodes, the potential produced is temperature dependent, except at the *isopotential point* (which varies depending on the type of electrode). It is therefore important that temperature compensation is used. A reference electrode is also needed so that the varying potential of the ISE can be compared with the steady value provided by the reference electrode (Section 10.1.3). If either potassium or chloride ions are being measured, then a double junction reference electrode is needed to prevent contamination of the solution by the internal solution of the reference electrode.

10.3.2 Uses of ion-selective electrodes

Ion-selective electrodes are easy to use, economical, easily transportable, capable of continuous monitoring without hazard and require little power. Because of all these advantages, they are widely used as shown in Table 10.2. Miniaturised electrodes called *spearhead microelectrodes* have been manufactured and are used to determine the ion contents of single cells, muscles and nerves.

Table 10.2 Uses of ion-selective electrodes (ISE).

Ion or gas detected	Application
Na^+	Analysis of seawater, serum, soil, skin
K^+	Analysis of serum (often combined with Na^+ electrode)
Cl^-	Quick test for cystic fibrosis; food analysis
Ca^{2+}	Analysis of serum, beer
NH_3	Analysis of solutions produced by Kjeldahl digestion of proteins (Section 3.2.2)
NO_3^-	Analysis of drinking water, fertilisers, microbial growth
Nitrogen oxides	Air pollution checks

10.4 Oxidation-reduction (redox) potentials

10.4.1 Principles

As was discussed in Section 10.1.2, compounds capable of existing in an oxidised and a reduced form can be represented by the general equation:

$$\text{reduced form} \rightleftharpoons \text{oxidised form} + ne^-$$

A mixture of the reduced and oxidised form of a substance (e.g. Fe^{2+}/Fe^{3+}; $NADH/NAD^+$) is known as a *redox couple*. If an inert electrode (such as platinum) is put into a solution of a redox couple, the metal will become charged, and the potential difference set up between the metal and the solution can be compared with the steady potential produced by an external reference electrode.

The scale of oxidation-reduction (redox) potentials is based on values obtained if the standard hydrogen electrode were to be used as reference and compared with the standard potential (E_0), produced by platinum dipping into equal concentrations of the oxidised and reduced forms of a substance at unit concentration (Section 10.1.3). Redox couples have positive or negative redox potentials depending on whether they are more oxidising or reducing than the standard hydrogen electrode. Thus ferri/ferrocyanide is more oxidising and has a redox potential of 0.36 V whilst $NAD^+/NADH$ is more reducing and has a redox potential of -0.32 V. The experimental potential measured depends on the ratio of oxidised to reduced forms and frequently on pH but does not depend to a great extent on the actual concentration involved. The measured redox potential, E_h, is related to the standard potential, E_0, by equation 10.7 which is a version of the Nernst equation (equation 10.2). This form of the equation applies if H^+ is not involved in the change from oxidised to reduced form:

$$E_h = E_0 + 2.303\frac{RT}{nF}\log\frac{[\text{oxid}]}{[\text{red}]} \qquad 10.7$$

where E_h = the measured redox potential of a couple of known composition (e.g. a mixture of 0.03 M oxidised form and 0.1 M reduced form)

E_0 = the standard redox potential with components at 1 M concentration at pH 0.

If H^+ is involved in the equation (i.e. the couple generates a pH change) then the equation becomes:

$$E_h = E_0 + 2.303\frac{RT}{nF}\log\frac{[\text{oxid}]}{[\text{red}]} + 2.303\frac{RT}{nF}\log\,[H^+]^a \qquad 10.8$$

where a = the number of protons involved in the reaction

In practice, E_0' values are most useful for biologists: they are similar to E_0

but the pH is not 0, and if not stated otherwise, is taken to be 7. A redox couple can theoretically be oxidised by a couple with a more positive E_0' than itself and it will oxidise a couple with a more negative E_0'. The final equilibrium is determined by the difference between the two potentials. The free energy charge in a coupled oxidation-reduction reaction is related to the difference in the redox potential of the couple and to the number of electrons involved (equation 10.1). It is difficult to predict the outcome of an interaction in a living cell of two redox couples whose E_0' are very similar since this may depend on the conditions within the cell. Factors which may influence the interaction are pH, relative concentrations of the molecules or ions and the presence of chelating agents.

10.4.2 Potentiometric titrations of oxidation-reduction reactions

The reduced form of a redox couple can be oxidised by the addition of a suitable oxidising agent, and the potential difference measured. A graph of potential difference against percentage oxidised form gives the relationship shown in Fig. 10.5. A titration curve of the reduction of the oxidised form would give a mirror image of Fig. 10.5. The actual shape of the graphs depends on whether a one electron change (curve A) or a two electron change (curve B) is taking place. Sometimes two electrons are donated in two separate steps e.g. quinones which have a stable intermediate, called semiquinones.

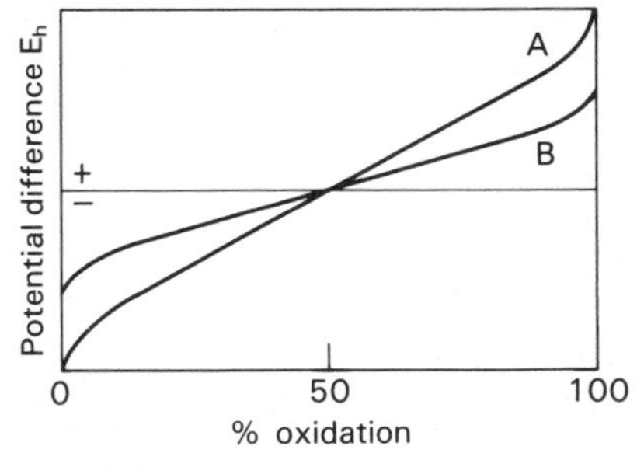

Fig. 10.5 Potentiometric titration curves – oxidation of the reduced form. Curve A is of a one electron change; curve B of a two electron change.

10.4.3 Oxidation-reduction indicators (redox dyes) and their uses

Most oxidative reduction indicators are brightly coloured when oxidised and colourless when reduced, exceptions are the tetrazoliums and viologens. Examples are shown in Table 10.3. These indicators can be used to detect the redox potential of a particular solution or may be used as electron donors or

Table 10.3 Redox indicators (artificial electron acceptors).

Indicator	E_0' value (V)	Disadvantages	Special uses
Methyl viologen	− 0.45	Autooxidises	Used to measure the rate of electron transport in chloroplast PSI
Benzyl viologen	−0.36	Autooxidises	As methyl viologen
TTC (2,3,5-triphenyl tetrazolium chloride)	−0.08	Reduction very pH dependent Light sensitive	Used to measure the rate of mitochondrial and yeast respiratory electron transport. Also in histology, used to locate respiratory electron transport
Methylene blue	+ 0.01	Autooxidises	Determination of bacterial contamination of milk
PMS (phenazine methosulphate)	+ 0.08	Rates depend on dye concentration. Can inhibit. Autooxidises; light sensitive. Interacts at several places in electron transport	Used to stimulate cyclic photophosphorylation in chloroplasts
DCPIP (2,6-dichlorophenol indophenol)	+ 0.22	Changes colour below pH 4	Used to measure the rate of electron transport in chloroplast PSII
Potassium ferricyanide	+ 0.36	Interacts at several places in electron transport.	Used to measure the rate of electron transport in chloroplast PSII, and to measure NADH dehydrogenase and succinate dehydrogenase activities in enzyme preparations

acceptors, in which case the rate of a reaction (an oxidation or a reduction) can be followed.

The rate of reduction of dye as measured in a spectrophotometer can also be used as the basis of an enzyme assay (Section 3.4.2), for example succinate dehydrogenase isolated from mitochondria. The electron transport processes of chloroplasts, mitochondria and bacteria can also be studied using indicator dyes. Provided they have an appropriate oxidation-reduction potential the electrons are accepted by the indicators, instead of being passed to the next electron carrier in the chain.

Unfortunately the majority of indicator dyes are not very specific and may receive electrons from several points in the electron transport chain. Moreover, some reduced dyes become readily reoxidised if oxygen is present; this is called *autooxidation*, and means that such dyes can only be used in anaerobic conditions, e.g. in Thunberg tubes. Care must be taken to interpret correctly the results of experiments using redox dyes because the dyes have been known to influence the nature of the reactions taking place and can inhibit enzymes or act as poisons for microorganisms. pH changes can also cause a change in colour or in the ease of reduction. There is also the problem that the dye may not be readily able to cross membranes and may, therefore, not be able to reach the appropriate subcellular site. However, in the case of organelles or artificial vesicles the fact that many dyes cannot penetrate the membrane can be exploited. Thus the extent of interaction of an electron transport component within the membrane and the external solution containing redox dye gives an indication as to which surface of the membrane the component is situated on. The experiment can then be repeated with inverted vesicles, so the other face of the membrane is then exposed to the dye (Section 10.1.2). The sub-cellular location of enzymes can be ascertained by using carefully prepared tissue slices and staining them with dyes (a histochemical technique). Tetrazolium chloride is frequently used for this purpose because it gives an insoluble precipitate which does not readily diffuse from its site of formation.

Use of Artificial Electron Donors These are useful for investigating electron transport in that they can bypass the site of action of inhibitors or restart electron transport when an essential component has been removed. The site of action of the inhibitor or the component can then be accurately pinpointed. The reduced form of a redox couple can be used as an electron donor. For example in studies with chloroplasts, manganous ions can replace water as the electron donor to chloroplasts. Similarly ascorbate is often used as an artificial donor, but is usually used in conjunction with another compound. In chloroplast work, for example, it has been found that ascorbate alone or ascorbate with phenylenediamine can replace water as the electron donor to PSII. In contrast, ascorbate with 2,6-dichlorophenol indophenol donates electrons just before PSI, i.e. at a totally different site. In mitochondrial work, ascorbate is often used with cytochrome *c* or with *N,N,N′,N′*-tetramethyl-*p*-phenylenediamine (TMPD): in both cases electrons are donated to cytochrome *c*.

10.5 The oxygen electrode

10.5.1 Principles

The reduction of oxygen at a cathode gives rise to a current which is proportional to the oxygen tension in the solution, provided that a voltage of 0.5 to 0.8 V is applied across the electrodes. Using a platinum or silver electrode in conjunction with a silver-silver chloride anode, four electrons are generated at the anode which are used to reduce a molecule of oxygen at the cathode. The oxygen tension at the cathode then drops to zero and this acts as a sink so oxygen diffuses towards it to make up the deficit. The reactions that occur are:

$$O_2 + 2H_2 + 2e^- \longrightarrow H_2O_2 + OH^-$$
$$H_2O_2 + 2e^- \longrightarrow 2OH^-$$

10.5.2 Types of oxygen electrode

Commercial oxygen electrodes differ considerably in their design but basically there are four types in use.

Open Electrode Consists of a platinum or gold wire exposed to the reaction mixture. In some forms, the sensitivity is increased by making the electrode rotate or vibrate, such electrodes being particularly useful for the measurement of very rapid oxygen changes during enzymic reactions. However, these electrodes are readily poisoned by such chemicals as ferricyanide, cyanide, ascorbate and indophenol dyes which are frequently used in experiments on respiration.

Recessed Microelectrode Consists of a platinum wire set in glass or epoxy resin such that only its end is exposed to the reaction mixture, which is present in a small funnel-shaped recess in the end of the glass or resin. A deep recess protects the electrode but makes it slow to respond. It has therefore been mainly replaced by the *Clark electrode*. Since the area of exposed electrode is small, the consumption of oxygen is low, but due to the nature of the recess, the electrode is only suitable for the measurement of oxygen concentrations which are not rapidly changing such as those in blood or plasma.

Clark Electrode Normally consists of a platinum cathode and a silver anode both immersed in the same solution of concentrated potassium chloride and separated from the test solution by a membrane. Oxygen diffuses from the solution in the reaction vessel through the membrane into the electrode compartment where it gives rise to a current. This type of electrode is the one most commonly used to study biochemical reactions, since the membrane prevents the electrodes from being contaminated by chemicals present in the test solution. It thereby overcomes one of the limitations of the open electrode. It has the slight disadvantage that the presence of the membrane

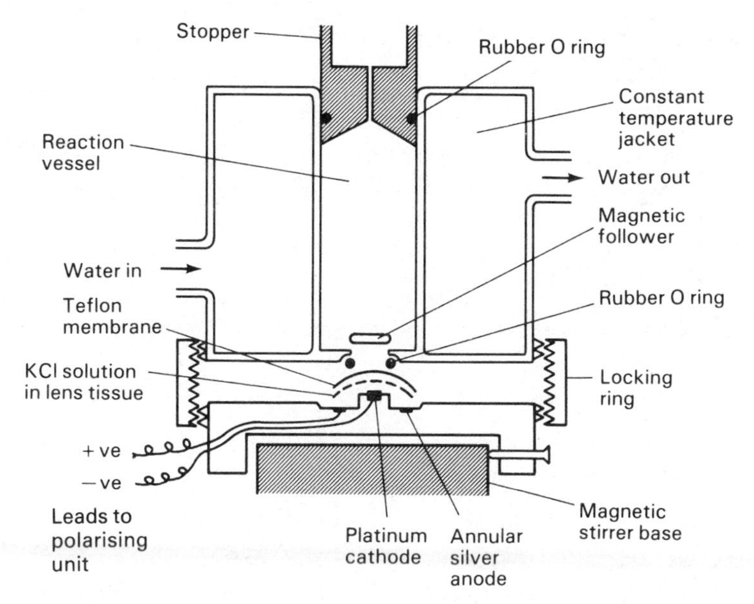

Fig. 10.6 Diagram of a dismantled Rank electrode.

lengthens the response time. The current produced depends on the cathode area and the amount of oxygen reaching the platinum cathode. There are many variations of the Clark electrode, the *Rank electrode* being one of the most commonly used (Fig. 10.6); probe-type oxygen electrodes are also used (Fig. 10.7).

Galvanic Probe Such a probe as a *Hersch cell* has a lead anode and a silver cathode, and unlike the other types of oxygen electrode, gives rise to an e.m.f. It is therefore a self driven probe and does not need an externally imposed voltage. The following reactions occur:

$$\text{Anode} \quad 2Pb + 6OH^- \longrightarrow 2PbO_2H^- + 2H_2O + 4e^-$$

$$\text{Cathode} \quad O_2 + 2H_2O + 4e^- \longrightarrow 4OH^-$$

This type of probe is readily sterilised and so is useful in industry, especially in monitoring fermentations, but it has the disadvantage of responding more slowly than a Clark electrode.

10.5.3 Operation of a Clark electrode (Rank electrode)

Teflon (12μm thick) is the usual choice for the membrane, but any other oxygen-permeable material may be used, including Cellophane, polythene, silicone rubber and 'Cling Film'. Care must be taken to ensure that the membrane does not become contaminated, for example it should not be touched by hand; creasing or twisting can also cause problems. Thinner

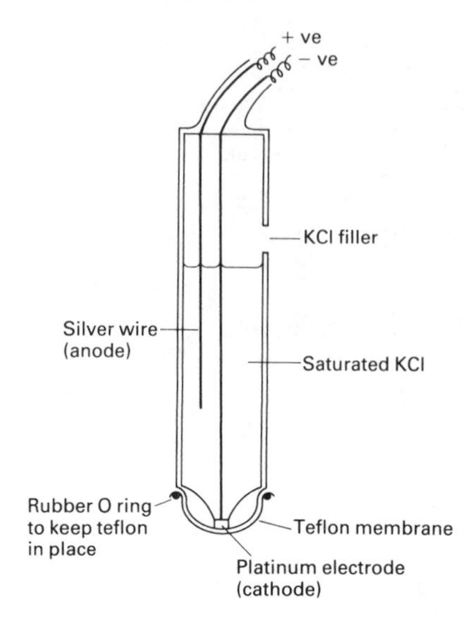

Fig. 10.7 A probe-type Clark oxygen electrode.

membranes give a quicker response but are more fragile. The membrane covers the electrodes and allows oxygen to diffuse into them while preventing other reaction ingredients from poisoning them. The electrodes are in electrical continuity with one another via potassium chloride solution, present in a piece of lens tissue, which holds the solution in place and which also makes its easier to exclude air bubbles from the electrode compartment. Electrodes should be clean and bright and if they are contaminated they can be cleaned with dilute ammonium hydroxide. When the membrane has been changed, several minutes should be allowed for the electrode to give a steady response.

The oxygen electrode is mounted on a stirring motor and a magnetic flea inserted into the reaction vessel. It is important that the solution is stirred as the platinum electrode uses up oxygen, and an artificially low reading will result if the stirring should stop. (Response time can be checked by restarting the stirrer and noting the time required for the current to return to its previous level.) Temperature control of the reaction vessel is necessary, so that reactions can be carried out in the correct conditions. The solubility and rate of diffusion of oxygen are both temperature dependent; also the action of the magnetic stirrer may cause a temperature rise. Water from a temperature controller is therefore pumped through the outer compartment of the oxygen electrode. During experiments, care should be taken to ensure that this temperature is reached by the reagents before the experiment is commenced; if ice cold solutions are pipetted into the reaction vessel and insufficient time allowed for the reagent to reach the required low temperature, incorrect results will be obtained. Calibration should be carried out at the same temperature as the experiment.

It is essential to have some mechanism of setting the correct voltage between the electrodes and also for adjusting the current for 0% and 100% oxygen value. In the case of the Rank oxygen electrode, this can be done by a unit called a polarising module. A recorder is also needed so that a continuous trace of oxygen content can be obtained. One hundred per cent oxygen can be set with a gain (sensitivity) control using distilled water which has stood in air at the temperature of the reaction vessel for several hours. The concentration of oxygen present at that temperature and pressure, which must be known for subsequent calculations, can be found in scientific tables, (e.g. *Handbook of Chemistry and Physics*, Weast, 1984). A more accurate calibration of the instrument can be obtained by using mitochondrial fragments to oxidise a known amount of NADH, the amount of this having been accurately determined by spectrophotometric means (Section 3.5). Zero per cent oxygen can be achieved either by adding a few crystals of sodium dithionite which removes oxygen from solution, or, biologically, by using yeast respiring sugars, or physically by bubbling nitrogen gas through the solution. At 0% oxygen, there should be no current flowing, but small faults in insulation of the platinum electrode may mean that a leakage current is present.

The calibration solutions are removed from the reaction vessel (usually by a suction pump) and the experimental samples pipetted in. The stopper is then put in position, so that oxygen from the air cannot enter. No air bubbles should be present. While the experiment is in progress and any change in oxygen content is being recorded, small additions (e.g. of inhibitor) can be made to the reaction vessel by a Hamilton syringe through the small hole in the stopper. Many chemicals are adsorbed onto the surface of the membrane and reaction vessel; hence it is important that the apparatus is thoroughly cleaned after each experiment. In addition, care must be taken when organic solvents are present in the reaction vessel since they may give an incorrect response. If the oxygen electrode is going to be reused without changing the membrane, then it is essential to leave distilled water in the reaction vessel to prevent the equipment drying out.

10.5.4 Applications of oxygen electrodes

Due to their ability to give a continuous trace, oxygen electrodes have largely replaced manometric techniques (Section 1.10) for studying reactions involving oxygen uptake and evolution.

Mitochondrial Studies The study of respiratory control and the effect of various inhibitors on mitochondrial respiration and the measurement of ADP/O ratios are best measured by the oxygen electrode (Section 10.1.2). As shown in Fig. 10.8 a slow rate of respiration occurs in tightly coupled mitochondria until ADP is added which starts ATP production and speeds up the rate of electron transport. (The slope of the oxygen electrode trace increases.) The phosphorylating mitochondria are then said to be in state 3 (an active state) and in the diagram, this is expressed as rate X. When the

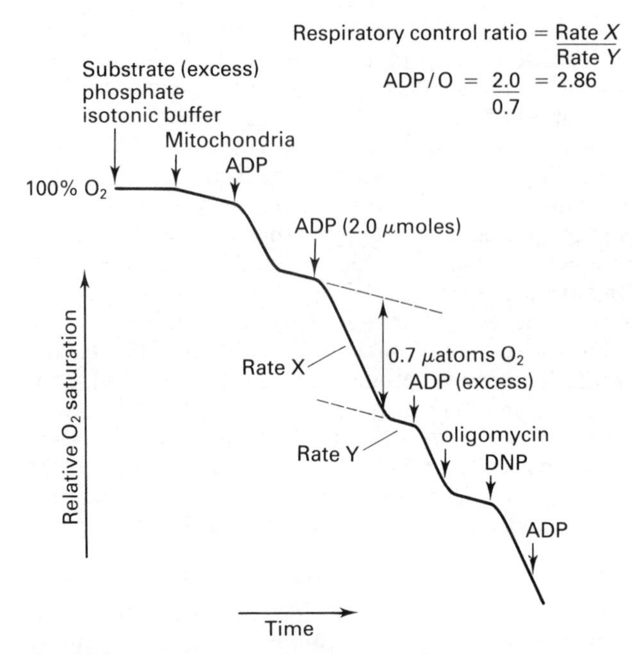

Fig. 10.8 A typical experimental trace of oxygen consumption for intact mitochrondria obtained using an oxygen electrode.

mitochondria have used up the ADP they can no longer phosphorylate and this causes a reduction in the rate of electron transport (rate Y). Mitochondria are then said to be in state 4 (an inactive state). The respiratory control ratio (shown as ratio Rate X/Rate Y in Fig. 10.8) is a measure of the extent of coupling of respiration and phosphorylation. If an inhibitor of phosphorylation, e.g. oligomycin, is added when ADP is present, then the rate of respiration slows down as shown in Fig. 10.8. If an uncoupler, e.g. 2,4-dinitrophenol (DNP), is added the rate of respiration speeds up and ADP has no effect in increasing the rate further (Fig. 10.8).

Inhibitors of electron transport can also be studied using an oxygen electrode. When used with electron donors such as succinate or tetramethylphenylenediamine (TMPD), the sites of inhibition can be determined (Fig. 10.1). A typical result is shown in Fig. 10.9.

Chloroplast Studies Oxygen evolution in cyanobacteria, algae, chloroplasts and chloroplast fractions enriched in photosystem II can be studied using a suitably illuminated Clark oxygen electrode. The oxygen content of the suspension medium is normally reduced below 100% oxygen by bubbling through nitrogen. This ensures that the oxygen produced stays in solution and is recorded.

Clinical Uses An early clinical use of oxygen electrodes was in monitoring heart-lung machines during open heart surgery. Because of their speed of

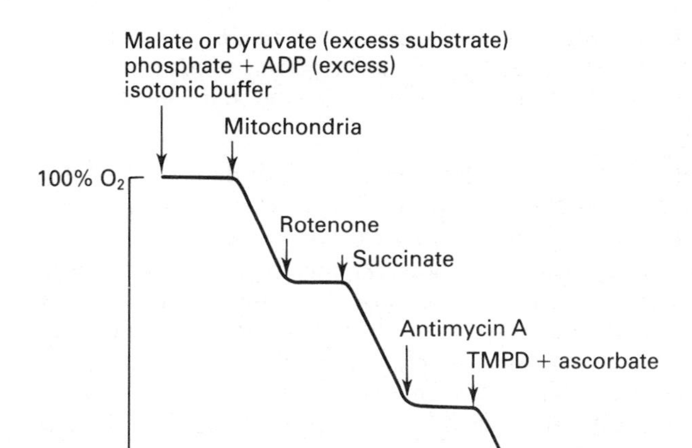

Fig. 10.9 Oxygen electrode trace showing the effect of inhibitors of electron transport and electron donors on mitochondrial respiration.

response and ease of operation, oxygen electrodes have been used for testing patients who are being treated with oxygen. Some oxygen electrodes have been specially modified to be small enough to be inserted into a blood vessel, but where possible, this is to be avoided because of the danger of infection or of blood clots forming. Small samples of blood from a warmed earlobe or a fingertip are used instead. Small Clark-type oxygen electrodes (usually called PO_2 electrodes) are used for the determination of oxygen content of blood.

Measurement of Oxygen in Bulk Liquids Oxygen concentrations are routinely monitored in fermentation processes, sewage and industrial waste treatment and in inland or ocean waters. This may involve use of a galvanic probe such as a Hersch cell (Section 10.5.2) or a variant of the Clark electrode called a *flush top sensor*. This has a large cathode which gives a high current (but a negligible current at zero oxygen concentration). The equipment is rugged and easily manufactured but to eliminate stirring effects the fluid must flow over it. Oxygen solubility is different in fresh water and salt water, so instruments have an adjustment for the degree of salinity in natural waters.

Enzyme Assays Enzyme assays are readily studied using a Clark oxygen electrode providing oxygen is involved in the reaction (Section 3.4.7). Glucose oxidase, D amino acid oxidase and catalase are examples of enzymes whose properties can be studied this way.

Microorganism Studies Bacteria which use oxygen as a terminal electron acceptor can be studied using an oxygen electrode, and the effect of electron inhibitors determined. Yeast respiration can be studied using different sugars. The most readily used sugars will speed up the rate of respiration in starved yeast which will give a steep slope on the oxygen electrode trace. This result shows that the sugar can enter the organism and can also be metabolised by respiration.

Multiparameter Studies Recent developments in the study of subcellular biochemistry include the combined use of an oxygen electrode with other methods such as pH and/or ion-selective electrodes (ISE) or spectrophotometric techniques. Detailed correlations can then be made, particularly if microprocessors are used to record and analyse the results. In such studies however, it must be remembered that oxygen electrodes tend to respond more slowly than other electrodes. As an example, the energy-linked mitochondrial uptake of ions can be followed in this way. The amount of oxygen uptake can be measured with an oxygen electrode, pH changes monitored using a pH electrode, mitochondrial swelling measured by spectrophotometric measurements at 546 nm and an ISE for calcium used to record ion changes.

10.6 Biosensors

10.6.1 Introduction and principles

Tissues, organelles, bacteria, yeast, enzymes, algae and antibodies can all be used as the basis of *biosensors*. These, by definition, consist of immobilised biological material in contact with a *transducer* which will turn the biochemical signal into an electrical signal. (A transducer turns one form of energy into another.) The generalised structure of one kind of biosensor is shown in Fig. 10.10. The types of transducer currently used are shown in Table 10.4. There is sometimes a problem in identifying a suitable transducer for use with a particular biological system.

Biosensors are already of importance and their significance is likely to increase as the technology develops. This is because they can be made to respond specifically and at high sensitivity to a wide range of molecules,

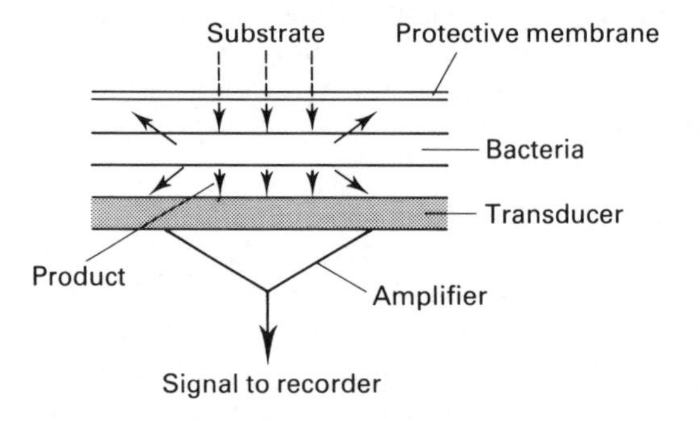

Fig. 10.10 Diagrammatic representation of a biosensor.

Table 10.4 Types of transducer used in biosensors.

Type	Operating principles	Reactions or molecules detected
Amperometric	A voltage is set and if the molecule is present, a current flows	O_2 (using Pt at −0.8V) (Section 10.5) H_2O_2 (using Pt at +0.8V) I_2, NADH
Ion-selective electrode/ pH electrode	Potential depends on ion concentration (Sections 10.2, 10.3)	H^+; Na^+; Cl^-
Gas-sensing electrode	Potential depends on gas concentration (Section 10.3)	CO_2; NH_3
Photomultipliers with fibre optics for bioluminescence	Light emission detected	If luciferase present, ATP detected (Section 3.4.4)
Photomultipliers with photodiode	Light absorption detected, e.g. pH changes cause a change in colour of a dye	Penicillin; urea
Thermistor	Heat of reaction detected	Almost universal (Section 3.4.9)
Piezoelectric crystal	Mass absorbed detected	Reactions involving volatile gases and vapours
Miniature field effect transistor (FET)	Microelectronic device	Can be made into an ion-selective field effect transistor (ISFET) to detect ions and a chemically sensitive field effect transistor (CHEMFET) to detect molecules
Non-specific ionic conductance	Measures increase in total number of ions	Many reactions, e.g. urease

including those of industrial and clinical significance. Pollution by pesticides can be monitored this way. They can also be used for industrial quality control and there are agricultural applications. In the clinical area, one of the most important research areas is the assay of glucose for the management of diabetes.

10.6.2 Enzyme electrodes

It is possible to exploit the specificity and diversity of enzymes to estimate the concentration of many different substrates. Some examples are given in Table 10.5. Enzyme electrodes can either be made to dip into a solution (this type is called a *bioprobe*) or can be made into a column (or sometimes a porous pad) and the solution passed through (forming a *bioreactor* or a flow system).

In bioprobes, the enzyme is immobilised by entrapment in a gel (e.g.

Table 10.5 Examples of biosensors.

Compound detected	Biological material used	Sensor	Immobilisation	Stability	Response time
Alcohol	Alcohol oxidase	O_2	Glutaraldehyde	2 weeks	1–2 min
β-glucose	β-Glucose oxidase	O_2	Chemical	3 weeks	1 min
Glutamate	Glutamate decarboxylase	CO_2	Glutaraldehyde	1 week	10 min
Penicillin	Penicillinase	pH	Polyacrylamide	2 weeks	15–30 sec
Urea	Urease	NH_4^+	Polyacrylamide	19 days	20–40 sec
Arginine	*Streptococcus faecium*	NH_3	Physically entrapped	20 days	20 min
Cholesterol	*Nocardia erythropolis*	O_2	Physically entrapped	4 weeks	35–70 sec
Nitrate	*Azotobacter vinelandii*	NH_3	Physically entrapped	2 weeks	7–8 min
NAD^+	NADase + *Escherichia coli*	NH_3	Dialysis membrane	1 week	5–10 min

polyacrylamide), by encapsulation, by ionic, adsorptive or covalent binding to carriers or by cross-linking either to another molecule of the same enzyme or to another protein, e.g. serum albumin. Glutaraldehyde, a bifunctional reagent, can be used to form cross-links or to give covalent binding, e.g. to a Teflon or nylon membrane. The substance to be measured which is usually the substrate for the enzyme, diffuses to the enzyme where a product is formed which then affects the electrode. For example, a penicillin electrode uses penicillinase which degrades penicillin and causes a pH change that is recorded using a pH electrode:

```
       R
       |
       C=O                                          R
       |                                            |
      HN                          H2O               C=O
       |  H   S      CH3           \                |
   H—C—C    \   C                   \              HN        S
       |  |      |   CH3    ————————————>           |  H /      \    CH3   + H+
       C—N______C           penicillinase       H—C—C          C
       ||      H \                                  |  |       | CH3
       O          COO-                            O=C  N_______C
                                                    |  H       H  COO-
                                                    O-

        penicillin                                    penicilloic acid
```

Enzyme electrodes normally have to be kept refrigerated to keep the lifetime as long as possible; even so, few have a lifetime of over a month. With some electrodes there is a problem of choosing the operating pH because the optimum pH for the enzyme reaction may be different from the optimum pH for detecting the product. A compromise pH may have to be selected.

As shown in Table 10.5, enzyme electrodes do not give an instant response. This is because the substrate has to diffuse through the membrane, and the product then has to diffuse to the transducer (sensing device). However, the response using immobilised enzymes is usually faster than that using immobilised cells where the cell membrane forms an additional barrier.

10.6.3 Bacterial electrodes (cell-based biosensors)

Some of the disadvantages of enzyme electrodes may be overcome by using bacterial electrodes. For example, bacterial electrodes are less sensitive to inhibition by solutes and are more tolerant of suboptimal pH and temperature values than are enzyme electrodes. Bacterial electrodes also tend to have a longer lifetime than enzyme electrodes (20 days or more compared with an average of 14 days). They are cheaper because an active enzyme does not have to be isolated. However, cells contain many enzymes and care has to be taken to ensure selectivity, e.g. by optimising storage conditions or adding specific enzyme or transport inhibitors to stop undesirable enzyme reactions. Mutant bacteria lacking enzymes can be used. Bacterial electrodes do have some disadvantages. For example, some have a longer response time than enzyme electrodes, but a more serious problem is the time taken for cell-based sensors to return to a baseline potential after use.

When bacterial electrodes were first developed, they involved the use of rather harsh techniques for the immobilisation such as polyacrylamide gel, or used cells whose permeability had been increased. Thus the cells were generally not viable. However, the enzymes within them were still active. More recent immobilisation techniques have tended to use more gentle physical methods so that viability is retained. The advantage of this is that such cells may be involved in converting substrate into a product via a multienzyme pathway, without having to immobilise each of the enzymes and then provide them with expensive coenzymes.

More than one kind of bacterium can be incorporated into one electrode which increases the number of potential applications. Thus biochemical oxygen demand (BOD) due to organic matter in waste water is detected by a mixed culture of bacteria obtained from soil since a single microorganism species would be unable to use all the organic compounds likely to be found in the sample.

It is also possible to combine an enzyme preparation with a micro-organism. In Table 10.5, an example of this is given; it is a detector for NAD^+. *Escherchia coli* cells provide the enzyme nicotinamide deaminase and it is combined with NADase from *Neurospora crassa*. The following reaction then occurs:

$$\text{NAD}^+ + \text{H}_2\text{O} \xrightarrow{\text{NADase}} \text{nicotinamide} + \text{ADP} - \text{ribose}$$

$$\text{nicotinamide} + \text{H}_2\text{O} \xrightarrow[\text{deaminase}]{\text{nicotinamide}} \text{nicotinic acid} + \text{NH}_3$$

and the ammonia released is detected by a gas-sensing electrode.

The detection of colesterol is of great clinical importance, and can be carried out by using *Nocardia erythropolis* immobilised in polyacrylamide on agar on an oxygen electrode. The reaction carried out is:

$$\text{cholesterol} + \text{O}_2 \xrightarrow[\text{oxidase}]{\text{cholesterol}} \text{cholest-4-en-3-one} + \text{H}_2\text{O}_2$$

The oxygen electrode measures the rate of oxygen uptake and this can be related to the cholesterol content of a biological sample (plasma).

Like enzymes electrodes, bacterial electrodes can be made into bioprobes or into bioreactors. Bioreactors are very suitable for the commercial production of metabolites, but can also be used as biosensors. In the future it may be possible to increase the life span of bacterial electrodes by stimulating growth of fresh cells on the membrane surface.

10.6.4 Enzyme-immunosensors

Enzyme-immunosensors of several kinds have been developed. They combine the molecular recognition properties of antibodies with the high sensitivity of enzyme based analytical methods. The enzyme is used as a marker as it reacts with its substrate, giving a change which can be detected by a transducer of some kind (Table 10.4). There is a similarity between this method and ELISA (Section 4.6).

The enzyme-immunosensor for IgG consists of anti-IgG antibody bound to a membrane and linked to an oxygen electrode. Catalase is bound to the IgG, and so forms a label for it. This complex is then mixed with the sample that contains an unknown amount of unlabelled IgG. The labelled and unlabelled IgG compete for the antibody on the membrane. After exposure to test an IgG solution, the sensor is rinsed to remove any non-specifically associated IgG and is then immersed in a H_2O_2 solution as a substrate for the catalase. The more unlabelled IgG present, the lower the amount of catalase present and therefore the lower the rate of oxygen evolution. A similar assay can be carried out for human chorionic gonadotropin (HCG), a diagnostic hormone for pregnancy. Catalase is again used to label the HCG, and oxygen evolution is followed as before.

In attempts to construct enzyme-immunosensors, bioluminescence, chemiluminescence and fluorescence principles are being explored, because of their great sensitivity. A luminescent immunoassay with catalase has been used to detect human serum albumin at 1 ng cm^{-3}.

10.7 Electrochemical detectors

Many substances can be relatively easily oxidised or reduced by applying a suitable potential to generate the necessary energy to drive the electrochemical reaction. In practice, the applied potential must exceed the reaction potential of the compound, but once the reaction potential has been exceeded and the reaction initiated, a further increase in potential does not increase the observed rate of reaction. The reaction potential, of course, varies from compound to compound and this forms the basis for selective detection. The technique is most commonly exploited for the detection of solutes in the effluent from chromatography columns especially HPLC.

The principle of the electrochemical detector is illustrated in Fig. 10.11 in which the amount of current (which is related to the rate of oxidation or reduction) is plotted as a function of the applied potential. The whole curve is referred to as an *electrochemical wave*.

At low potential (between A and B) the current is too low to start the reaction. At point B, the reaction starts. Point C (the *half wave potential*) is where the wave has reached half its final height and because this potential is characteristic of a compound, it can be used as the basis of identification. At point D, further changes in the applied potential do not alter the rate of the electrode reaction. The difference in height between B and D is related to the

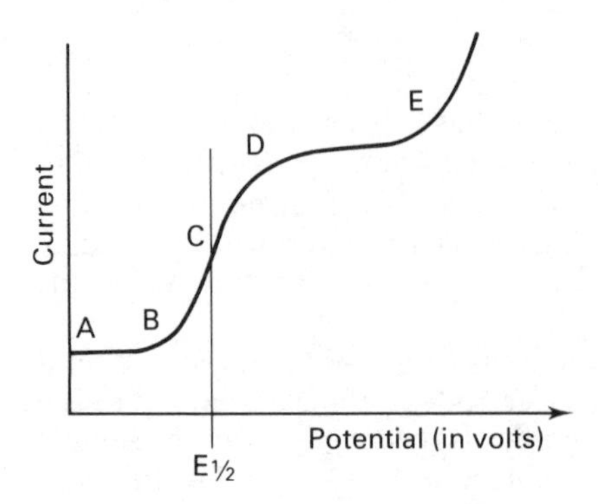

Fig. 10.11 An electrochemical wave.

concentration of the substance. At point E, a second and unrelated electrochemical reaction commences.

Normally in an electrochemical detector, the potential is first chosen so that it is sufficiently high to cause the reaction to occur. It is then maintained at that level and a trace of current against time is recorded. As the effluent containing the solute comes off the end of the HPLC column it flows through the detector and a peak is recorded. The trace produced may look similar to those produced by UV or fluorescence detectors but the sensitivity will be different.

Not all compounds in the liquid give rise to a current at the particular potential chosen. This means that if two compounds have not been chromatographically separated and one alone gives a current at that potential, then the electrochemical detector will give one peak only, and it will be easy to make quantitative measurements despite the presence of the second substance.

The detector requires an electrolyte in the mobile phase so that a current may flow. This means that reverse phase and ion-exchange chromatography (which both use aqueous solvents) can be coupled to electrochemical detectors if sufficient electrolytes are present. Electrochemical detectors are very sensitive to flow changes and can only be used with pulse-free pumps (Section 6.8.2).

In a *wall jet electrochemical detector*, the electrodes are mounted in a fluorocarbon block. The liquid flows under pressure up to the electrodes and through a jet to impinge on the surface of the working electrode and this is why this type of electrochemical detector is called a wall jet electrode (Fig. 10.12). A high current efficiency ensues, and also the working electrode stays clean, as the products of the electrochemical reaction are washed off by the flow. As well as the working electrode (which is glassy carbon) there are two other electrodes present. One is a counter or auxilliary electrode which is stainless steel and forms part of the inlet. Any deviation from the preset voltage of the working electrode can be corrected by the auxilliary electrode. The reference electrode silver/silver chloride gives a steady potential, and is in contact with the liquid as it leaves the wall jet flow cell.

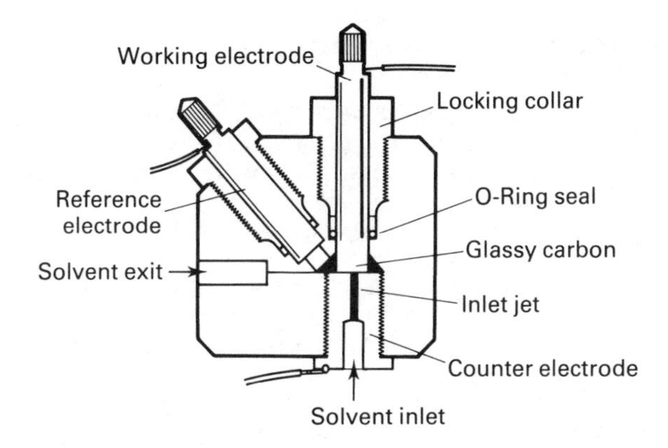

Fig. 10.12 A wall jet electrochemical detector.

Chemicals can be detected by either oxidation, in which the working electrode is kept at a positive potential, or reduction in which the working electrode is negative. At very high positive or high negative potentials the current becomes large due to the solvent reacting. These are referred to as the *anodic and cathodic limits of solvent.*

Most commonly the molecule is converted into its oxidised form releasing electrons giving the current (Fig. 10.13). The oxidised form is often unstable (e.g. a free radical) and it then reacts irreversibly to give a stable product. Sometimes the reactions that occur depend on pH. Aromatic phenols, aromatic amines, heterocyclic nitrogen atoms and sulphur-containing molecules can be detected by oxidation, i.e. positive potentials are used. Thus drugs such as aspirin, paracetamol, morphine, nicotine and caffeine can be analysed by the oxidation mode.

Oxygenating compounds such as quinones, peroxides and amides can be detected by reduction. This approach is also suitable for aromatic nitro and nitroso groups and halogen compounds. From an electrochemical point of view, however, reductions are more difficult to carry out than oxidations as it is commonly difficult to obtain ideal experimental conditions. Thus solvents must be degassed, as oxygen will undergo electrochemical reactions, and meticulous care has to be taken to ensure that oxygen cannot re-enter the solution. At voltages more negative than -0.8 V, water will evolve hydrogen, and so lower voltages have to be used.

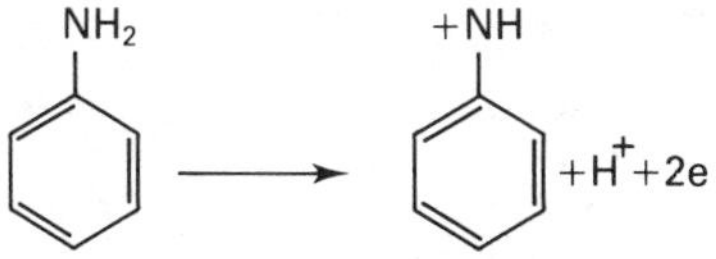

$$C_6H_5NH_2 \longrightarrow C_6H_5\overset{+}{N}H + H^+ + 2e$$

Fig. 10.13 Oxidation of an aromatic amine to give a free radical.

10.8 Suggestions for further reading

Buck, R.P. (1984). *Electrochemistry of Ion Selective Electrodes.* In *Comprehensive Treatise of Electrochemistry Vol. 8. Experimental Methods in Electrochemistry* (White, R.E., Bockris, J.O'M., Conway, B.E. and Yeager, E. Eds). Plenum Press, N.Y. (Ion-selective electrodes.)

Corcoran, C.A. and Rechnitz, G.A. (1985). *Trends in Biotechnology*, **3**, 92–6. (Cell based biosensors.)

Gronow, M. (1984). *Trends in Biochemical Sciences*, **9**, 336–40. Biosensors. (Enzyme and bacterial electrodes.)

Harris, D.C. (1982). *Quantitative Chemical Analysis.* W.H. Freeman and Co., USA. (pH and ion selective electrodes, redox.)

Lessler, M.A. (1982). *Adaptation of Polarographic Oxygen Sensor for Biochemical Assays.* In *Methods of Biochemical Analysis*, **28**, 175–99 (Glick, D., Ed.). Wiley-Interscience, New York. (Oxygen electrodes.)

Morf, W.E. (1981). *The Principles of Ion Selective Electrodes and Membrane Transport.* In *Studies in Analytical Biochemistry* Vol. 2. Elsevier, Amsterdam. (Ion-selective electrodes.)

Nicholls, D.G. (1982). *Bioenergetics: An Introduction to the Chemiosmotic Theory.* Academic Press, London. (For measurement of membrane potential and pH differences.)

Orion Research (1982). *Handbook of Electrode Technology.* Orion Research Incorporated. (Ion selective electrodes.)

Rottenberg, H. (1979). *Methods in Enzymology*, **55**, 547–69. (Fleischer, S., Packer L., Eds). Academic Press, New York. (For measurement of membrane potential and pH differences.)

Weast, R.C. (1984). *Handbook of Chemistry and Physics*, 65th edition. The Chemical Rubber Co., Cleveland, Ohio. (For details of oxygen in solution.)

Westcott, C.C. (1978). *pH Measurements.* Academic Press, London.

Index